Multiple 3-Phase Fault Tolerant Permanent Magnet Machine Drives

IEEE Press
445 Hoes Lane
Piscataway, NJ 08854

Multiple 3-Phase Fault Tolerant Permanent Magnet Machine Drives

Design and Control

Bo Wang
Southeast University
China

Jiabin Wang
The University of Sheffield
United Kingdom

IEEE*Press*
WILEY

Published by John Wiley & Sons, Inc., Hoboken, New Jersey.
Published simultaneously in Canada.

Library of Congress Cataloging-in-Publication Data Applied for:

Hardback ISBN: 9781394252015

Cover Design: Wiley
Cover Image: © Bo Wang

Set in 9.5/12.5pt STIXTwoText by Straive, Chennai, India

To all our colleagues and collaborators, past and present, who contributed to and supported the research of fault-tolerant machine drives.

Contents

About the Authors

Bo Wang received a Bachelor's degree and a Master's degree, in Electrical Engineering, from Nanjing University of Aeronautics and Astronautics, Nanjing, China, in 2009 and 2012, respectively, and a PhD in Electronic and Electrical Engineering from the University of Sheffield, Sheffield, United Kingdom (UK), in 2018.

From 2012 to 2014, he was a Senior Engineer at Delta Electronics Co. Ltd, Nanjing, China. From 2017 to 2018, he was a Research Associate at the Department of Electronic and Electrical Engineering, University of Sheffield, Sheffield, UK. Since 2018, he has joined the School of Electrical Engineering, Southeast University, Nanjing, China, and now he is a full professor. From 2020 to 2022, he joined Hong Kong Polytechnical University, Hong Kong, China under the Hong Kong Scholar program.

He received several awards from the Xiaomi Young Scholarship, Hong Kong Scholarship, and Zhishan Young Scholarship funding programs. He is, currently, an editor of several technical journals, including *IET Power Electronics, Power Electronics and Drives, Chinese Journal of Electrical Engineering, CES Transactions on Electrical Machines and Systems*. His research interests include permanent magnet machine drives, electric traction, and fault-tolerant systems.

Jiabin Wang graduated with a Bachelor of Engineering degree in 1982 and a Master of Engineering degree in 1986 from Jiangsu University, Zhengjiang, China, and a PhD from the University of East London, London, United Kingdom (UK), in 1996, all in electrical and electronic engineering.

Since February 2023, he has been an emeritus professor in Electrical Engineering at the University of Sheffield, Sheffield, UK. From 1986 to 1991, he was with the Department of Electrical Engineering at Jiangsu University, where he was appointed a lecturer in 1987 and an associated professor in 1990. He was a postdoctoral research associate at the University of Sheffield, Sheffield, UK, from 1996 to 1997, and a senior lecturer at the University of East London from 1998 to 2001. In 2002 he joined the Electrical Machines and Drives Research group at the University of Sheffield as a senior lecturer, became a reader in 2007, and a professor in Electrical Engineering in 2010.

His research encompasses novel rotary and linear electrical machines and drives, advanced control techniques for electrical drives, electrical power trains, and "more-electric" technologies, with a particular focus on high integrity, fault-tolerant, and high-efficiency electric drives and associated condition-monitoring techniques for applications in aerospace, automotive, and renewable energy systems.

He is a fellow of the Institute of Engineering and Technology, UK, and a senior member of the Institute of Electrical and Electronics Engineers, USA.

Preface

Due to their high efficiency, zero-emission, good controllability, simple and flexible physical layout, electrical machine drives are increasingly used in transportation electrification, such as electric vehicles, railway traction, marine and aerospace propulsion, and their actuations. However, electronic and electrical components and cable interconnections are susceptible to adverse operation conditions and environmental stresses, such as high temperatures, and high levels of mechanical vibration and chemical erosion. The failures are usually abrupt and may lead to catastrophic consequences and considerable economic losses. Thus, high reliability and fault tolerance are essential for electrical machine drives in safety-critical applications.

To achieve high reliability and desirable fault tolerance, many novel machine topologies, such as multiphase machines and multiple 3-phase machines with innovative stator and rotor configurations and power electronic control, have been developed to cope with various electric faults, such as open-circuit faults, short-circuit faults, and winding interturn short-circuit faults as a result of insulation breakdown. Comprehensive and computationally efficient fault modeling techniques have also evolved to gain an in-depth understanding of various fault behaviors and to facilitate the development of fault detection and postfault control strategies. In this regard, effective fault detection is essential to identify the fault type, isolate it, and enable an appropriate mitigation action for continued postfault operations.

However, very few materials exist in the literature, which present a comprehensive and integrated treatment of the design and control of fault-tolerant machine drives, including the selection of fault-tolerant machine topology, fault modeling, fault detection, and post-fault control considering failures in both the machine and power inverter. The existing body of knowledge on fault-tolerant machine drives is fragmented, mainly in numerous technical journals and conference papers, which makes it difficult to apply them effectively in engineering designs and further developments. Thus, this book aims to fill this knowledge gap and address the pertinent issues on the design, operation, modeling, and control of fault-tolerant machine drives, including fault detection and mitigation, in a comprehensive and integrated manner.

While various fault-tolerant machine drive topologies and technologies are reviewed and discussed, this book focuses on the design and control of multiple 3-phase fault-tolerant permanent magnet machine drives. The conventional overlapped distributed 3-phase winding is reconfigured as multiple nonoverlapped 3-phase winding modules, each driven by a standard 3-phase inverter. Due to the segregated winding configuration and independent drive for each 3-phase module, physical, electrical, and thermal isolations are obtained in the multiple 3-phase subsystems of the electrical drive. As a result, a failure in one 3-phase subsystem does not significantly affect the operation of other healthy 3-phase subsystems. The modular approach minimizes the risk of fault propagation to the other subsystems, providing tolerance to various electric faults. By exploiting reluctance torque in a permanent magnet machine with a multilayer rotor topology, high torque density and the ability to tolerate short-circuit faults can be achieved. This book introduces the fault-tolerant machine drive system, analyses various fault behaviors, and presents comprehensive design and fault modeling techniques, including fault detection, postfault control, and alternative winding configurations for further reduction in short-circuit currents. The methods and approaches described in the book also apply to developing other fault-tolerant machine drives.

The authors have worked on developing fault-tolerant machine drive technologies at Southeast University and the University of Sheffield for more than 20 years and published over 50 SCI/EI-indexed papers on this topic. This book organizes a selected collection of the research outcomes and is arranged into seven chapters as outlined below.

Chapter 1 introduces the background knowledge and current state-of-the-art fault-tolerant machine drives.

Chapter 2 describes the multiple 3-phase permanent magnet-assisted synchronous reluctance machine with segregated windings and its ability to tolerate various electric faults. It also analyzes the mutual magnetic coupling among the 3-phase subsystems in the machine.

Chapter 3 presents the global design optimization process of the fault-tolerant machine drive to achieve performance requirements in healthy operations and the ability to tolerate worst-case fault scenarios, namely inter-turn short circuits, which lead to the largest possible fault current.

Chapter 4 establishes a general modeling technique for operations in healthy and various fault conditions and outlines the key characteristics of fault behaviors.

Chapter 5 develops fast and reliable fault detection techniques, particularly focusing on the detection of inter-turn short circuits at low speed.

Chapter 6 devises post-fault control strategies for fault-tolerant machine drives to minimize torque ripple and fault current in worst-case conditions.

Chapter 7 proposes several alternative winding configurations for the multiple 3-phase machine drive to reduce fault current in worst-case conditions further.

The authors thank professors Cheng Ming, Hua Wei, Deng Fujin, Antonio Griffo, and Zhao Wenxiang for their valuable suggestions on this book. The authors also thank postgraduate researchers Mr. Feng Xiaobao and Mr. Wei Zihan for their support in preparing this book.

Finally, the authors thank their families for their love and support in the research and writing of this book.

Bo Wang
Nanjing, China
Jiabin Wang
Sheffield, United Kindom

1

Introduction to Fault-Tolerant Machine Drives

1.1 Background of Fault-Tolerant Machine Drives

Advanced electric drive systems are increasingly being used in a wide range of applications from industrial automation, household appliances, and transportation to oil and gas, mining, and renewable energy industries, where efficient and reliable electric-to-mechanical energy conversion or vice versa is essential. Extensive research activities on electrical drives have been undertaken in both academic and industrial organizations [1]. A typical electric drive is composed of a power converter, a control unit, and an electric motor, generally known as an electric machine, as shown in Fig. 1.1. The power converter contains typically power electronic devices (i.e. Insulate Gate Bipolar Transistor (IGBT), Metal-Oxide-Semiconductor Field-Effect Transistor (MOSFET), SiC, and diode), gate drives, and passive components (capacitor and damping resistors, etc.), which are responsible for driving the motor. The control unit consists mainly of microprocessor and its associated electronic circuitry in addition to various sensors. The electric motor delivers controllable torque to a mechanical payload by converting electrical power into mechanical power or vice versa.

As advances emerge fast in the areas of materials, electric machine design and manufacturing, power electronics, microprocessors, and sensing, electrical drives are capable of delivering desirable features such as high-power density, high efficiency, low emission, and good controllability, compared to other counterparts, namely, mechanical, hydraulic, or pneumatic drive/actuation systems [2]. Aircraft employing electrical actuators, electrical propulsion, and power generation in the form of more electric, hybrid, and full-electric aircraft can leverage the merits of weight saving, economical fuel consumption, low CO_2 emission, increased functionality, and less maintenance [3]. Another emerging example is the electric vehicle (EV) replacing traditional internal combustion engine (ICE) for low CO_2 and low harmful pollutant emissions [4, 5]. However, high reliability

Multiple 3-Phase Fault Tolerant Permanent Magnet Machine Drives: Design and Control, First Edition.
Bo Wang and Jiabin Wang.

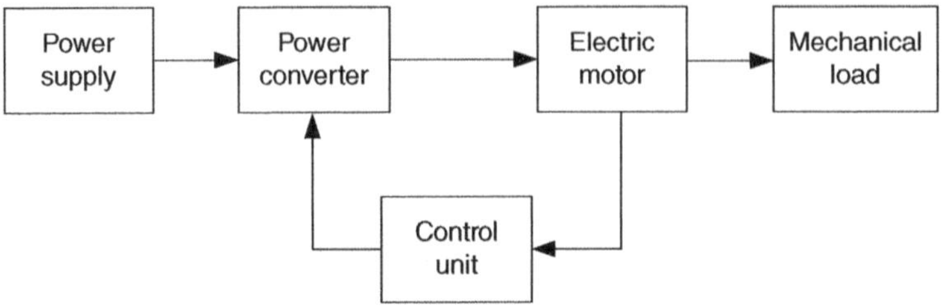

Figure 1.1 Illustration of a typical electric drive system.

is also an essential requirement for these safety critical applications, which should be addressed at the system design stage [6].

In the aforementioned safety critical applications, the electrical drives are expected to continue operation if a fault occurs, or at least being fail-safe without catastrophic damage [7]. Otherwise, the unexpected fault may cause casualties and huge economic losses [8]. Thus, fault tolerance should be considered to attain the reliability requirement for the targeted applications.

Fault tolerance means that the system is capable of performing at a satisfactory level of operation in the presence of fault. It is a common requirement that has been investigated in various areas, such as fault-tolerant computing systems [9], distributed power systems [10], and high availability internet servers. In the scope of electrical drives, the fault tolerance mainly means it is capable of maintaining the original or an acceptable output torque or power level after a fault. The acceptable level defines the minimum output, which should be considered at the primary design stage of such systems.

1.2 Frequent Faults in Electric Drives

An electrical drive is a complex electromechanical system composed of an electronic controller, a power converter, an electric motor, and sensors. These components are exposed to electrical, thermal, mechanical, and environmental stresses as well as chemical corrosion. Fault may occur in each of these components. Studies in [11, 12] have been carried out to investigate the failure distribution of electric machines. The results of the survey show the bearing faults account for the majority of the failures, as much as 51%, followed by stator winding faults, up to 25%. Other faults such as rotor bars and end rings in induction machines, shafts, and other unidentified failures take up the remaining percentage in Fig. 1.2(a). The investigation data in [13] also illustrates that electrical winding failures amount to a failure rate of 1.4×10^{-7} failures per hour in military-grade machines and 1.0×10^{-6} in industrial machines. Since these surveys are mainly focused on induction machines, permanent magnet (PM) failure in permanent magnet synchronous machines (PMSMs) is not included.

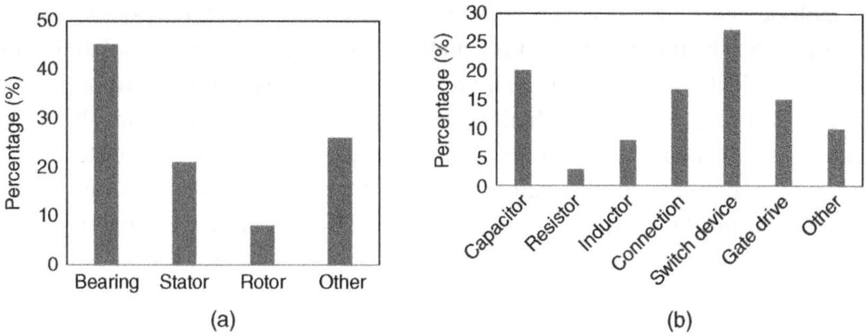

Figure 1.2 Fault distribution in electrical drives: (a) machine side and (b) converter side.

In fact, partial demagnetization is a frequent fault for PM machines due to a strong armature reaction field, overheating, and excessive mechanical stress and vibration [14].

A similar industry survey was conducted on failures in converters in [15]. The survey indicates that the most vulnerable component is the switching devices, followed by capacitors and gate drive circuitry. Open circuit in one phase due to device and connection failures is also a frequent fault. Failures associated with resistors and inductors are quite rare and only observed in a few applications, as shown in Fig. 1.2(b). The survey result shows most IGBT/MOSFET device failures result from thermal and power cycling, with a typical failure rate of 2.78×10^{-6} failures per hour. Additionally, the controller and sensors may also experience faults during operation. Nevertheless, it should be noted that the probability of the microcontroller and sensor faults is much lower.

As mentioned above, many potential faults may occur in the system. In this chapter, the principal device and electromagnetic faults under consideration that may occur within an electric drive are shown in Table 1.1.

On the machine side, the winding insulation degrades gradually due to electrical, thermal, and mechanical stresses and finally develops into open-circuit or short-circuit failure. The short-circuit failure can be classified as interphase and

Table 1.1 Potential faults occurring in an electric drive.

Machine side	Drive side
Winding open circuit	Switch device open circuit
Winding interphase short circuit	Switch device short circuit
Winding intraphase short circuit	DC-link capacitor failure
Demagnetization	Controller/sensor failure
Uncontrolled generation failure at high speed	

intraphase short circuit, which occur between phases or within a single phase, respectively. The intraphase fault is usually caused by turn-to-turn insulation failure. In particular, an intraphase fault involving a few turns, also known as a turn fault, is reported as the worst-fault scenario since only a few turns are short circuited. The resultant fault current is massive and the excessive hotspot temperature may lead to catastrophic failure. Partial demagnetization is another common fault in PM machines due to the excessive armature reaction field, overheating, and a high level of mechanical stress and vibration. It may cause torque reduction and increased torque ripple, etc.

On the drive side, the switch device is also subjected to open-circuit and short-circuit failure due to electric and thermal–mechanical stress during repeated switching on and off operations. DC-link capacitor is exposed to combined electrical and thermal stress during inverter operation and hence contributes to a considerable failure rate in electric drives [16]. Gate drive failure gives rise to similar consequences of switching device failures and may be incorporated into the switch device fault mechanism.

Besides, another possible fault is the uncontrolled generation, particularly for PM machines. If the power converter fails when the machine is rotating at high speed, the electromotive force (emf) may be much higher than the DC-link voltage and consequently cause uncontrolled rectification via the freewheeling diodes in the power converter. This may damage the DC-link components if the generated power is excessive and cannot be absorbed [17, 18].

So far, most of the fault-tolerant electrical drives focus on the faults described above [8, 13], since the most frequent bearing failure can be significantly reduced by regular maintenance, online monitoring, and replacement, whereas the controller and sensor faults are less likely.

1.3 Design Requirements of Fault-Tolerant Machine Drives

The requirements for the fault-tolerant systems in distributed power systems have been investigated in [10]. The methodology for fault-tolerant electrical drives follows relatively similar principles. The principal guideline is one fault in the system should be isolated in a subunit and has limited effect on the remaining healthy part, which can be in place to maintain uninterrupted operation. As extensively discussed in literature, four design criteria for fault-tolerant electrical drives are summarized.

A) **Partitioning and Redundancy:** A fundamental specification for fault-tolerant system is that a single fault would not disrupt the whole system.

Therefore, the fault must be confined to a relatively independent subsystem. This implies the system should be partitioned into several subunits. A fault would only disable the fault module, which would not cause malfunction of the whole system. Then, the remaining healthy modules could continue operation to meet the output requirement in the faulty condition. The trade-off between the redundancy number and cost is often debated at the design stage [3].

B) **Fault Isolation:** Partitioning alone is not enough to prevent the system breakdown from a faulty module, since certain types of fault may affect the remaining healthy subunits or even propagate to the whole system. It may lead to more severe consequences. Thus, many measures are employed to achieve fault isolation. Specifically, magnetic, electrical, thermal, and physical isolations are required for ideal fault-tolerant electrical drives to accommodate fault [19].

C) **Fault Detection:** On one hand, fault detection can help the electrical drive to isolate fault by taking appropriate fault mitigation action while on the other hand, continued operation needs information on fault type and location to perform appropriate postfault control. Fault must be detected before severe consequences occur. Indeed, fault detections are core techniques for fault-tolerant machine drives and require extensive and in-depth investigations.

D) **Postfault Control:** After a fault, the machine drive changes its physical behaviors and the control law under healthy conditions may lose its effectiveness. For example, new voltage harmonics and excessive fault current may arise, and consequently, the current tracking quality with conventional control techniques deteriorates significantly, resulting in excessive torque pulsation, etc. Thus, a new postfault control strategy should be in place to protect the system and maintain its operation for continued functionality.

In summary, the above four requirements should be taken into consideration to design an integrated fault-tolerant electric machine drive system. However, current research on the fault-tolerant machine drive is quite fragmented, and most literatures only focus on one of them.

1.4 Current State-of-the-Art Techniques of Fault-Tolerant Machine Drives

To achieve an integrated fault-tolerant machine drive, a number of effective techniques have been developed, which include fault-tolerant machine drive topology, fault modeling, fault detection, and postfault control. These techniques are essential to accommodate various failures by transferring the machine drive from healthy operation to postfault operation in a reliable manner. The fault-tolerant machine drive topology enables the fault isolation and postfault operation while the fault modeling technique assesses machine performance in both healthy and faulty conditions and also aids in fault detection. Fault detection can identify and classify the faults and consequently trigger appropriate mitigation actions. After the fault, the previous control law becomes invalid since the machine has changed its physical behaviors. It requires a new postfault control strategy to accommodate the fault and suppress excessive and undesirable fault current if present. Subsequently, these key techniques are reviewed and the main challenges are highlighted.

1.4.1 Fault-Tolerant Machine Drive Topologies

Various fault-tolerant machine topologies have been reported and investigated in literature. The most straightforward approach is to adopt two or more redundant machine drive modules either in series or in parallel [7, 20]. In case of a failure in one module, the fault is isolated and the other module can continue its operation. However, the use of multiple machine modules for redundant operation occupies large space and necessitates additional accessories to guarantee operation, resulting in low power density and bulky size. Therefore, this approach becomes less attractive.

Alternatively, some level of fault tolerance may be achieved on a typical 3-phase machine drive by employing a neutral connection of the 3-phase winding to the midpoint of the DC link or to a fourth inverter leg as shown in Fig. 1.3(a) and (b) [21], respectively. It can cope with an open-circuit fault either in the inverter or in the 3-phase windings. By employing the neutral connection, the two remaining phase currents can be controlled independently. Zero-sequence current is utilized to generate the equivalent rotating magneto-motive force (MMF) in the machine if one phase is open circuited.

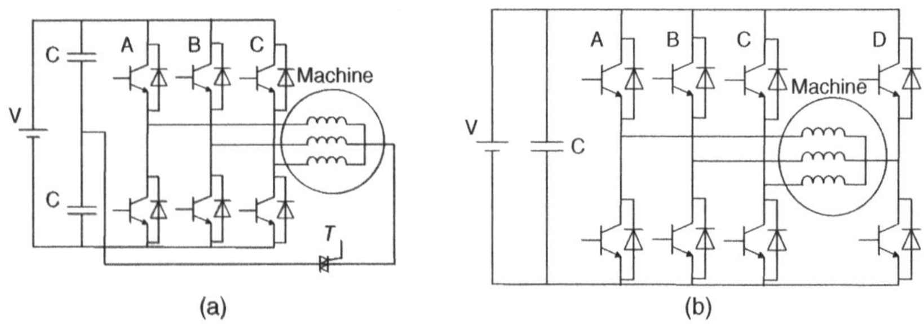

Figure 1.3 3-Phase fault-tolerant machine drive: (a) full bridge with DC midpoint and (b) four-leg inverter.

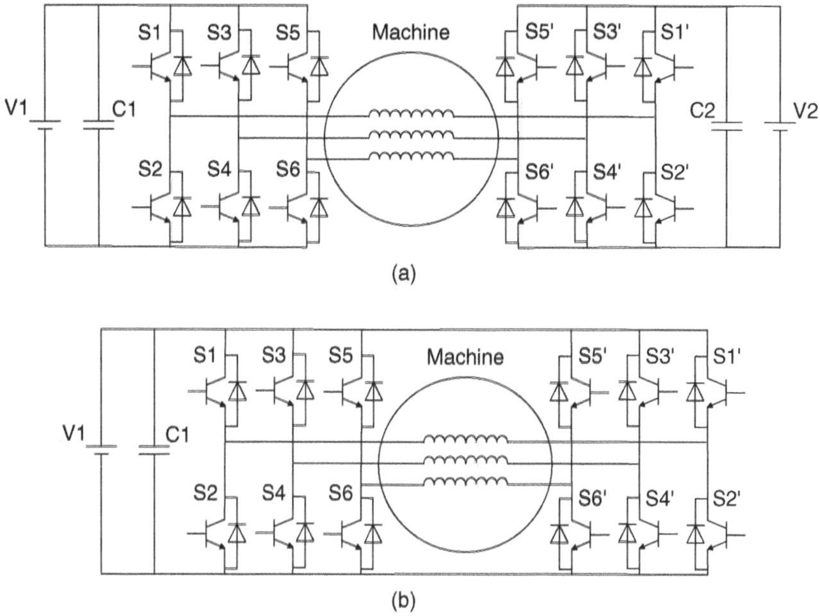

Figure 1.4 3-Phase machine using open-end winding drive (a) with independent power supplies and (b) with single power supply.

Another common fault-tolerant approach for 3-phase machines is to employ open-end winding drives as shown in Fig. 1.4. Both terminals of a 3-phase winding are available (opened) for connections to dual 3-phase inverters. The dual drives can share one DC link or use independent power supplies [22]. In addition to better fault-tolerant capability, they also bring the merits of 3-level voltage output, lower DC-link voltage for a given rated voltage of the machine and high efficiency. In case of a switch failure in one inverter, drive operation can be sustained by appropriate control of the healthy side inverter.

The access to the neutral point in Fig. 1.3 can be eliminated by employing more than three phases in a single machine as shown in Fig. 1.5 [23, 24]. Depending on the number of phases, a multiphase machine is capable of continuous operation when one or more than one phase has failed. Multiphase machine is initially investigated for high-power applications. By dividing the required power between multiple phases, higher power levels can be realized with specific rated power electronic converter modules. It can also reduce torque pulsation and lower DC-link harmonic current. As described above, a neutral point connection is required for a 3-phase machine after one phase is open circuited so that the current in the remaining two phases can be controlled independently. However, it is possible to take advantage of the additional degrees of freedom in a multiphase

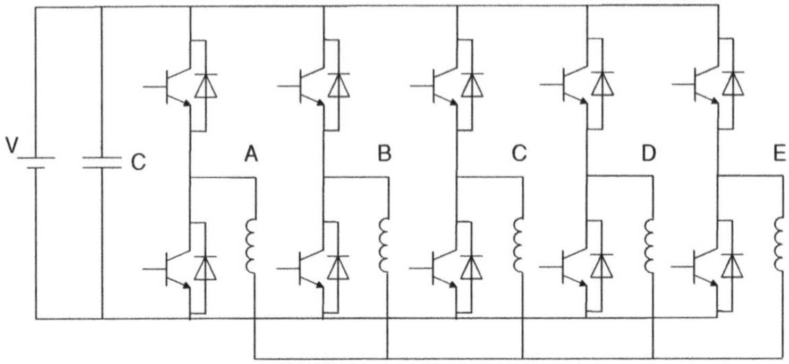

Figure 1.5 Multiphase (5-phase for illustration) fault-tolerant machine drive.

machine where the number of phases is greater than 3. With appropriate control of the power converter, a multiphase machine can continue its operation in case of an open-circuit fault without connecting the neutral point. Five-phase, six-phase, and seven-phase machines are most extensively investigated for both induction machines and PM machines [25–27]. Additionally, various control algorithms have been proposed to achieve the maximum attainable torque or minimum torque ripple [28–31] under fault conditions.

One common disadvantage of the 3-phase and multiphase fault-tolerant machine mentioned above is the complex postfault control since the machine no longer operates in a balanced manner after a fault. To alleviate this problem, the machine windings can be configured as multiple 3-phase windings with each driven by an independent inverter [32] as shown in Fig. 1.6. If a fault occurs in one set of 3-phase winding, the whole set will be taken out of service. The remaining sets can continue their balanced operation. The benefits of such fault-tolerant drives are that the postfault control strategy can be much simpler as the operation of the remaining balanced 3-phase windings can be easily maintained. It should be noted, however, that the strong and undesirable magnetic coupling may still exist in different 3-phase winding sets. Consequently, the short-circuit current may still be excessive and torque ripple may increase due to the mutual coupling. This concept has been realized in both induction machines, synchronous machines, and PM machines [25, 26, 33–35]. However, the majority of the above solutions cannot cope with excessive current in the event of short-circuit fault.

To accommodate short-circuit fault in power electronic switches or phase windings, alternative fault-tolerant drives are developed. Owing to its rugged, magnet-free rotor structure and concentrated windings, the switched reluctance machine (SRM) is inherently fault tolerant as shown in Fig. 1.7 [36, 37]. Each phase winding can be magnetically, thermally, and electrically isolated. Due to low mutual coupling between phases, the short-circuit current can be limited in a

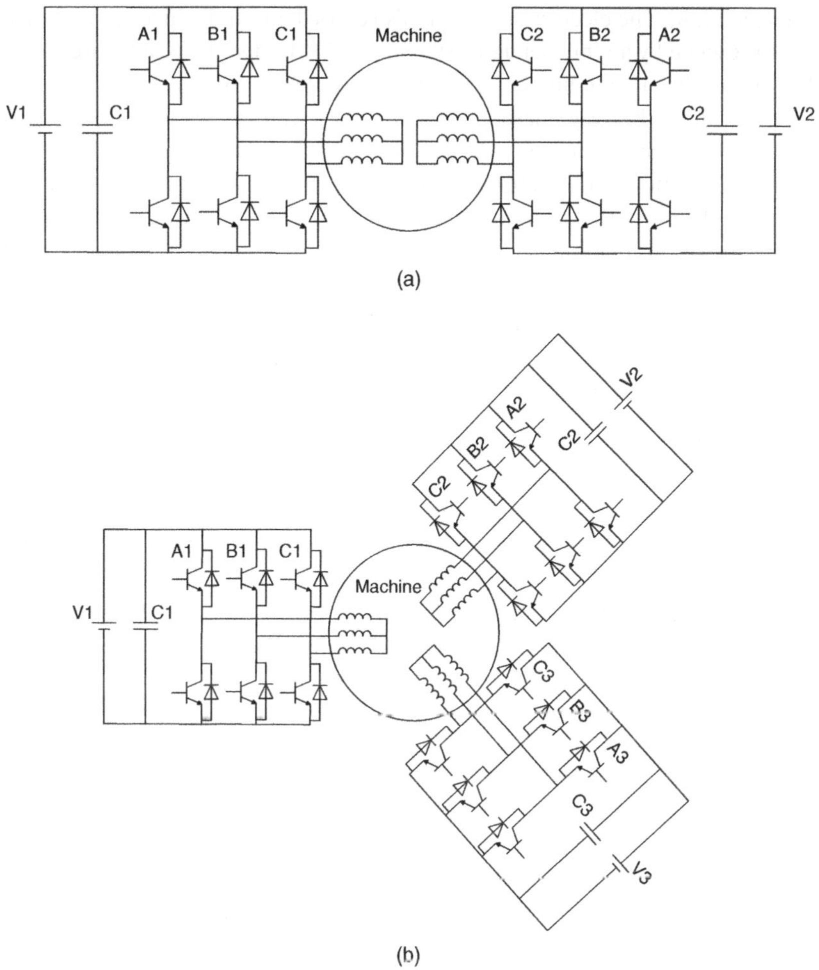

Figure 1.6 Multiple 3-phase fault-tolerant machine drive: (a) dual 3-phase and (b) triple 3-phase.

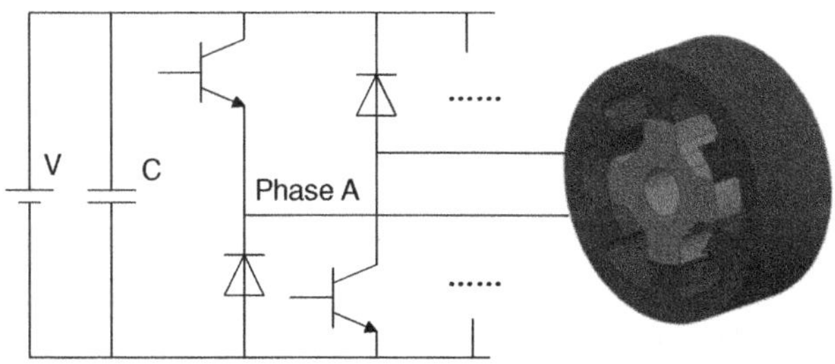

Figure 1.7 SRM machine drive.

safe region [38]. The electric isolation is also carried out in the inverter, and each phase is excited by an independent asymmetrical H-bridge. Thus, a failure in one phase either in the winding or in the bridge would not influence the operation of the remaining healthy phases. The faulty phase can be deactivated simply by disabling the active switches. Additionally, the active switching devices in the bridge are in series with a winding, so in case of switch misfiring, there is a delay in the rise of current that prevents the shoot-through fault of the bridge. However, an SRM drive exhibits high torque ripple, undesirable noise and vibration, and inferior torque density/efficiency. These undesirable attributes are even worse in fault condition. Furthermore, the SRM drive may lose its starting capability due to a phase absence.

To enhance the torque performance of a fault-tolerant drive, a single layer, fractional-slot concentrated winding (FSCW) PM machine is developed as shown in Fig. 1.8 [39], which facilitates electrical, magnetic, thermal, and physical isolations between phases. Each winding is wound on every alternative tooth, so each slot only contains one phase coil, facilitating thermal and physical isolation between phases. With the help of surface-mounted magnet rotor and the unwound stator teeth, the magnetic isolation between phases is also largely realized. The tooth tip and slot opening are specially designed by controlling their depth and width to obtain per-unit self-inductance in each phase in such a way that under a short-circuit fault, the fault current is limited. Further, the modular design philosophy is extended to the inverter with each phase excited by a single-phase H-bridge, as shown in Fig. 1.8; thus, each phase is also electrically isolated. The machine is fault tolerant to both winding open or short circuit and switch open or short circuit. In an event of an interturn short-circuit fault, the short-circuit current can be limited to an acceptable level by applying a terminal short circuit (TSC) on the fault phase via H-bridge control. This methodology has also been extended to switched flux, also known as flux switching, PM machines in [40].

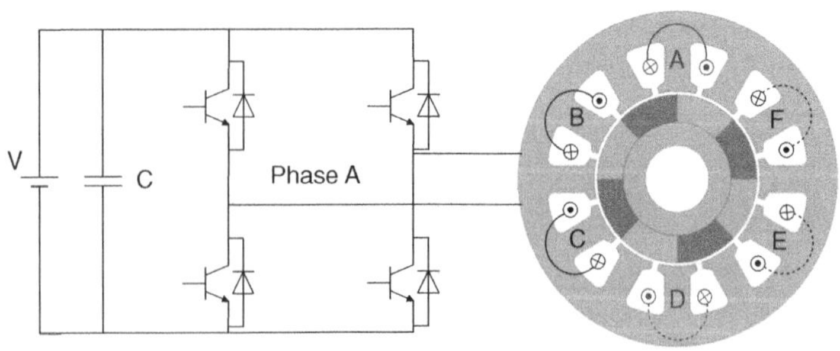

Figure 1.8 Multiphase FSCW fault-tolerant machine drive.

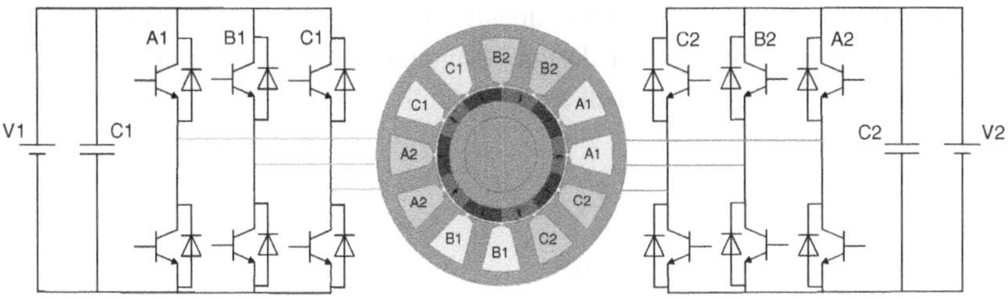

Figure 1.9 Multiple 3-phase FSCW fault-tolerant machine drive.

Similarly, multiple 3-phase configuration can be employed for the FSCW fault-tolerant machine as shown in Fig. 1.9, which facilitates the fast integration of standard 3-phase inverters and simple control [41]. Each 3-phase set winding is driven by an independent 3-phase inverter and can tolerate open-circuit, short circuit, and interturn fault.

However, due to the concentrated windings, very little reluctance torque can be exploited. The output torque of an FSCW PM machine purely relies on the PM field and, consequently, an increase in torque capability leads to higher flux linkage. The presence of a strong PM field poses a safety hazard to the machine as it cannot be turned off in the event of a fault [42]. The back emf may be much higher than the DC-link voltage at high speed and, consequently, uncontrolled rectification via the diodes of a power converter may occur in case of an inverter failure. Excessive regenerative power that flows to the DC-link capacitor may cause catastrophic failure to the drive system. Significant braking torque may also be imposed on the rotor causing excessive stress to the mechanical system or a sudden reduction in speed that may induce an accident. Thus, the maximum back emf should be limited [43]. However, this is in conflict with the requirements for torque production. Further, the use of concentric windings also gives rise to sub- and high-order MMF space harmonics, which produce extra rotor loss. Hence, the magnets need to be segmented axially or circumferentially to avoid excessive rotor temperature [44]. The magnets are also subject to the risk of demagnetization in high temperature and in case of a drive failure [45]. The demagnetization leads to torque reduction and increase in machine currents for closed-loop controlled drive under the same load condition. The vicious cycle may continue, causing further increases in temperature and demagnetization.

According to the foregoing review, it can be concluded that the conventional 3-phase induction machine and PMSM drives are only fault tolerant to open-circuit failure while SRM and FSCW PM machine could tolerate open-circuit, short-circuit, and interturn fault. Multiphase machine drive normally has better postfault torque capability compared to multiple 3-phase counterpart albeit requires more complex postfault control. PM machine has the merits of high efficiency and high torque density; however, the high back emf also poses challenges to the system design. In general, no existing topology could achieve high performance as well as good fault tolerance in a simple and cost-effective manner.

1.4.2 Fault Modeling Techniques

For fault-tolerant machine drive systems, the postfault behavior is notably changed and should be carefully assessed to understand the fault behavior. For example, maximum torque capability and torque ripple in postfault operation should be assessed [46]. New voltage harmonics would appear that pose an extra constraint on voltage if field weakening operation is required [17]. In addition, in order to optimize the output capability in fault condition, accurate fault modeling is necessary to guide machine design and to facilitate postfault control in order to achieve high efficiency and high-power density [7] in both healthy and postfault operations. Especially for interturn fault, which causes excessive fault current and results in local hotspots, its behavior should be captured to facilitate the development of diagnostic techniques and corresponding mitigation action [47]. Thus, various fault modeling techniques are being actively investigated.

Lumped parameter-based modeling is commonly employed to describe the machine in both healthy and fault conditions. Lumped parameters, inductance values in particular may be extracted from analytical derivations [48], finite element (FE) simulations [49], or experimental tests [50]. For example, based on closed-form voltage equations established from this process, the responses of an interior permanent magnet (IPM) machine under single-phase open circuit [51] and symmetrical and asymmetrical short-circuit faults [52] were investigated. An induction machine with one-phase open circuit was modeled and its control behavior was accessed in [53]. The fault modeling technique has been extended to multiphase induction machines by using the inductance matrices [54]. With derived lumped parameters, circuit-oriented method can also be employed to investigate induction machine behavior [55] with faults in both stator and rotor sides, such as broken rotor bar. A five-phase synchronous reluctance machine with open-circuit faults was modeled in the synchronous frame in [56] when the current was controlled in the synchronous frame with zero steady-state error. A novel field reconstruction method was used to analyze the machine performance and vibration [57] for an FSCW PM machine. This method utilizes the fields produced by the current in a single slot and the field from the magnets to reconstruct the field distribution along the air gap with arbitrary current excitation.

It should be noted that most of the modeling methods above consider the machine as a linear system, and hence compromise the model accuracy. The postfault behavior can also be evaluated by FE simulations [58]. However, the FE method is quite time consuming, and not suitable for developing fault detection strategies, control methods and performing comprehensive performance assessment over a wide operation region.

Accurate interturn fault modeling is very important to improve the understanding of fault behavior, which is essential for the development of fault detection algorithms [59]. A survey of stator turn fault modeling techniques was presented in [60]. In [61], a transient model for an induction machine with interturn fault was derived using reference frame transformation theory. In [62], magnetic saturation was considered by using modulation of air gap permeance at twice the fundamental frequency. Stator turn fault in surface-mounted PM machines was discussed in [49, 63–66]. The inductance values were derived by turn ratio (the ratio of partially short-circuited turns to the total number of turns in a coil), permeance network, and FE simulation. In addition, back emf harmonics were considered in [50] to improve the model accuracy. An analytical approach was proposed in [48] to capture the inductance and PM flux. Particularly, variation of slot leakage inductances under interturn fault, which has a significant impact on the turn fault current, was also investigated. It is shown that the leakage inductance is strongly dependent on the position of the fault turns in a slot, which will finally affect thc fault current.

An important fact in a turn fault condition is that the winding distributions are no longer symmetrical. To represent the winding distribution, and the resultant MMF in such fault conditions, the winding function proves to be a powerful tool. The inductances for the fault turns, fault phases, and other healthy phases can be obtained by integrating the product of the representative turn function, winding function, and airgap permeance as described in [67, 68], assuming linear magnetic characteristics. The theory has been applied in a salient pole synchronous machine with coils in series and in parallel connections [69, 70]. However, for an IPM machine, the magnetic saturation should be considered since it has a crucial influence on the field distribution. In [59], a semi-analytical model of IPM with turn fault was derived by using the healthy *dq* flux maps to deduce the flux in a fault condition. The magnetic nonlinearity was considered in the healthy flux map. Further, a high-fidelity model was developed in [71] based on the inverse of a four-dimensional (4D) table extracted from FE simulation that exhibits high accuracy. However, it can only deal with the turn fault with fixed turns and fixed position (coil location and slot position). The investigations in [47, 72, 73] show that both the slot location and coil location of a turn faul affect the fault current. This location dependency causes a significant difference in the fault current and consequently affects fault detection and fault-tolerant machine design.

It follows that although various modeling tools are available for fault behavior prediction, the accuracy, flexibility, and computational efficiency are still limited. Particularly for interturn fault, the impact of fault turn location on fault behavior is worth investigation in order to assess the worst-case scenario in terms of fault current and detectability.

1.4.3 Fault Detection Techniques

Fault detection is of key importance for fail-safe and fault-tolerant machine drives in order to take appropriate fault mitigation actions. Timely and reliable detection is essential to avoid false alarms and associated unnecessary downtime. All possible failures in a machine drive, namely, winding or device open-circuit, short-circuit, stator turn fault, etc., should be well detected, classified, and located. Among them, detection of device open-circuit in power converters and interturn short-circuit fault in machine windings is most actively investigated due to their high failure rate. Thus, they are reviewed and will be addressed subsequently in this chapter.

Open-Circuit Fault Detection

Open-circuit faults in motor drives often arise from overloading, overheating, and aging [74–76]. In general, they disrupt current flow in one phase, resulting in a rapid reduction in output torque [77, 78]. Furthermore, postfault actions may lead to increased voltage or current stress on other power switches, thereby posing a substantial safety risk to the motor drives and potentially triggering secondary faults [79]. Thus, it is imperative to explore effective methods for open-circuit fault detection and take appropriate mitigation measures.

Open-circuit faults in machine drive systems can be broadly categorized into two primary types, namely, inverter open-circuit fault and winding open-circuit fault [80], as illustrated in Fig. 1.10(a) and (b), respectively. The detection of winding open-circuit faults is relatively straightforward through current sensors typically used in a drive system. Hence, existing research mainly focuses on the detection of open-circuit fault of a switching device in an inverter (typically, the active switch is open circuited while the diode is normal). Currently, there are two predominant approaches for detecting open-circuit fault, namely, signal-based detection methods and model-based detection methods [81].

Signal-based detection methods typically rely on changes in current or voltage signals following a fault occurrence to identify open-circuit fault. The most apparent characteristics of an open-circuit fault is that the phase current associated with the fault device during the response interval becomes zero. Due to this feature, literature [82] utilizes variations in the current vector slope to achieve open-circuit fault detection and localization. However, current signals are susceptible to disruptions from load variations. To mitigate this challenge,

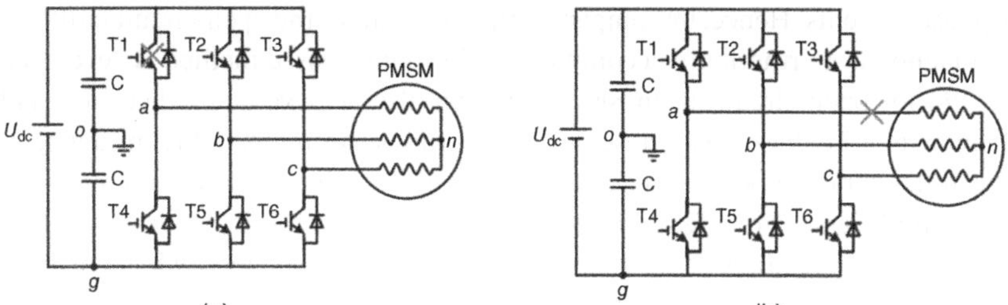

Figure 1.10 Open-circuit fault illustration: (a) switch open-circuit fault and (b) winding open-circuit fault.

Peuget *et al.* [83] introduces the normalization of Park vectors to reduce the impact of load fluctuations in the fault detection. Similarly, literature [84] employs the average of normalized 3-phase currents to detect open-circuit faults. Since the asymmetry of phase currents after the fault can introduce harmonics components, literature [85] employs the harmonic components of phase currents for open-circuit fault detection. Additionally, depending on the installation positions of current sensors, DC-link current [86] is also utilized for open-circuit fault detection. The acquisition of current signals is relatively simple; however, it is prone to load disturbances and long detection cycle.

By contrast, voltage signals normally offer faster detection. Literatures [87–89] analyze open-circuit detection by examining changes in phase voltages, line voltages, neutral point voltages, and the voltage between power device collector and emitter. Reference [90] achieves μs-level detection of open-circuit faults by monitoring the midpoint voltage of 3-phase bridge arm with respect to the ground. Literature [91] uses zero-sequence voltage for single-phase open-circuit detection and discrimination between inverter and winding. However, measuring voltage signals usually requires additional voltage sensors and high sampling rate hardware circuits.

Model-based methods detect the fault by analyzing the difference between model-predicted signals and actual signals [92]. In literature [93–95], the reference current of the motor is calculated using the reference voltage output from the *dq*-axis controller, while the actual current is obtained through physical sensors. During normal operation, the reference current agrees well with the actual signal. However, when an open-circuit fault occurs in a motor drive, a small interval with zero current will result due to the fault. Therefore, a significant difference arises between the reference and actual current, which can be utilized for open-circuit fault detection. Similarly, an approach based on voltage residuals is employed in [96] and [97]. The reference voltage is obtained from the *dq*-axis controller while the actual voltage is calculated from a model with measured

3-phase currents. Hence, by comparing the two signals, the open-circuit fault can be diagnosed. Furthermore, a common mode voltage-based method is developed in [98] for open-circuit diagnosis, which can also achieve quick detection and localization. Model-based fault detection methods can mitigate the influence of controller parameters and transient processes on fault detection, as both reference and actual signals are affected by these factors. However, it is worth noting that this approach requires more knowledge of motor parameters, which may be subjected to variations due to changes in operation conditions.

Interturn Short-Circuit Fault Detection

Interturn short-circuit fault, denoted as turn fault for short notation, as a result of turn-to-turn insulation failure, is known as severer fault while it is more challenging to detect. It usually only involves a few turns in a machine winding, resulting in a benign fault signature as seen from machine terminals and consequently low signal-to-noise ratio for detection. However, excessive fault current flows in the fault turns, which requires fast detection and mitigation action. This necessitates a rapid and computationally efficient detection within a few tens of milliseconds before developing into a catastrophic failure.

In terms of turn fault monitoring, a few standard offline methods including the insulation resistance measurement, direct current test, alternating current test, surge comparison test, and partial discharge test have been discussed in [99, 100]. Online detecting schemes would be more preferred since they provide continuous stator insulation monitoring and enable appropriate mitigation actions. Benefiting from the features of noninvasive and no additional sensors, most of the online turn fault detection techniques are based on motor current signal analysis (MCSA) by inspecting the stator current spectrum to spot the degradation [101]. In [99, 102], the induction machine insulation failure was detected by tracking the negative sequence current and impedance, which was insensitive to the machine slip. In fact, the methods in [103, 104] based on Park's Vector Approach were also employed to monitor the negative sequence current. However, it was shown in [61] that the negative sequence component was not a reliable fault indicator since it does not change in a predictable manner when exposed to the existing asymmetry or supply unbalance. Hence, a more robust fault indicator was developed by making use of the off-diagonal term of the sequence component in the impedance matrix [105, 106]. The disadvantage of this method was that it is required to store two sets of test data under different voltage unbalance conditions. An alternative way was introduced in [107] by exploiting feedforward neural network to compensate for the arbitrary supply voltage and nonidealities in the machine or instrumentation.

Alternatively, model-based approaches were also investigated in [108–111] for interturn fault by monitoring the estimation errors. It mainly follows the same approach as the open-circuit fault detection by comparing the measured signals and predicted values. This method required an accurate online model, which was usually achieved by adaptive state observer or online learning. In [112], the interturn short-circuit fault was alarmed by tracking the back emf estimator for a PMSM. The temperature influence on the winding resistance was compensated by a transient thermal model; however, the influence on the PM flux variation was not considered, which may compromise detection accuracy.

The feasibility of turn fault detection by injecting high-frequency signal was analyzed in [113, 114] by measuring the high-frequency negative sequence current and further investigated in [115] by monitoring the incremental inductances. It should be noticed that high-frequency injection introduces additional noise in the current and may reduce voltage operating range at high speed. As a result, an advanced method was proposed in [116] by comparing the inherent high-frequency PWM current ripple between different phases, where the switching inverter was exploited as a natural source of high-frequency signal injection into the motor. Therefore, there is no need to inject additional high-frequency component, avoiding the undesirable effects.

It should be noted that for the fault detection techniques reviewed above, the respective fault indicators vary with speed and load and this variation should be compensated in order to improve sensitivity and robustness of the detection. On the other hand, the speed and load transients usually cause false alarm in harmonic-based fault detection techniques. It is important that such false alarm or false positive is eliminated. Last but not least, the classification and location identification of different faults are of paramount importance in order to apply appropriate mitigation action.

1.4.4 Postfault Control Strategies

After a fault, the internal physical behavior of a machine drive is significantly changed. Hence, the original control law is no longer applicable. Further, once a fault has been detected and classified, some forms of reconfiguration in the power converter and its connection to the motor may be required in order to limit fault propagation and facilitate postfault operation. Therefore, it becomes essential to adopt fault-tolerant control strategy to achieve reliable operation. According to different objectives and fault-tolerant schemes employed, the existing fault-tolerant control strategies can be mainly divided into three categories, namely, output torque control, postfault current tracking, and short-circuit current suppression.

At present, output torque control is most extensively investigated. The basic approach is to control the healthy phase currents to maintain a constant rotating MMF after an open-circuit fault. Fu and Lipo [26] proposed a general current calculation method for multiphase machines under open-circuit fault condition [26] and presented a reduced-order fault-tolerant control with decoupling in [117], which maintains the MMF on the fundamental plane by compensating the asymmetry caused by the fault. It is realized by controlling the special harmonic components in the stator currents. However, the reduced order decoupling matrix introduces asymmetric synchronous rotation voltage, which requires feedforward compensation. An optimal torque control is proposed in [28] for five-phase PM machines. The objective function is constructed for torque demand as a constraint with the minimum power consumption, and the optimal currents for postfault operation are obtained by the Lagrange algorithm. However, the current commands contain high-order harmonic components, which pose difficulties for current tracking at high speed. Parsa and Toliyat [33] introduced a fault-tolerant operation strategy based on instantaneous power theory for five-phase PM machine with third harmonic in back emf. Wang *et al.* [118] studied a 15-phase induction motor with nonsinusoidal power supply for integrated power system (IPS) in ship propulsion. The stator windings are composed of three sets of five-phase windings. In such designs, the utilization of the stator MMF can be enhanced by injecting the third harmonic and hence the torque density is increased. The fault-tolerant control strategy is studied in [119] under symmetric and asymmetric open-circuited phase operation. Direct torque control of multiphase motor under fault condition is also investigated in [120].

It should be pointed out that during postfault operation, voltage harmonics will appear due to the presence of fault and asymmetric operation. The voltage harmonics will deteriorate current tracking performance and lead to current distortion. At high-speed operation, voltage harmonics will further reduce the available voltage margin and affect the stability of the system operation. Therefore, it is necessary to study the control algorithm of current tracking in postfault operation. The current commands generated from a fault-tolerant control algorithm can be tracked by the well-known hysteresis control for its fast response speed and strong robustness. However, this method has the disadvantage of variable switching frequency. The work reported in [121] proposed a fault-tolerant control technique for a seven-phase motor under open-circuit fault, and the current tracking errors were reduced by multiple PI controllers operating in the synchronous rotating coordinate system of different harmonic spaces, as shown in Fig. 1.11. However, due to the presence of the voltage harmonics and limited bandwidth of the PI controller, the tracking performance for the AC current commands may not be satisfactory. The authors in [24] studied model predictive control for the postfault operation of a multiphase induction motor, and the switching state is selected for minimum current error based on a model predictive algorithm. However, the model predictive control heavily relies on accurate mathematical model and requires intensive online computation for the optimal solution, which increases the cost and complexity of the digital processing

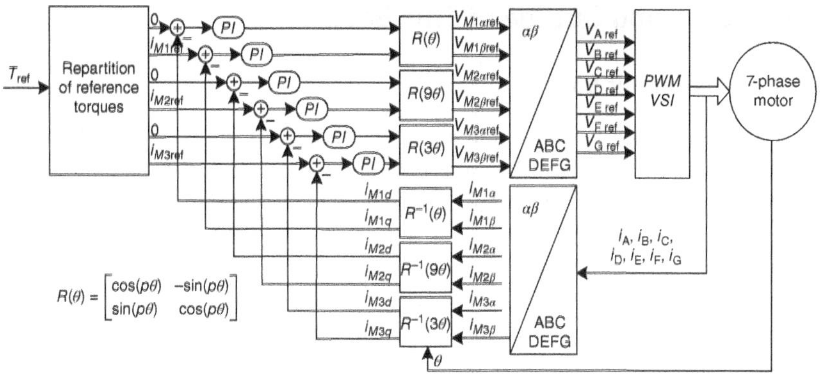

Figure 1.11 Fault-tolerant control diagram of seven-phase motor. Source: Tao *et al.* [122]/Proceedings of the Chinese Society of Electrical Engineering.

units. In addition, field weakening control algorithms are rarely investigated for postfault operation of machine drives at high speed.

In case of a short-circuit fault, especially interturn short-circuit fault, it is essential to adopt timely control to suppress the excessive fault current and avoid further failures. TSC realized via the inverter is usually applied to the whole phase winding where the fault turns are located to reduce the fault current. However, its effectiveness in limiting the fault current is dependent on the location of the short-circuit turns in the winding. Therefore, literature [123] studies a short-circuit current suppression method based on current injection into the healthy part of the winding, which cancels the flux linkage of the faulty turns and consequently reduce the short-circuit current. However, this method requires accurate information on the fault condition, such as the number of short-circuited turns and their physical position in the winding, which is practically impossible to determine. For the 3-phase machine with open winding drive as shown in Fig. 1.4, the flux linkage of the short-circuit turns can be actively reduced by controlling the currents in the healthy phases as described in [124]. As a result, the fault current is effectively reduced. Jiang *et al.* [125] proposed a technique for interturn short-circuit current suppression for PM starter/generator in motoring and generating modes, respectively. With this scheme, however, the torque capability of the drive is significantly reduced.

In summary, although technical issues pertinent to design and operation of fault-tolerant drive systems, such as fault-tolerant machine drive topologies, fault modeling, fault detection, fault mitigation, and postfault operation, have been extensively investigated in literature, some key problems are still remaining and worth further study. Further, the techniques for addressing these issues should be considered at the system level and well integrated to establish a complete machine drive system capable of fault-tolerant operation.

1.5 Scope and Outline of This Book

1.5.1 Scope of this Book

This book is devoted to presenting a systematic and integrated approach to the design and control of fault-tolerant machine drives, including fault-tolerant machine topologies, design optimization, fault modeling, fault detection, fault mitigation, and postfault control algorithm. For each of these topics, the recent advances and current state-of-the-art will be reviewed and the needs for further research in order to develop the technology to sufficient level of maturity will be discussed. The development of advanced fault-tolerant machine drive topologies is a key step to provide a desired level of fault tolerance without significant compromise on the machine drive performance in healthy conditions. Appropriate machine design optimization should be taken to ensure that the machine drive can meet the requirement for performance targets in both healthy and postfault operating conditions. This will be facilitated by a generic and comprehensive fault modeling technique for predicting the machine fault behavior in a computationally efficient manner. The advanced fault modeling technique will also provide a useful tool for the development of fault detection techniques and postfault control strategies. Effective and reliable fault detection techniques are indispensable for monitoring the machine drive operations and for undertaking appropriate mitigation measures when a fault occurs. Suitable postfault control strategy will finally realize the fault-tolerant operation after a fault. Thus, all these technical issues should be addressed, and suitable solutions developed and implemented in order to realize an effective fault-tolerant machine drive system.

In this book, the process of design, control, and development of high reliability/availability fault-tolerant machine drives is described through a specific example by incorporating all the above technical aspects for safety critical applications. Generally, different fault-tolerant machine drive topologies may be preferred for diversified applications. Depending on the machine drive topology employed, the design optimization, modeling, detection, and postfault control techniques may vary accordingly. However, the general process and internal logic are the same for different types of fault-tolerant machine drives. This book offers a generic methodology and systematic approach to the design and control of a typical fault-tolerant machine drive. It also intends to stimulate further developments of the core technologies pertinent to fault-tolerant machine drives.

1.5.2 Outline of this Book

This book is arranged with seven chapters. Each chapter is briefly introduced as follows:

Chapter 1 introduces the background knowledge and recent advances of fault-tolerant machine drives. The frequent failure modes in electrical drives are explained, and the measures and requirements to accommodate these faults as well as to facilitate postfault operations are discussed. Current state-of-the-art techniques for machine topology, fault modeling, fault detection, fault mitigation, and postfault control are reviewed to provide a comprehensive understanding and insight into the fault-tolerant machine drive technology.

To establish an integrated fault-tolerant machine drive, a novel multiple 3-phase PM machine with segregated windings is introduced in **Chapter 2**. The key merits of the fault-tolerant drive topology are assessed against other key state-of-the-art candidates. In particular, its performance is compared against the conventional 3-phase drives in healthy operations, and its ability to tolerate various faults is evaluated and discussed. The machine fault behavior and mutual coupling mechanism are explained from the viewpoint of MMF distribution.

Based on the proposed machine topology in Chapter 2, the design optimization process of a 40 kW machine is described as a case study in **Chapter 3**. An integrated design procedure is developed to consider both the healthy and fault modes of operation while satisfying electrical, thermal, and mechanical constraints. The optimized machine is prototyped and extensively tested for different operation modes in healthy and under fault conditions.

A general modeling technique is developed in **Chapter 4** considering the healthy operation, open-circuit, short-circuit, and unbalanced operation conditions. The model can also be adapted for turn fault with varying numbers of faulted turns and fault location. The accuracy of the model is verified by FE prediction and experimental tests.

Fast and reliable detection techniques are developed in **Chapter 5** for the most common faults, including device open-circuit and winding interturn short-circuit fault. The open-circuit fault is detected by using the relative β-axis residual voltage signals of the three phases. The second harmonics in the instantaneous active power and reactive power are employed as fault indicators for turn fault detection. The effectiveness of the detection methods is evaluated with the proposed general modeling technique and on prototype tests.

Chapter 6 describes optimal torque control strategy for fault tolerant multiphase FSCW PM machines and develops a turn fault mitigation strategy to minimize the fault current for the multiple 3-phase PMA-SynRM drive. Optimal current control strategy is employed to deliver smooth output torque during post fault operation for both constant torque and constant power operating region. By making use of the additional degree of freedom from the multiple 3-phase windings, appropriate phase currents are injected to the fault 3-phase set to reduce the residual flux of the fault turns, leading to a much lower turn fault current and enhanced fault tolerance.

Winding configuration has a dominant impact on the machine performance and its ability to tolerate faults. To further suppress potentially excessive turn fault current, four new segregated winding configurations, namely, delta-connected winding, star-delta-connected winding, mixed-pitch winding, and concentric winding, are proposed in **Chapter 7** to reduce the residual flux linkage of the fault turns. With appropriate mitigation actions, the fault current is reduced by changing the winding connection or modifying the winding layout leading to more cost-effective solutions.

References

1 Y. Song and B. Wang, "Survey on reliability of power electronic systems," *IEEE Transactions on Power Electronics*, vol. 28, no. 1, pp. 591–604, 2013.

2 J. A. Tegopoulos, "Electrical machines for electromotion and their design," in *The 21st IEEE Convention of the Electrical and Electronic Engineers in Israel*. 2000, pp. 209–212.

3 W. Cao, B. C. Mecrow, G. J. Atkinson, J. W. Bennett, and D. J. Atkinson, "Overview of electric motor technologies used for more electric aircraft (MEA)," *IEEE Transactions on Industrial Electronics*, vol. 59, no. 9, pp. 3523–3531, 2012.

4 K. T. Chau, C. C. Chan, and C. Liu, "Overview of permanent-magnet brushless drives for electric and hybrid electric vehicles," *IEEE Transactions on Industrial Electronics*, vol. 55, no. 6, pp. 2246–2257, 2008.

5 J. Wang, Z. P. Xia, and D. Howe, "Three-phase modular permanent magnet brushless Machine for torque boosting on a downsized ICE vehicle," *IEEE Transactions on Vehicular Technology*, vol. 54, no. 3, pp. 809–816, 2005.

6 A. Mohammadpour, S. Mishra, and L. Parsa, "Fault-tolerant operation of multiphase permanent-magnet machines using iterative learning control," *IEEE Journal of Emerging and Selected Topics in Power Electronics*, vol. 2, no. 2, pp. 201–211, 2014.

7 N. Bianchi and S. Bolognani, "Design of a fault-tolerant IPM motor for electric power steering," in *IEEE 36th Power Electronics Specialists Conference, 2005. PESC '05,* 2005, p. 2873.

8 A. Boglietti, A. Cavagnino, A. Tenconi, and S. Vaschetto, "The safety critical electric machines and drives in the more electric aircraft: A survey," in *35th Annual Conference of IEEE Industrial Electronics, 2009. IECON '09,* 2009, pp. 2587–2594.

9 D. A. Rennels, "Fault tolerant computing concepts and examples," *IEEE Transactions on Computers*, vol. C-33, no. 12, pp. 1116–1129, 1984.

10 R. V. White, "Fault tolerance in distributed power systems," in *IV IEEE International Power Electronics Congress, 1995. Technical Proceedings. CIEP 95*, 1995, pp. 121–128.

11 MOTOR RELIABILITY WORKING GROUP. "Report of large motor reliability survey of industrial and commercial installations, part I," *IEEE Transactions on Industry Applications*, vol. IA-21, no. 4, 1985, pp. 853–864.

12 O. V. Thorsen and M. Dalva, "A survey of faults on induction motors in offshore oil industry, petrochemical industry, gas terminals and oil refineries," in *Petroleum and Chemical Industry Conference, 1994. Record of Conference Papers., Institute of Electrical and Electronics Engineers Incorporated Industry Applications Society 41st Annual*, 1994, pp. 1–9.

13 B. C. Mecrow, D. J. Atkinson, A. G. Jack, S. Green, J. A. Haylock, and J. Coles, "The need for fault tolerance in an aeroengine electric fuel control system," in *IEE Colloquium on Electrical Machines and Systems for the More Electric Aircraft (Ref. No. 1999/180)*, 1999, pp. 9/1–9/5.

14 A. Sarikhani and O. A. Mohammed, "Demagnetization control for reliable flux weakening control in PM synchronous machine," *IEEE Transactions on Energy Conversion*, vol. 27, no. 4, pp. 1046–1055, 2012.

15 S. Yang, A. Bryant, P. Mawby, D. Xiang, L. Ran, and P. Tavner, "An industry-based survey of reliability in power electronic converters," *IEEE Transactions on Industry Applications*, vol. 47, no. 3, pp. 1441–1451, 2011.

16 P. Sun, C. Gong, X. Du, Q. Luo, H. Wang, and L. Zhou, "Online condition monitoring for both IGBT module and DC-link capacitor of power converter based on short-circuit current simultaneously," *IEEE Transactions on Industrial Electronics*, vol. 64, no. 5, pp. 3662–3671, 2017.

17 N. Bianchi, S. Bolognani, and M. Zigliotto, "Analysis of PM synchronous motor drive failures during flux weakening operation," in *27th Annual IEEE Power Electronics Specialists Conference, 1996. PESC '96 Record.*, 1996, vol. 2, pp. 1542–1548.

18 A. M. E. Refaie, M. R. Shah, and K. K. Huh, "High-power-density fault-tolerant PM generator for safety-critical applications," *IEEE Transactions on Industry Applications*, vol. 50, no. 3, pp. 1717–1728, 2014.

19 A. G. Jack, B. C. Mecrow, and J. A. Haylock, "A comparative study of permanent magnet and switched reluctance motors for high-performance fault-tolerant applications," in *IEEE Transactions on Industry Applications*, vol. 32, no. 4, pp. 889–895, July–Aug. 1996.

20 B. Vaseghi, N. Takorabet, J. P. Caron, B. Nahid-Mobarakeh, F. Meibody-Tabar, and G. Humbert, "Study of different architectures of fault-tolerant actuator using a two-channel PM motor," *IEEE Transactions on Industry Applications*, vol. 47, no. 1, pp. 47–54, 2011.

21 M. B. de Rossiter Correa, C. B. Jacobina, E. R. C. da Silva, and A. M. N. Lima, "An induction motor drive system with improved fault tolerance," *IEEE Transactions on Industry Applications*, vol. 37, no. 3, pp. 873–879, 2001.

22 B. Wang *et al.*, "Overall assessments of dual inverter open winding drives," in *2015 IEEE International Electric Machines & Drives Conference (IEMDC)*, 2015, pp. 1029–1035.

23 N. Bianchi, S. Bolognani, M. D. Pre, and E. Fornasiero, "Post-fault operations of five-phase motor using a full-bridge inverter," in *Power Electronics Specialists Conference, 2008. PESC 2008. IEEE*, 2008, pp. 2528–2534.

24 M. Bermudez, I. Gonzalez-Prieto, F. Barrero, H. Guzman, M. J. Duran, and X. Kestelyn, "Open-phase fault-tolerant direct torque control technique for five-phase induction motor drives," *IEEE Transactions on Industrial Electronics*, vol. 64, no. 2, pp. 902–911, 2017.

25 E. Levi, R. Bojoi, F. Profumo, H. A. Toliyat, and S. Williamson, "Multiphase induction motor drives - a technology status review," *Electric Power Applications, IET*, vol. 1, no. 4, pp. 489–516, 2007.

26 J.-R. Fu and T. A. Lipo, "Disturbance-free operation of a multiphase current-regulated motor drive with an opened phase," *IEEE Transactions on Industry Applications*, vol. 30, no. 5, pp. 1267–1274, 1994.

27 A. Tani, M. Mengoni, L. Zarri, G. Serra, and D. Casadei, "Control of multiphase induction motors with an odd number of phases under open-circuit phase faults," *IEEE Transactions on Power Electronics*, vol. 27, no. 2, pp. 565–577, 2012.

28 J. Wang, K. Atallah, and D. Howe, "Optimal torque control of fault-tolerant permanent magnet brushless machines," *IEEE Transactions on Magnetics*, vol. 39, no. 5, pp. 2962–2964, 2003.

29 S. Dwari, L. Parsa, and T. A. Lipo, "Optimum control of a five-phase integrated modular permanent magnet motor under normal and open-circuit fault conditions," in *Power Electronics Specialists Conference, 2007. PESC 2007. IEEE*, 2007, pp. 1639–1644.

30 N. Bianchi, S. Bolognani, and M. D. Pre, "Strategies for the fault-tolerant current control of a five-phase permanent-magnet motor," *IEEE Transactions on Industry Applications*, vol. 43, no. 4, pp. 960–970, 2007.

31 H. Zhou, W. Zhao, G. Liu, R. Cheng, and Y. Xie, "Remedial field-oriented control of five-phase fault-tolerant permanent-magnet motor by using reduced-order transformation matrices," *IEEE Transactions on Industrial Electronics*, vol. 64, no. 1, pp. 169–178, 2017.

32 L. Alberti and N. Bianchi, "Experimental tests of dual three-phase induction motor under faulty operating condition," *IEEE Transactions on Industrial Electronics*, vol. 59, no. 5, pp. 2041–2048, 2012.

33 L. Parsa and H. A. Toliyat, "Fault-tolerant interior-permanent-magnet machines for hybrid electric vehicle applications," *IEEE Transactions on Vehicular Technology*, vol. 56, no. 4, pp. 1546–1552, 2007.

34 Y. Fan, W. Zhu, X. Zhang, M. Cheng, and K. T. Chau, "Research on a single phase-loss fault-tolerant control strategy for a new flux-modulated permanent-magnet compact in-wheel motor," *IEEE Transactions on Energy Conversion*, vol. PP, no. 99, pp. 1–9, 2016.

35 M. J. Duran, I. G. Prieto, M. Bermudez, F. Barrero, H. Guzman, and M. R. Arahal, "Optimal fault-tolerant control of six-phase induction motor drives with parallel converters," *IEEE Transactions on Industrial Electronics*, vol. 63, no. 1, pp. 629–640, 2016.

36 S. Gopalakrishnan, A. M. Omekanda, and B. Lequesne, "Classification and remediation of electrical faults in the switched reluctance drive," *IEEE Transactions on Industry Applications*, vol. 42, no. 2, pp. 479–486, 2006.

37 Y. Hu, C. Gan, W. Cao, W. Li, and S. J. Finney, "Central-tapped node linked modular fault-tolerance topology for SRM applications," *IEEE Transactions on Power Electronics*, vol. 31, no. 2, pp. 1541–1554, 2016.

38 B. Lequesne, S. Gopalakrishnan, and A. M. Omekanda, "Winding short circuits in the switched reluctance drive," *IEEE Transactions on Industry Applications*, vol. 41, no. 5, pp. 1178–1184, 2005.

39 B. C. Mecrow, A. G. Jack, J. A. Haylock, and J. Coles, "Fault-tolerant permanent magnet machine drives," *IEE Proceedings – Electric Power Applications*, vol. 143, no. 6, pp. 437–442, 1996.

40 J. T. Chen and Z. Q. Zhu, "Comparison of all- and alternate-poles-wound flux-switching PM machines having different stator and rotor pole numbers," *IEEE Transactions on Industry Applications*, vol. 46, no. 4, pp. 1406–1415, 2010.

41 J. Chai, J. Wang, K. Atallah, and D. Howe, "Performance comparison and winding fault detection of duplex 2-phase and 3-phase fault-tolerant permanent magnet brushless machines," in *Conference Record of the 2007 IEEE Industry Applications Conference, 2007. 42nd IAS Annual Meeting*, 2007, pp. 566–572.

42 J. Dusek, P. Arumugam, C. Brunson, E. K. Amankwah, T. Hamiti, and C. Gerada, "Impact of slot/pole combination on inter-turn short-circuit current in fault-tolerant permanent magnet machines," *IEEE Transactions on Magnetics*, vol. 52, no. 4, pp. 1–9, 2016.

43 S. H. Han, T. M. Jahns, M. Aydin, M. K. Guven, and W. L. Soong, "Impact of maximum back-EMF limits on the performance characteristics of interior permanent magnet synchronous machines," in *Conference Record of the 2006 IEEE Industry Applications Conference Forty-First IAS Annual Meeting*, 2006, vol. 4, pp. 1962–1969.

44 J. Wang, K. Atallah, R. Chin, W. M. Arshad, and H. Lendenmann, "Rotor eddy-current loss in permanent-magnet brushless AC machines," *IEEE Transactions on Magnetics*, vol. 46, no. 7, pp. 2701–2707, 2010.

45 V. I. Patel, J. Wang, and S. S. Nair, "Demagnetization assessment of fractional-slot and distributed wound 6-phase permanent magnet machines," *IEEE Transactions on Magnetics*, vol. 51, no. 6, pp. 1–11, 2015.

46 W. Wang, M. Cheng, B. Zhang, Y. Zhu, and S. Ding, "A fault-tolerant permanent-magnet traction module for subway applications," *IEEE Transactions on Power Electronics*, vol. 29, no. 4, pp. 1646–1658, 2014.

47 D. C. Patel and M. C. Chandorkar, "Modeling and analysis of stator interturn fault location effects on induction machines," *IEEE Transactions on Industrial Electronics*, vol. 61, no. 9, pp. 4552–4564, 2014.

48 B. Sen and J. Wang, "Analytical modelling of stator turn fault in surface mounted permanent magnet machines," in *Energy Conversion Congress and Exposition (ECCE), 2013 IEEE*, 2013, pp. 4445–4452.

49 O. A. Mohammed, Z. Liu, S. Liu, and N. Y. Abed, "Internal short circuit fault diagnosis for PM machines using FE-based phase variable model and wavelets analysis," *IEEE Transactions on Magnetics*, vol. 43, no. 4, pp. 1729–1732, 2007.

50 L. Romeral, J. C. Urresty, J. R. Riba Ruiz, and A. Garcia Espinosa, "Modeling of surface-mounted permanent magnet synchronous motors with stator winding interturn faults," *IEEE Transactions on Industrial Electronics*, vol. 58, no. 5, pp. 1576–1585, 2011.

51 B. A. Welchko, T. M. Jahns, and S. Hiti, "IPM synchronous machine drive response to a single-phase open circuit fault," *IEEE Transactions on Power Electronics*, vol. 17, no. 5, pp. 764–771, 2002.

52 B. A. Welchko, T. M. Jahns, W. L. Soong, and J. M. Nagashima, "IPM synchronous machine drive response to symmetrical and asymmetrical short circuit faults," *IEEE Transactions on Energy Conversion*, vol. 18, no. 2, pp. 291–298, 2003.

53 A. M. S. Mendes and A. J. M. Cardoso, "Fault-tolerant operating strategies applied to three-phase induction-motor drives," *IEEE Transactions on Industrial Electronics*, vol. 53, no. 6, pp. 1807–1817, 2006.

54 R. Kianinezhad, B. Nahid-Mobarakeh, L. Baghli, F. Betin, and G. A. Capolino, "Modeling and control of six-phase symmetrical induction machine under fault condition due to open phases," *IEEE Transactions on Industrial Electronics*, vol. 55, no. 5, pp. 1966–1977, 2008.

55 A. Pantea *et al.*, "Six-phase induction machine model for electrical fault simulation using the circuit-oriented method," *IEEE Transactions on Industrial Electronics*, vol. 63, no. 1, pp. 494–503, 2016.

56 R. Hyung-Min, K. Ji-Woong, and S. Seung-Ki, "Synchronous-frame current control of multiphase synchronous motor under asymmetric fault condition due to open phases," *IEEE Transactions on Industry Applications*, vol. 42, no. 4, pp. 1062–1070, 2006.

57 Z. Weidong, B. Fahimi, and S. Pekarek, "A field reconstruction method for optimal excitation of permanent magnet synchronous machines," *IEEE Transactions on Energy Conversion*, vol. 21, no. 2, pp. 305–313, 2006.

58 M. Dai, A. Keyhani, and T. Sebastian, "Fault analysis of a PM brushless DC Motor using finite element method," *IEEE Transactions on Energy Conversion*, vol. 20, no. 1, pp. 1–6, 2005.

59 B. Sen, J. Wang, and P. Lazari, "A detailed transient model of Interior Permanent Magnet motor accounting for saturation under stator turn fault," in *2013 IEEE Energy Conversion Congress and Exposition (ECCE)*, 2013, pp. 3548–3555.

60 A. Gandhi, T. Corrigan, and L. Parsa, "Recent advances in modeling and online detection of stator interturn faults in electrical motors," *IEEE Transactions on Industrial Electronics*, vol. 58, no. 5, pp. 1564–1575, 2011.

61 R. M. Tallam, T. G. Habetler, and R. G. Harley, "Transient model for induction machines with stator winding turn faults," *IEEE Transactions on Industry Applications*, vol. 38, no. 3, pp. 632–637, 2002.

62 S. Nandi, "A detailed model of induction machines with saturation extendable for fault analysis," *IEEE Transactions on Industry Applications*, vol. 40, no. 5, pp. 1302–1309, 2004.

63 I. Jeong, B. J. Hyon, and K. Nam, "Dynamic modeling and control for SPMSMs with internal turn short fault," *IEEE Transactions on Power Electronics*, vol. 28, no. 7, pp. 3495–3508, 2013.

64 B. Vaseghi, B. Nahid-mobarakh, N. Takorabet, and F. Meibody-Tabar, "Inductance identification and study of PM motor with winding turn short circuit fault," *IEEE Transactions on Magnetics*, vol. 47, no. 5, pp. 978–981, 2011.

65 N. Leboeuf, T. Boileau, B. Nahid-Mobarakeh, N. Takorabet, F. Meibody-Tabar, and G. Clerc, "Inductance calculations in permanent-magnet motors under fault conditions," *IEEE Transactions on Magnetics*, vol. 48, no. 10, pp. 2605–2616, 2012.

66 F. Wu, P. Zheng, and T. M. Jahns, "Analytical modeling of inter-turn short circuit for multiphase fault-tolerant PM machines with fractional-slot concentrated windings," in *Energy Conversion Congress and Exposition (ECCE), 2015 IEEE*, 2015, pp. 6970–6977.

67 K. Kyung-Tae, P. Jun-Kyu, H. Jin, and K. Byeong-Woo, "Comparison of the fault characteristics of IPM-type and SPM-type BLDC motors under inter-turn fault conditions using winding function theory," *IEEE Transactions on Industry Applications*, vol. 50, no. 2, pp. 986–994, 2014.

68 T. Kim, H. W. Lee, and S. Kwak, "The internal fault analysis of brushless DC motors based on the winding function theory," *IEEE Transactions on Magnetics*, vol. 45, no. 5, pp. 2090–2096, 2009.

69 X. Tu, L. A. Dessaint, M. E. Kahel, and A. O. Barry, "A new model of synchronous machine internal faults based on winding distribution," *IEEE Transactions on Industrial Electronics*, vol. 53, no. 6, pp. 1818–1828, 2006.

70 X. Tu, L. A. Dessaint, N. Fallati, and B. D. Kelper, "Modeling and real-time simulation of internal faults in synchronous generators with parallel-connected windings," *IEEE Transactions on Industrial Electronics*, vol. 54, no. 3, pp. 1400–1409, 2007.

71 B. Sen, J. Wang, and P. Lazari, "A high-fidelity computationally efficient transient model of interior permanent-magnet machine with stator turn fault," *IEEE Transactions on Industrial Electronics*, vol. 63, no. 2, pp. 773–783, 2016.

72 Z. Sun, J. Wang, D. Howe, and G. Jewell, "Analytical prediction of the short-circuit current in fault-tolerant permanent-magnet machines," *IEEE Transactions on Industrial Electronics*, vol. 55, no. 12, pp. 4210–4217, 2008.

73 P. Arumugam, "Design optimization on conductor placement in the slot of permanent magnet machines to restrict turn-turn short-circuit fault current," *IEEE Transactions on Magnetics*, vol. 52, no. 99, pp. 1–8, 2016.

74 M. Alavi, D. Wang, and M. Luo, "Short-circuit fault diagnosis for three-phase inverters based on voltage-space patterns," *IEEE Transactions on Industrial Electronics*, vol. 61, no. 10, pp. 5558–5569, 2014.

75 S. Yang, D. Xiang, A. Bryant, P. Mawby, L. Ran, and P. Tavner, "Condition monitoring for device reliability in power electronic converters: a review," *IEEE Transactions on Power Electronics*, vol. 25, no. 11, pp. 2734–2752, 2010.

76 B. Lu and S. K. Sharma, "A literature review of IGBT fault diagnostic and protection methods for power inverters," *IEEE Transactions on Industry Applications*, vol. 45, no. 5, pp. 1770–1777, 2009.

77 X. Zhou, J. Sun, P. Cui, Y. Lu, M. Lu, and Y. Yu, "A fast and robust open-switch fault diagnosis method for variable-speed PMSM system," *IEEE Transactions on Power Electronics*, vol. 36, no. 3, pp. 2598–2610, 2021.

78 H. Tang, W. Li, and Z. Wu, "Influence of inverter open circuit fault on multiple physical quantities in the PMSM," *IEEE Transactions on Power Electronics*, vol. 38, no. 1, pp. 901–916, 2023.

79 X. Wu *et al.*, "A fast and robust diagnostic method for multiple open-circuit faults of voltage-source inverters through line voltage magnitudes analysis," *IEEE Transactions on Power Electronics*, vol. 35, no. 5, pp. 5205–5220, 2020.

80 J. Hang, J. Zhang, M. Cheng, and S. Ding, "Detection and discrimination of open-phase fault in permanent magnet synchronous motor drive system," *IEEE Transactions on Power Electronics*, vol. 31, no. 7, pp. 4697–4709, 2016.

81 C. Song, X. Zhao, X. Sun, N. Diao, and J. Zhao, "Fast in-loop single-tube open-circuit fault diagnosis for voltage source inverter based on duality of current-time pairs," *IEEE Transactions on Industrial Electronics*, vol. 70, no. 9, pp. 9572–9581, 2023.

82 H. Mesai-Ahmed, I. Jlassi, A. J. M. Cardoso, and A. Bentaallah, "Multiple open-circuit faults diagnosis in six-phase induction motor drives using stator current analysis," *IEEE Transactions on Power Electronics*, vol. 37, no. 6, pp. 7275–7285, 2022.

83 R. Peuget, S. Courtine, and J. P. Rognon, "Fault detection and isolation on a PWM inverter by knowledge-based model," *IEEE Transactions on Industry Applications*, vol. 34, no. 6, pp. 1318–1326, 1998.

84 H. Yan, Y. Xu, F. Cai, H. Zhang, W. Zhao, and C. Gerada, "PWM-VSI fault diagnosis for a PMSM drive based on the fuzzy logic approach," *IEEE Transactions on Power Electronics*, vol. 34, no. 1, pp. 759–768, 2019.

85 J. O. Estima and A. J. M. Cardoso, "A new approach for real-time multiple open-circuit fault diagnosis in voltage-source inverters," *IEEE Transactions on Industry Applications*, vol. 47, no. 6, pp. 2487–2494, 2011.

86 J. H. Choi, S. Kim, D. S. Yoo, and K. H. Kim, "A diagnostic method of simultaneous open-switch faults in inverter-fed linear induction motor drive for reliability enhancement," *IEEE Transactions on Industrial Electronics*, vol. 62, no. 7, pp. 4065–4077, 2015.

87 H. Yan, Y. Xu, J. Zou, Y. Fang, and F. Cai, "A novel open-circuit fault diagnosis method for voltage source inverters with a single current sensor," *IEEE Transactions on Power Electronics*, vol. 33, no. 10, pp. 8775–8786, 2018.

88 R. L. d. A. Ribeiro, C. B. Jacobina, E. R. C. d. Silva, and A. M. N. Lima, "Fault detection of open-switch damage in voltage-fed PWM motor drive systems," *IEEE Transactions on Power Electronics*, vol. 18, no. 2, pp. 587–593, 2003.

89 Z. Li, H. Ma, Z. Bai, Y. Wang, and B. Wang, "Fast transistor open-circuit faults diagnosis in grid-tied three-phase VSIs based on average bridge arm pole-to-pole voltages and error-adaptive thresholds," *IEEE Transactions on Power Electronics*, vol. 33, no. 9, pp. 8040–8051, 2018.

90 M. A. Rodríguez-Blanco *et al.*, "A failure-detection strategy for IGBT based on gate-voltage behavior applied to a motor drive system," *IEEE Transactions on Industrial Electronics*, vol. 58, no. 5, pp. 1625–1633, 2011.

91 Q. T. An, L. Z. Sun, K. Zhao, and L. Sun, "Switching function model-based fast-diagnostic method of open-switch faults in inverters without sensors," *IEEE Transactions on Power Electronics*, vol. 26, no. 1, pp. 119–126, 2011.

92 M. Salehifar, R. S. Arashloo, M. Moreno-Eguilaz, V. Sala, and L. Romeral, "Observer-based open transistor fault diagnosis and fault tolerant control of five-phase permanent magnet motor drive for application in electric vehicles," *IET Power Electron*, vol. 8, no. 1, pp. 76–87, 2015.

93 Q. T. An, L. Sun, and L. Z. Sun, "Current residual vector-based open-switch fault diagnosis of inverters in PMSM drive systems," *IEEE Transactions on Power Electronics*, vol. 30, no. 5, pp. 2814–2827, 2015.

94 I. Jlassi, J. O. Estima, S. K. E. Khil, N. M. Bellaaj, and A. J. M. Cardoso, "A robust observer-based method for IGBTs and current sensors fault diagnosis in voltage-source inverters of PMSM drives," *IEEE Transactions on Industry Applications*, vol. 53, no. 3, pp. 2894–2905, 2017.

95 I. Jlassi, J. O. Estima, S. K. E. Khil, N. M. Bellaaj, and A. J. M. Cardoso, "Multiple open-circuit faults diagnosis in back-to-back converters of PMSG drives for wind turbine systems," *IEEE Transactions on Power Electronics*, vol. 30, no. 5, pp. 2689–2702, 2015.

96 S. M. Jung, J. S. Park, H. W. Kim, K. Y. Cho, and M. J. Youn, "An MRAS-based diagnosis of open-circuit fault in PWM voltage-source inverters for PM synchronous motor drive systems," *IEEE Transactions on Power Electronics*, vol. 28, no. 5, pp. 2514–2526, 2013.

97 N. M. A. Freire, J. O. Estima, and A. J. M. Cardoso, "A voltage-based approach without extra hardware for open-circuit fault diagnosis in closed-loop PWM AC regenerative drives," *IEEE Transactions on Industrial Electronics*, vol. 61, no. 9, pp. 4960–4970, 2014.

98 Y. Cheng, Y. Sun, X. Li, H. Dan, J. Lin, and M. Su, "Active common-mode voltage-based open-switch fault diagnosis of inverters in IM-drive systems," *IEEE Transactions on Industrial Electronics*, vol. 68, no. 1, pp. 103–115, 2021.

99 J. Sottile and J. L. Kohler, "An on-line method to detect incipient failure of turn insulation in random-wound motors," *IEEE Transactions on Energy Conversion*, vol. 8, no. 4, pp. 762–768, 1993.

100 R. M. Tallam *et al.*, "A survey of methods for detection of stator-related faults in induction machines," *IEEE Transactions on Industry Applications*, vol. 43, no. 4, pp. 920–933, 2007.

101 S. Nandi, H. A. Toliyat, and X. Li, "Condition monitoring and fault diagnosis of electrical motors-a review," *IEEE Transactions on Energy Conversion*, vol. 20, no. 4, pp. 719–729, 2005.

102 J. Sottile, F. C. Trutt, and J. L. Kohler, "Experimental investigation of on-line methods for incipient fault detection" in *Conference Record of the 2000 IEEE Industry Applications Conference*, 2000, vol. 4, pp. 2682–2687.

103 S. M. A. Cruz and A. J. M. Cardoso, "Stator winding fault diagnosis in three-phase synchronous and asynchronous motors, by the extended Park's vector approach," *IEEE Transactions on Industry Applications*, vol. 37, no. 5, pp. 1227–1233, 2001.

104 A. J. M. Cardoso, S. M. A. Cruz, and D. S. B. Fonseca, "Inter-turn stator winding fault diagnosis in three-phase induction motors, by Park's vector approach," *IEEE Transactions on Energy Conversion*, vol. 14, no. 3, pp. 595–598, 1999.

105 L. Sang Bin, R. M. Tallam, and T. G. Habetler, "A robust, on-line turn-fault detection technique for induction machines based on monitoring the sequence component impedance matrix," *IEEE Transactions on Power Electronics*, vol. 18, no. 3, pp. 865–872, 2003.

106 S. Cheng, P. Zhang, and T. G. Habetler, "An impedance identification approach to sensitive detection and location of stator turn-to-turn faults in a closed-loop multiple-motor drive," *IEEE Transactions on Industrial Electronics*, vol. 58, no. 5, pp. 1545–1554, 2011.

107 R. M. Tallam, T. G. Habetler, and R. G. Harley, "Stator winding turn-fault detection for closed-loop induction motor drives," *IEEE Transactions on Industry Applications*, vol. 39, no. 3, pp. 720–724, 2003.

108 C. H. D. Angelo, G. R. Bossio, S. J. Giaccone, M. I. Valla, J. A. Solsona, and G. O. Garcia, "Online model-based stator-fault detection and identification in induction motors," *IEEE Transactions on Industrial Electronics*, vol. 56, no. 11, pp. 4671–4680, 2009.

109 M. Yi Lu, M. A. Masrur, C. ZhiHang, and Z. Baifang, "Model-based fault diagnosis in electric drives using machine learning," *IEEE/ASME Transactions on Mechatronics*, vol. 11, no. 3, pp. 290–303, 2006.

110 B. Sen and J. Wang, "A fast detection technique for stator inter-turn fault in multi-phase permanent magnet machines using model based approach," in *7th IET International Conference on Power Electronics, Machines and Drives (PEMD 2014)*, 2014, pp. 1–6.

111 S. Nadarajan, S. K. Panda, B. Bhangu, and A. K. Gupta, "Online model-based condition monitoring for brushless wound-field synchronous generator to detect and diagnose stator windings turn-to-turn shorts using extended Kalman filter," *IEEE Transactions on Industrial Electronics*, vol. 63, no. 5, pp. 3228–3241, 2016.

112 A. Sarikhani and O. A. Mohammed, "Inter-turn fault detection in PM synchronous machines by physics-based back electromotive force estimation," *IEEE Transactions on Industrial Electronics*, vol. 60, no. 8, pp. 3472–3484, 2013.

113 F. Briz, M. W. Degner, A. Zamarron, and J. M. Guerrero, "Online stator winding fault diagnosis in inverter-fed AC machines using high-frequency signal injection," *IEEE Transactions on Industry Applications*, vol. 39, no. 4, pp. 1109–1117, 2003.

114 F. Briz, M. W. Degner, P. Garcia, and A. B. Diez, "High-frequency carrier-signal voltage selection for stator winding fault diagnosis in inverter-fed AC machines," *IEEE Transactions on Industrial Electronics*, vol. 55, no. 12, pp. 4181–4190, 2008.

115 J. Arellano-Padilla, M. Sumner, and C. Gerada, "Winding condition monitoring scheme for a permanent magnet machine using high-frequency injection," *IET Electric Power Applications*, vol. 5, no. 1, pp. 89–99, 2011.

116 B. Sen and J. Wang, "Stator interturn fault detection in permanent-magnet machines using PWM ripple current measurement," *IEEE Transactions on Industrial Electronics*, vol. 63, no. 5, pp. 3148–3157, 2016.

117 Y. Zhao and T. A. Lipo, "Modeling and control of a multi-phase induction machine with structural unbalance," *IEEE Transactions on Energy Conversion*, vol. 11, no. 3, pp. 578–584, 1996.

118 D. Wang, X. Z. Wu, and S. W. Cheng, "Steady state performance analysis of multi-phase induction motor with non-sinusoidal supply," *Sci SinTech*, vol. 45, pp. 322–329, 2015.

119 X. Zheng, W. Ma, D. Wang, X. Yi, and H. Liu, "Fault-tolerant control of fifteen-phase induction machine under asymmetrical fault condition," *Proceedings of the CSEE*, vol. 38, no. 4, pp. 1221–1298, 2018.

120 Y. Zhou, X. Lin, and M. Cheng, "A fault-tolerant direct torque control for six-phase permanent magnet synchronous motor with arbitrary two opened phases based on modified variables," *IEEE Transactions on Energy Conversion*, vol. 31, no. 2, pp. 549–556, 2016.

121 F. Locment, E. Semail, and X. Kestelyn, "Vectorial approach-based control of a seven-phase axial flux machine designed for fault operation," *IEEE Transactions on Industrial Electronics*, vol. 55, no. 10, pp. 3682–3691, 2008.

122 T. Tao, Z. Wenxiang, C. Ming, and W. Zheng, "Review on fault-tolerant control of multi-phase machines and their key technologies," *Proceedings of the CSEE*, vol. 39, no. 2, pp. 316–326, 2019.

123 A. J. Mitcham, G. Antonopoulos, and J. J. A. Cullen, "Implications of shorted turn faults in bar wound PM machines," *IEE Proceedings – Electric Power Applications*, vol. 151, no. 6, pp. 651–657, 2004.

124 B. A. Welchko, J. Wai, T. M. Jahns, and T. A. Lipo, "Magnet-flux-ing control of interior PM machine drives for improved steady-state response to short-circuit faults," *IEEE Transactions on Industry Applications*, vol. 42, no. 1, pp. 113–120, 2006.

125 Y. Jiang, Z. Zhang, W. Jiang, W. Geng, and J. Huang, "Three-phase current injection method for mitigating turn-to-turn short-circuit fault in concentrated-winding permanent magnet aircraft starter generator," *IET Electric Power Applications*, vol. 12, no. 4, pp. 566–574, 2018.

2

Multiple 3-Phase Fault-Tolerant Machine Drive with Segregated Windings

2.1 Introduction

While electrical machine drives possess enhanced functionality, adaptability, and controllability compared with pneumatic, hydraulic, and mechanical-based solutions [1], their relatively low reliability becomes an apparent disadvantage for safety critical applications since an unexpected fault in the machine or inverter would cause serious consequences or economic losses. Thus, fault-tolerant machine drives are the key enabling techniques to enhance their reliability or availability in safety critical applications [2, 3].

The performance of a machine drive and its fault-tolerant capability are mainly determined by the machine topology and the associated drive configuration. While various solutions reported in literature have been reviewed in Chapter 1 to strike a good balance between the two aspects, the main challenge to achieve high performance and good fault tolerance with attainable cost remains. In this regard, induction machine-based drives are known as inferior in efficiency and power density and less tolerant to faults due to the overlapped windings. While switched reluctance machine drives are inherently tolerant to various electric faults in one phase owing to the concentrated winding, they suffer from relatively poor performance in torque controllability and high level of noise and vibration. For permanent magnet (PM)-based machine drives, the main challenge is to accommodate short-circuit fault, especially the turn fault in which only a few turns in a coil are short circuited, since the magnetic field produced by the PM cannot be turned off during the fault. Consequently, the excessive short-circuit current will be induced and cause a rapid localized temperature rise, leading to permanent damage to insulation and eventually catastrophic failures. By employing tooth-wound and segregated winding configurations, it is possible to limit the short-circuit current in the event of a terminal short circuit (TSC) by employing per-unit phase inductance design. However, it is very difficult to limit the fault current in a few short-circuited turns to an acceptable level because of

Multiple 3-Phase Fault Tolerant Permanent Magnet Machine Drives: Design and Control, First Edition.
Bo Wang and Jiabin Wang.

the presence of strong magnetic coupling between the rest part and faulted turns of the coil. Further, no significant reluctance torque is produced in this type of machine. Consequently, a high inverter VA rating is required with the per-unit phase inductance design and the drives also suffer from poor field weakening capability. Recently, synchronous reluctance machine-based drives have gained advantages over induction machine counterparts while it is relatively easier to manage fault current in the event of a turn fault because of the absence of PMs.

Thus, it is feasible to take advantages of high torque density, high efficiency of PM machines, and low short-circuit current characteristics of synchronous reluctance machines in the form of a PM-assisted synchronous reluctance machine (PMA-SynRM). This type of machine is usually equipped with distributed overlapping windings and a multilayer rotor with magnets inserted inside to maximize the reluctance torque. The PM torque usually accounts for about 1/3 of the total torque while the reluctance torque contributes to 2/3. Due to the combined torque production mechanism, the merits of high torque density, low PM usage, and low back electromotive force (bemf) with extended field weakening capability are maintained for the PMA-SynRM [4, 5], while the disadvantages of high bemf, and excessive fault current associated with conventional PM machines is eliminated. Especially, this machine features an inherent one per-unit characteristic current, which is a unique superiority for fault-tolerant machine design with a large safe margin. However, the distributed overlapping windings used in this machine yield strong mutual inductive coupling between phases. Windings of different phases are overlapped and bundled together in the overhang region, which may give rise to interphase short-circuit fault and facilitate fault propagation between phases and cause other negative impacts [6–8]. Thus, it would be desirable if the overlapped windings could be further separated to enhance fault tolerance.

Aiming for safety critical applications, this chapter is dedicated to develop a machine drive with high performance and good fault tolerance by employing a novel segregated, multiple 3-phase winding configuration for the PMA-SynRM topology. It eliminates overlapping in different 3-phase winding sets while having a modular structure to enhance fault tolerance and maintain high torque density. Detailed theoretical analysis and finite element (FE) simulation have been conducted to assess its performance under healthy and various fault conditions.

For the multiple phase or multiple 3-phase machine drives, mutual coupling between different phases or different 3-phase modules is inevitable. Such coupling is of key importance to understand the behaviors of a fault-tolerant machine under fault and postfault conditions since such knowledge forms the basis for fault-tolerant machine design, fault diagnostic, and development of postfault control strategy. For this purpose, the mutual coupling mechanism between different modules of the multiple 3-phase PMA-SynRM drive is investigated by analyzing the magneto-motive force (MMF) distribution of each module and their combined effect. The results provide analytical means to assess the machine's behavior under open-circuit and short-circuit fault conditions.

2.2 PMA-SynRM with Segregated Windings

The standard distribution of 3-phase windings in a PMA-SynRM (as seen in Fig. 2.1) produces a nearly sinusoidal MMF in the air gap, but it lacks fault tolerance. To achieve redundant operation, the windings can be arranged as three independent 3-phase sets, each driven by a standard 3-phase inverter. However, the windings of these 3-phase sets are overlapped and bundled together in the overhang region, which increases the risk of interphase short-circuit fault. Additionally, there is a strong magnetic coupling between the different 3-phase sets, and a fault in one 3-phase module can spread heat to the other healthy sets through the overlapped windings. Therefore, a fault in one part cannot be isolated and may potentially affect other healthy phases.

To enhance the fault tolerance of the PMA-SynRM, a segregated triple redundant 3-phase winding configuration is proposed in Fig. 2.2 [9]. This configuration achieves physical, electrical, and thermal isolations. Unlike the conventional overlapped windings, the triple redundant configuration divides the windings into three separate 3-phase winding sets, providing physical and thermal isolation by eliminating any physical contact between different sets. The electrical isolation is achieved by adopting an independent inverter for each 3-phase set.

As shown in Fig. 2.2 and compared with the winding distribution in Fig. 2.1, the positions of the coils in phases C, F, and I are reversed, but the induced voltages remain unchanged due to the periodic air gap flux density distribution. This results in each winding set forming a balanced 3-phase winding, maintaining the same MMF distribution and performance in healthy operations as the original configuration in Fig. 2.1. The PMA-SynRM's advantages, such as reduced magnet usage, low bemf, inherent large reluctance torque, high efficiency, and high torque density, are all preserved. Additionally, the segregated windings and independent inverters provide excellent fault tolerance by eliminating the risk of fault propagation between different 3-phase winding sets and offering redundancy for various faults [10].

In the event of an open-circuit fault in an inverter switch or windings, the faulty 3-phase winding set can be deactivated by switching off all the switches in that set. This allows the remaining two healthy 3-phase sets to continue operation in a balanced manner, albeit with a reduction in output torque or power. If a short circuit occurs in an inverter switch or 3-phase winding, TSC can be applied to the faulty 3-phase set by switching on the entire top or bottom switches of the

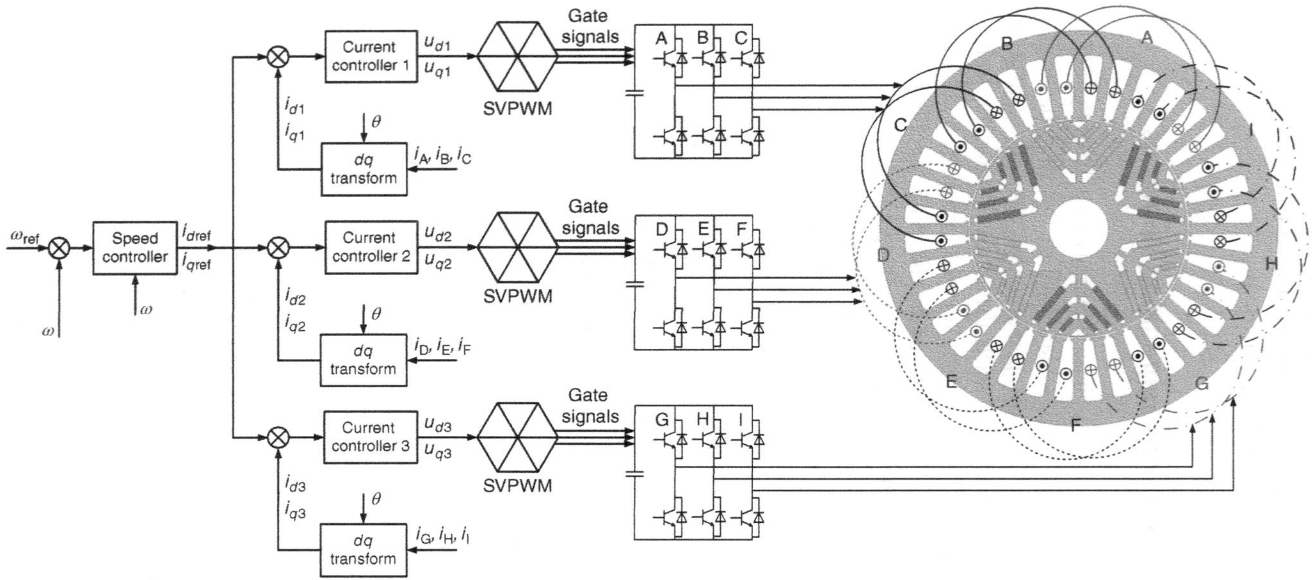

Figure 2.1 PMA-SynRM with conventional overlapped windings.

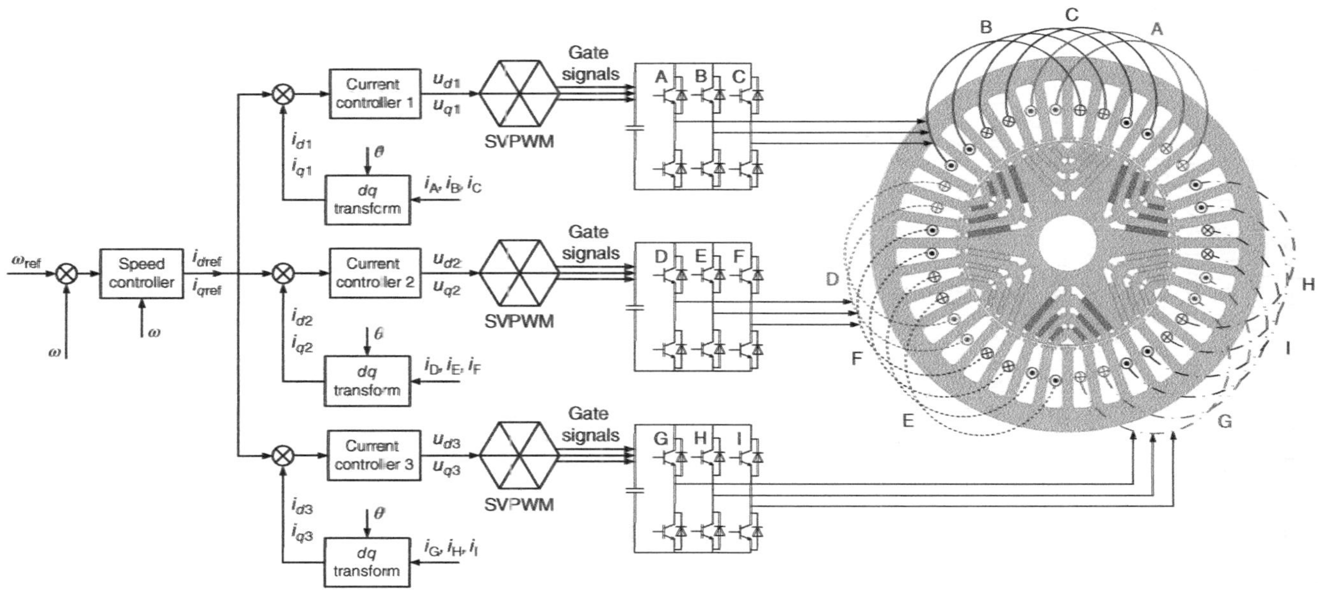

Figure 2.2 PMA-SynRM with segregated windings.

3-phase inverter. Due to the low PM flux in the machine, the short-circuit current is limited below or close to the rated value. In the case of an interturn short circuit, the current of the faulted turn can be alleviated by applying TSC to nullify the flux linkage of the faulted turns, and hence controlling the localized fault current to a thermally sustainable level. The separation of the windings enables the fault in one 3-phase system to be isolated so that the other two 3-phase systems can continue to operate and generate torque or power. The multilayer rotor core provides better protection for the magnets, and even if they are partially demagnetized, the reduction in torque capability will be insignificant since magnet torque only accounts for approximately 30% of torque in this type of machine. The machine's low bemf ensures that in the event of inverter or control failure, there is no uncontrollable rectification even at high speed, leading to the inverter fail-safe feature highly desirable in many applications.

The PMA-SynRM with segregated windings yields high performance and strong fault tolerance without much-added costs or penalties. The segregated windings require only a simple reconfiguration with a slight increase in overhang length to create multiple 3-phase winding sets, each driven by a standard 3-phase inverter. While the winding segregation is developed primarily for PMA-SynRM, it can also be applied to other machines with distributed windings, including synchronous reluctance machines, surface-mounted PM machines, synchronous wound field machines, and induction machines. The number of 3-phase modules can be adjusted to meet the specific requirements of the application.

2.3 Fault-Tolerant Capability Assessment

During machine and drive operation, it is important to be aware of the various faults that may occur. Understanding of these failure modes, potential mitigating measures, and corresponding fault-tolerant capabilities are crucial for developing a fault-tolerant drive system, including effective modeling, control, and fault detection techniques. This section investigates the fault-tolerant capabilities of the PMA-SynRM with segregated windings against various types of most likely faults. To quantitatively examine these failure modes and postfault operations, an example specification of the proposed PMA-SynRM is given in Table 2.1 for FE analysis. The gamma angle, which is defined as the angle between the current vector and the q-axis of the synchronous reference frame, will be taken into account in all simulations performed at a base speed of 4000 rpm with the rated current excitation, unless otherwise stated.

2.3.1 Healthy Operation

First, the health condition of the machine is evaluated. The machine is excited with the rated current of 120 A, and the optimal gamma angle for maximum torque

Table 2.1 Specifications of the PMA-SynRM.

Specification	Symbol	Value
Base speed	n_b	4000 rpm
Maximum speed	n_m	19,200 rpm
Rated power	P_r	35 kW
Rated current and gamma angle	I_{rated}	120 A (51°)
Nominal DC-link voltage	V_{dc}	270 V
Number of turns per coil	N	8

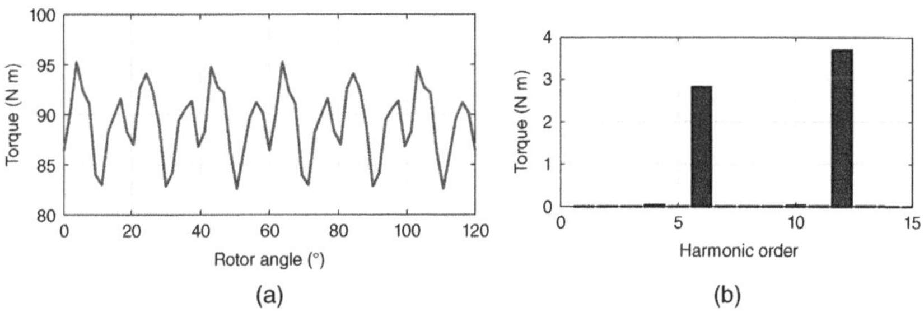

Figure 2.3 Torque and spectrum in healthy condition: (a) torque waveform and (b) torque spectrum.

per Ampere (MTPA) operation is determined to be 51°. The torque waveform obtained is depicted in Fig. 2.3(a), where the machine produces an average torque of 89.2 N m at the rated operating point. The torque ripple, mainly caused by the interaction of rotor saliency and stator slotting, is assessed to be 14.1%, with the 6th and 12th harmonics being the major contributors as shown in Fig. 2.3(b). The PM torque and reluctance torque contribute 30% and 70% of the total torque, respectively [11]. Despite the relatively low PM usage, the machine still achieves a high torque density of 31 (N m)/L (estimated based on active volume), comparable to similar Surface-mounted permanent magnet (SPM) and Interior-mounted permanent magnet (IPM) designs with similar cooling conditions, owing to the significant contribution of the reluctance torque component.

Figures 2.4 and 2.5 present the magnetic flux distribution and flux density under the rated conditions, respectively. It is noticeable that the distribution repeats three times periodically over 360°. Therefore, the flux linkages of the three 3-phase sets are identical in healthy conditions. For instance, Fig. 2.6 shows the flux linkage waveforms of set ABC, which exhibit 3-phase symmetry despite some distortions due to space harmonics and magnetic saturation.

Figure 2.7 provides an assessment of the efficiency map across the entire speed range under rated torque and power. The machine demonstrates superior

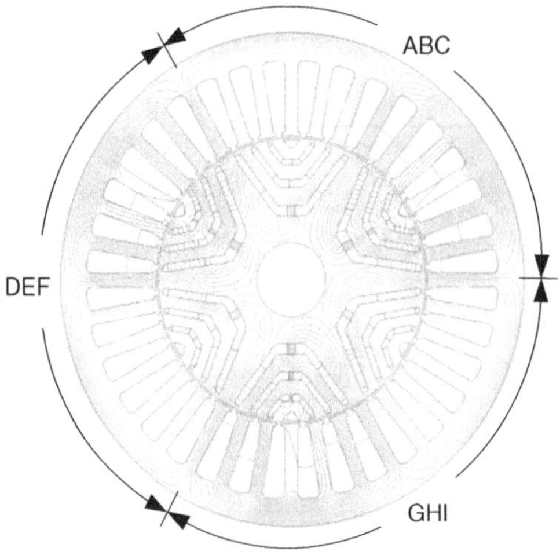

Figure 2.4 Flux distribution in healthy condition.

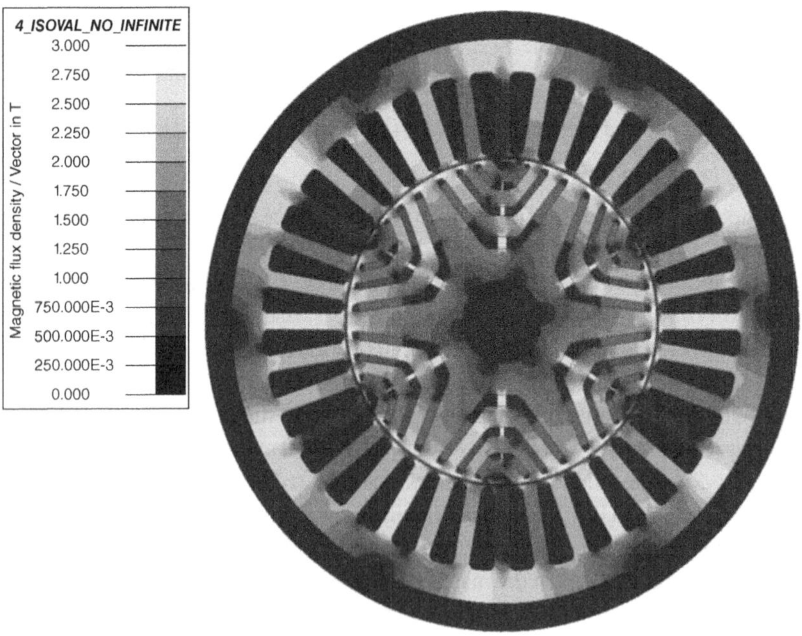

Figure 2.5 Flux density distribution in healthy condition.

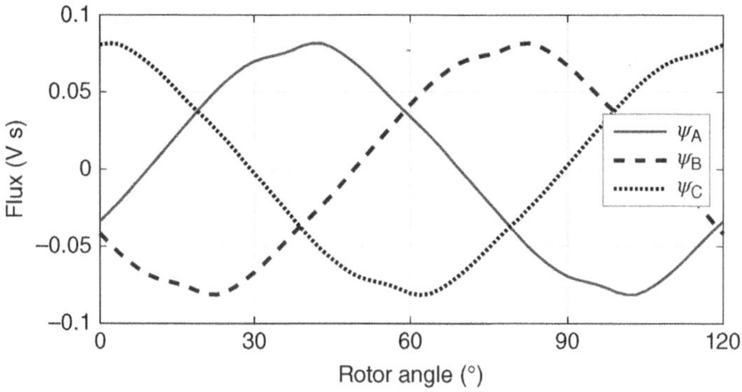

Figure 2.6 Phase flux linkages in healthy condition.

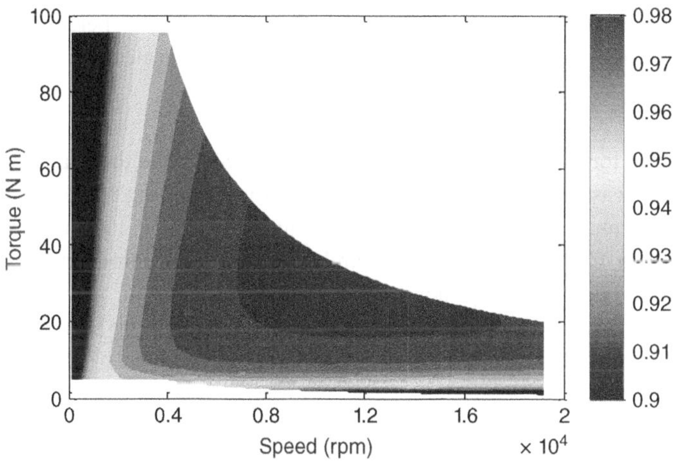

Figure 2.7 Efficiency map under rated power operation region.

efficiency in most areas, particularly at higher speeds owing to the combination of PM torque and reluctance torque. At the base speed and rated torque, the efficiency reaches 95.83%. This result confirms that the machine can maintain excellent performance when operating under healthy conditions.

2.3.2 One 3-Phase Set Open Circuit

If there is an open-circuit fault in any phase of a 3-phase winding set or in the power switching devices, all switches in the 3-phase inverter can be opened to deactivate the faulty 3-phase set. This will result in zero currents in the faulty set, leading to open-circuit failure mode for that set. The remaining two 3-phase

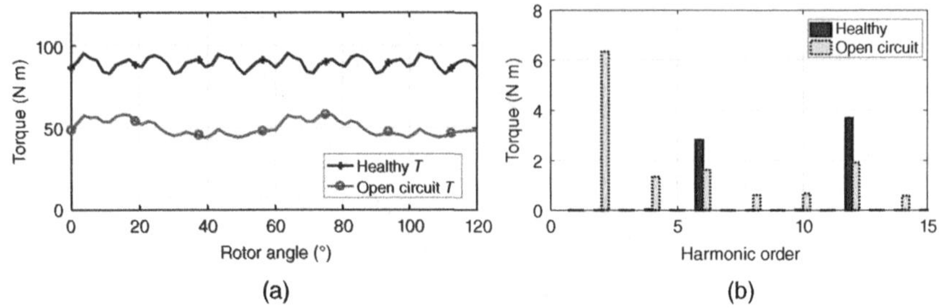

Figure 2.8 Torque with set ABC open-circuited (a) torque waveforms and (b) torque spectrums.

sets will still be excited by the same currents, and the resulting torque waveform is illustrated in Fig. 2.8. Under this condition, the machine's average torque is 50 N m. The open-circuit torque capability ρ_{oc}, which is defined as the percentage ratio of the open-circuit torque T_{oc} to the rated torque T_r, can be calculated as

$$\rho_{oc} = T_{oc}/T_r \times 100\% = 56\% \tag{2.1}$$

The reduction in torque is slightly higher than 1/3 pu. Since the thermal load is significantly decreased with one set open circuited, the remaining two healthy 3-phase sets can increase their currents to compensate for the loss of torque. Now the torque ripple is 15.8%. Besides the 6th and 12th harmonic torque ripples, the 2nd harmonic becomes the major torque ripple in the torque waveform. This phenomenon is caused by the electromagnetic imbalance in the remaining two healthy 3-phase sets, due to the open-circuit fault, as indicated by the flux distribution in Fig. 2.9.

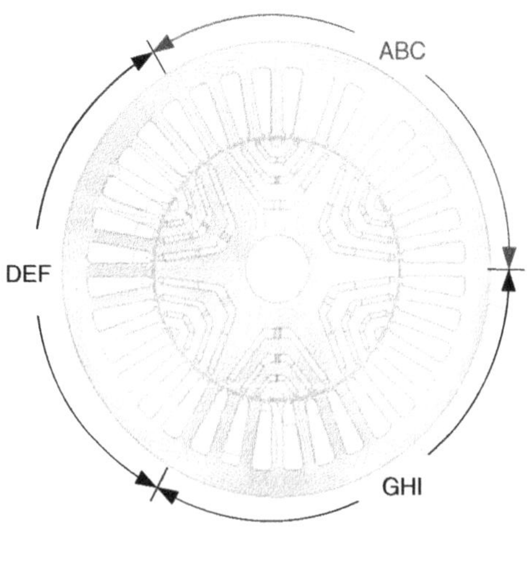

Figure 2.9 Flux distribution with set ABC open circuited.

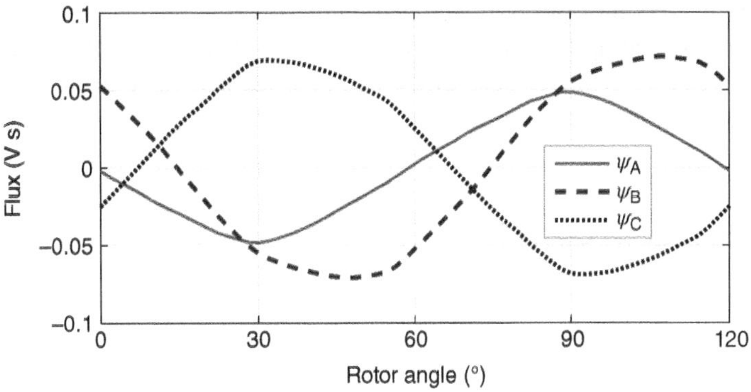

Figure 2.10 Faulty phase flux linkages with set ABC open circuited.

The resultant heavy distortion in the phase flux linkages of the ABC set is shown in Fig. 2.10. This is due to notable mutual magnetic coupling between the three 3-phase winding sets. Although the currents in the faulty set are zero, the currents in the remaining two healthy sets generate a non-zero MMF in the fault set region, leading to asymmetric MMF distribution in the air gap. Further details will be explained subsequently. Consequently, the flux linkages of all the 3-phase sets are no longer symmetrical under the combined influence of the rotor magnets and the currents in the two healthy sets.

Indeed, the flux linkage waveforms of the DEF and GHI sets, which are in healthy condition, as depicted in Fig. 2.11, are also no longer symmetrical. The distortion in the flux linkage waveforms resulting from the asymmetry differs for each phase. Phases D, E, F, G, and I exhibit relatively minor distortions, while more significant distortions are evident in phase H. This phenomenon is

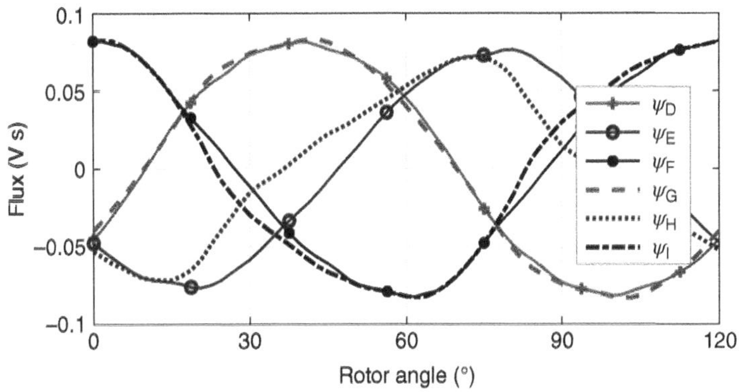

Figure 2.11 Healthy phase flux linkages with set ABC open circuited.

linked to the interaction of the asymmetric armature reaction field with that of the rotor magnets, leading to different levels of saturation seen by each phase. Hence, the level of distortion in each phase will be dependent on directions of rotation, current phase advance angle, and modes of operation (motoring or generating).

2.3.3 One 3-Phase Set Short Circuit

In the event of a switch in the inverter or phase winding being short circuited, the affected 3-phase set should have all its top or bottom switches turned on the inverter to create a TSC. However, the remaining healthy 3-phase sets will still be powered by their inverters for uninterrupted operation, similar to the one set open-circuit fault. The resulting short-circuit phase currents, as shown in Fig. 2.12, are much lower than the rated current due to the relatively low PM flux. Therefore, the faulty set will not produce excessive heat to jeopardize the machine safe operation. This low short-circuit current is an inherent feature of this type of machine, and there is no penalty in the machine design to achieve this advantage. However, the short-circuit phase currents are asymmetrical, which can be attributed to the mutual coupling between the faulty set windings and the healthy sets of the windings since they are not completely magnetically isolated.

Figure 2.13 shows the flux distribution during the short-circuit condition. It can be observed that the region occupied by the short-circuited ABC set experiences significantly lower flux density, as the short-circuit currents nearly nullify the flux. The resulting flux linkage waveforms of the fault set, plotted in Fig. 2.14, also indicate much lower flux linkages compared to the healthy condition. The residual flux linkages mainly consist of zero-sequence components, which are of

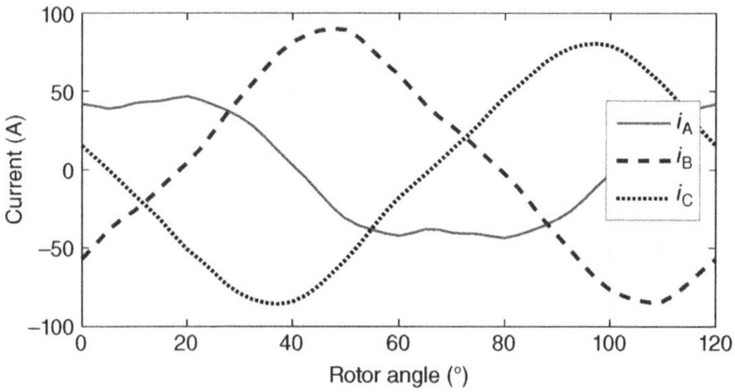

Figure 2.12 Short-circuit phase currents.

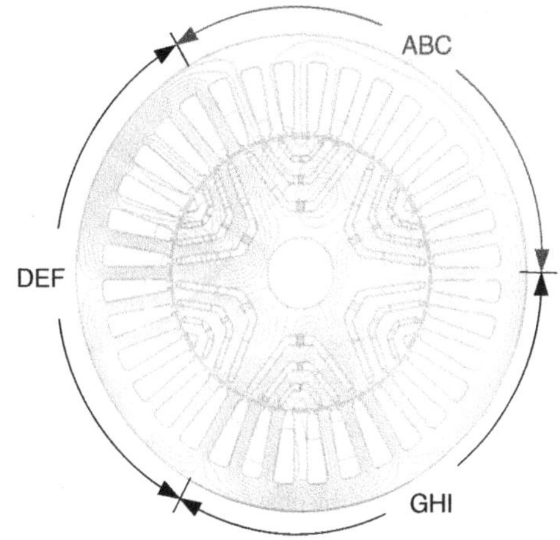

Figure 2.13 Flux distribution with set ABC short circuited.

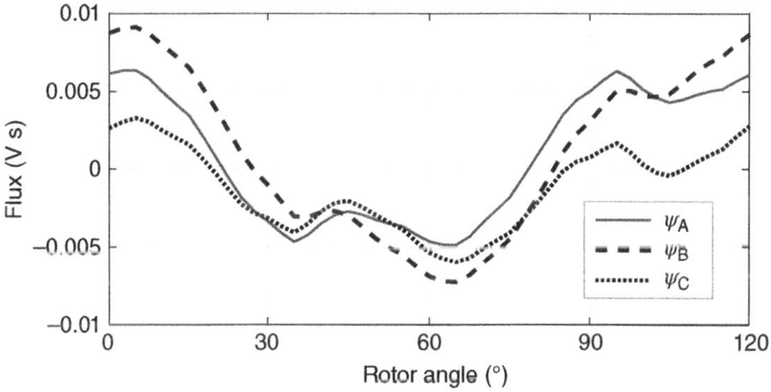

Figure 2.14 Faulty phase flux linkages with set ABC short circuited.

a similar waveform shape with very little phase shift between each other. This is because there are no zero-sequence currents in the star-connected 3-phase winding to nullify them. Phase B exhibits the highest flux linkage, which is consistent with the highest phase current observed in Fig. 2.12. In contrast, the flux-density distribution in set DEF and GHI regions appears to be almost normal, as evident from the flux-linkage waveforms shown in Fig. 2.15.

Under the short-circuit condition, the torque waveform is plotted in Fig. 2.16, with the average torque and torque ripple being 58 N m and 14%, respectively. The short-circuit torque capability ρ_{sc} is evaluated to be 65%, nearly two thirds of the rated torque and slightly higher than that in the open-circuit scenario. The torque ripple is slightly lower than that in the open-circuit case and mainly contains the 2nd and 12th harmonics. These results demonstrate that the machine is capable of accommodating the terminal short-circuit faults.

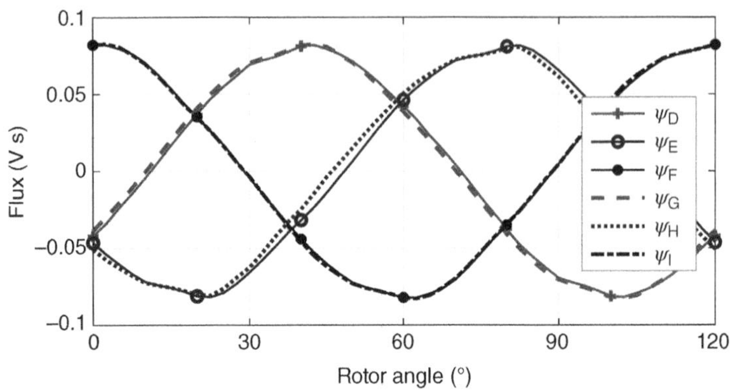

Figure 2.15 Healthy phase flux linkages with set ABC short circuited.

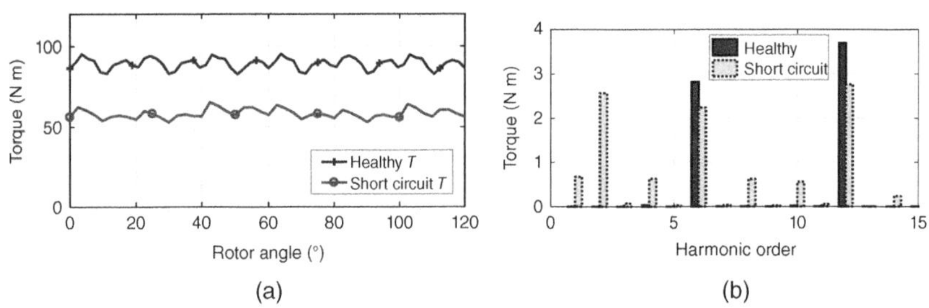

Figure 2.16 Torque with set ABC short-circuited (a) torque waveforms and (b) torque spectrums.

2.3.4 Intraphase Turn Fault

In mash or randomly wound machine windings, different turns of the same phase winding are in direct physical contact with each other, and their insulation relies mainly on very thin enamel coating of the magnet wire. As such, the turn insulation is most vulnerable under combined mechanical, thermal, electrical, and environmental stress. Hence, a short-circuit failure involving in a few turns is most likely among other types of short-circuit failure modes. Consequently, a turn insulation failure within a single phase will lead to a short-circuit fault, known as an intraphase turn fault, as illustrated in Fig. 2.17. Such a fault generates a substantial fault current, which increases as the number of short-circuited turns decreases, with the short circuit of a single turn located near the slot opening being the most critical case.

Assuming a generic case, a single turn situated at the top of the slot (close to the slot opening) within coil B2 of the ABC set is depicted in Fig. 2.18. All phase windings carry the rated load current of 120 A at a phase advancing angle of 51° at

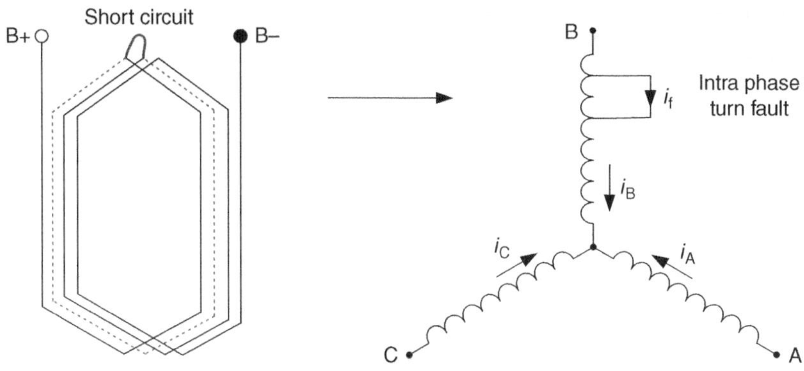

Figure 2.17 Illustration of intraphase turn fault.

Figure 2.18 Coil location of the fault turn.

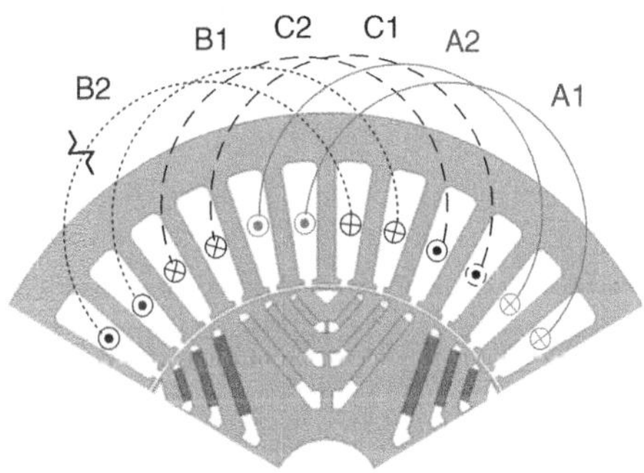

4000 rpm. FE simulation of this fault results in an 1100 A fault current i_f as shown in Fig. 2.19, whose rms value is 9.1 times greater than the rated. Immediate detection and action are therefore required to prevent a disastrous failure. Figure 2.20 illustrates the slightly reduced flux distribution across coil B2 under fault conditions when compared to healthy coils E2 and H2. The much greater fault current involving a small number of short-circuited turns can be understood by the fact that the fault current is roughly the PM flux linkage of the short-circuit turns divided by the inductance. The former is proportional to the number of turns while the latter is to the turns squared. Hence, the fault current is inversely proportional to the number of short-circuited turns. Such understanding also explains why a short circuit in a turn located close to a slot opening yields the largest fault current because it carries a larger PM flux but has lower leakage inductance.

To mitigate the fault current, TSC is deployed through the inverter connected to the affected 3-phase set. This introduces short-circuit currents within the healthy turns and coils of that set, which counters the external flux to the faulted turns generated by magnets and currents in the healthy sets. Therefore,

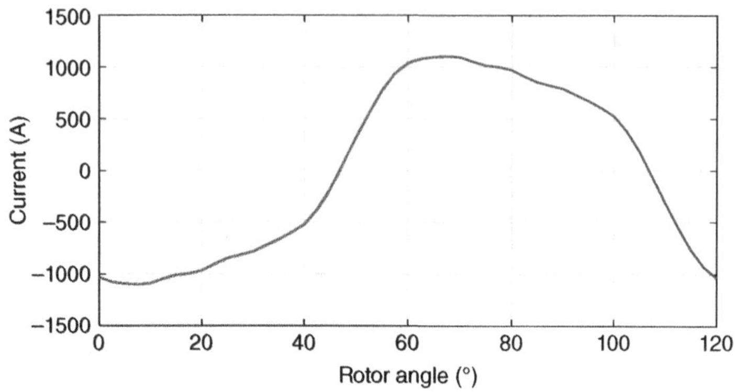

Figure 2.19 Turn fault current in coil B2 with 120 A load current at 4000 rpm.

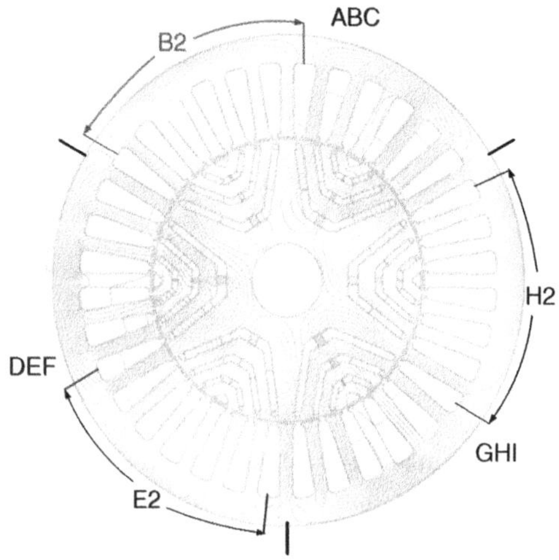

Figure 2.20 Flux distribution with turn fault in coil B2.

the flux encompassing the faulted turn diminishes significantly, resulting in much-reduced fault current.

Analysis [9, 12] indicates that the fault current magnitude is influenced by the coil location of the fault turn and the currents in the healthy 3-phase sets. The altered flux linkage of the faulted turn after TSC reflects combined influences from magnets and currents in both the faulty and healthy sets. The imbalanced currents in the fault set modify the flux linkage and result in differing fault currents at various coil locations. Given that one 3-phase set comprises six coils, as seen in Fig. 2.18, the fault is analyzed at six different coil locations with variation in currents in the healthy sets under MTPA conditions. Figure 2.21 displays how the RMS fault currents in six coils of the ABC set after TSC varies with the load currents in the healthy sets at 4000 rpm, indicating that the fault current in coil

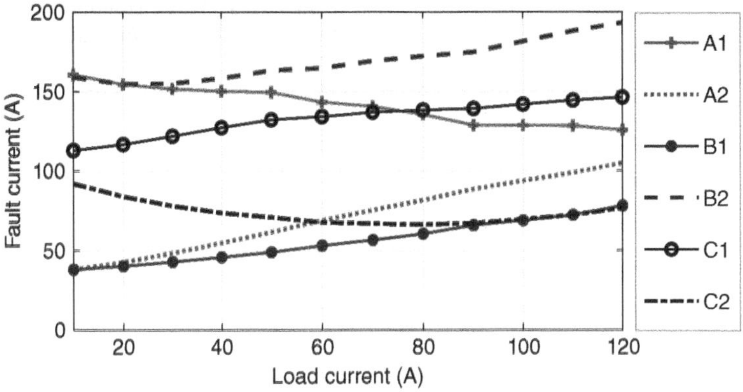

Figure 2.21 Variations of RMS turn fault currents in six coils with load current in healthy 3-phase sets in motoring mode.

B is the highest and rises with an increase in the load current. Thus, the most severe case is ascertained to be a single-turn fault in coil B2 with a 120 A current in the healthy sets, which should pay special attention during design optimization. It should be noted that if the currents in the healthy sets are greater than the rated, the fault current will also increase. However, such an operating condition is considered to occur only for a short period and hence not sustainable even in healthy operations.

Figure 2.22 presents the current waveforms for the faulted turn and the healthy turns in the ABC set following TSC application. The resulting phase currents are unbalanced but considerably lower than the rated value. The unbalanced currents, in conjunction with those from healthy sets, cause variable fault currents when the

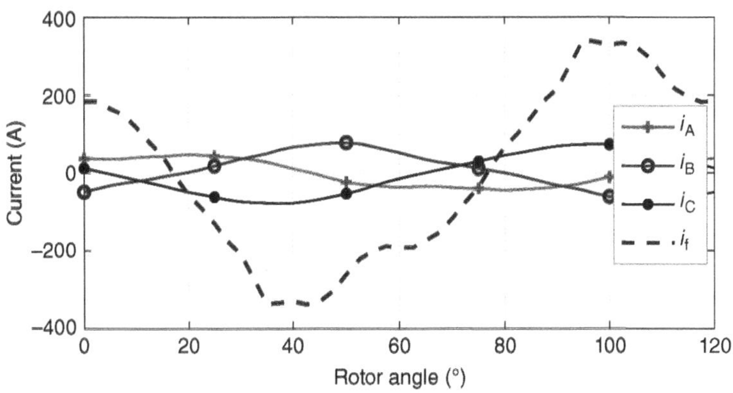

Figure 2.22 Turn fault current and phase currents in ABC with turn fault in coil B2 set after TSC.

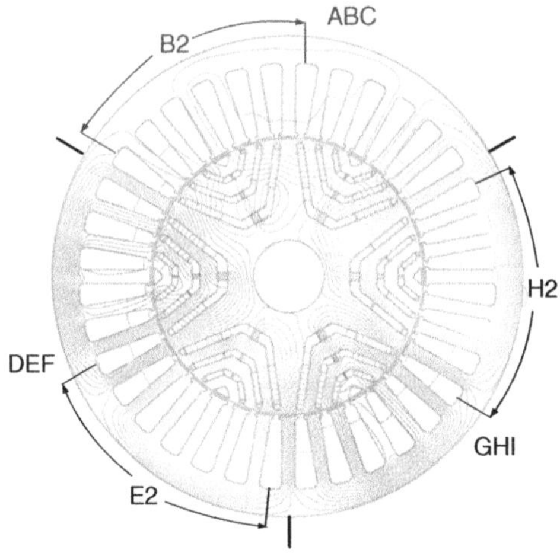

Figure 2.23 Flux distribution in turn fault condition after TSC.

fault occurs in different coils. Although the faulted turn current reaches 2.6 pu in the severest case, the machine can still dissipate the induced heat with the temperature in the fault turn below the safe limit because the total heat generated by the ABC set under the TSC application is lower than that of the rated. The flux distribution, as shown in Fig. 2.23, indicates that short-circuit phase currents have effectively reduced the external flux in the ABC region compared to the scenario in Fig. 2.20, thereby reducing the turn fault current as the residual flux diminishes. Figure 2.24 contrasts the torque waveforms under both the healthy and fault conditions, confirming the machine's ability to produce approximately two-thirds of the rated torque after the TSC application.

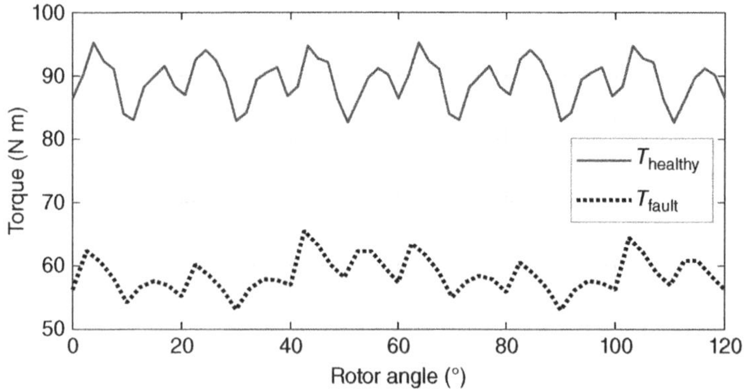

Figure 2.24 Comparison of torque waveforms under healthy and turn fault condition after TSC.

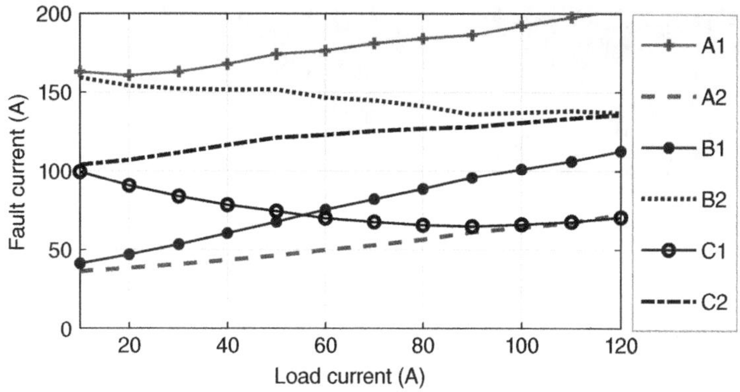

Figure 2.25 Variations of RMS turn fault currents in six coils with load current in healthy 3-phase sets in generating mode.

It should be noted that the highest turn fault current occurs in coil B2, which is the trailing coil of the ABC 3-phase set, when the machine operates in an anticlockwise direction under motoring conditions. Should the direction of rotation changes to clockwise direction, the worst case would occur in the trailing coil A1. Conversely, in generating mode, the severest turn fault occurs in coil A1 when rotating in the anticlockwise direction as illustrated in Fig. 2.25. In this operating condition, coil A1 is the leading coil of the ABC 3-phase set. The change to that seen in the motoring operation is owing to the reversed current sequences in the healthy sets in the generating mode. Therefore, depending on the mode of operation and rotation direction, the most critical fault current after TSC can be found in coils A1 or B2.

2.3.5 Interphase Turn Fault

Since the phases in one module of the triple 3-phase machine still overlap in the overhang region, a short circuit between two different phases may occur if the phase-to-phase insulation fails. In a mash wound machine, phase-to-phase insulation is usually provided by the insulation paper known as phase separator, as illustrated in Fig. 2.26. While this provides additional insulation protection, the insulation breakdown between two phases cannot be completely avoided due to high-level stresses in both the phase-to-phase voltages and mechanical vibration in the overhang region.

A short circuit involving a few turns across two different phases is termed as an interphase turn fault, as shown in Fig. 2.27. Short-circuit fault will be induced in the fault turns and affect the phase currents. Considering the extreme cases, the worst interphase turn fault would be a two-turn short circuit involving one turn in

Figure 2.26 Phase-to-phase insulation in the overhang region.

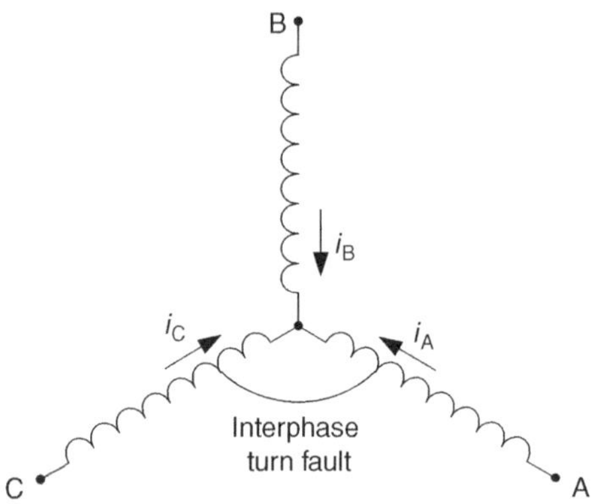

Figure 2.27 Interphase turn fault illustration.

each phase. As shown in Fig. 2.28, it may occur at the input terminal end or at the neutral point end. In the case of the former, excessive current will result and trigger current protection or even damage the inverter. This will also lead eventually to an open-circuit failure, which has been previously analyzed. It is therefore the interphase faults at the neutral end become the worst scenarios which is worthy of further investigation.

Since the interphase turn fault involves at least two turns in the short-circuit loop, it is less severe compared with the single-turn intraphase fault. Though the fault current is expected to be lower than the intraphase turn fault, the same mitigation measure, namely, application of TSC via the inverter, should be taken to limit the fault current to a sustainable level. It is therefore the interphase faults at the neutral end with TSC application as illustrated in Fig. 2.29 are simulated by FE analysis in order to assess the resultant fault current.

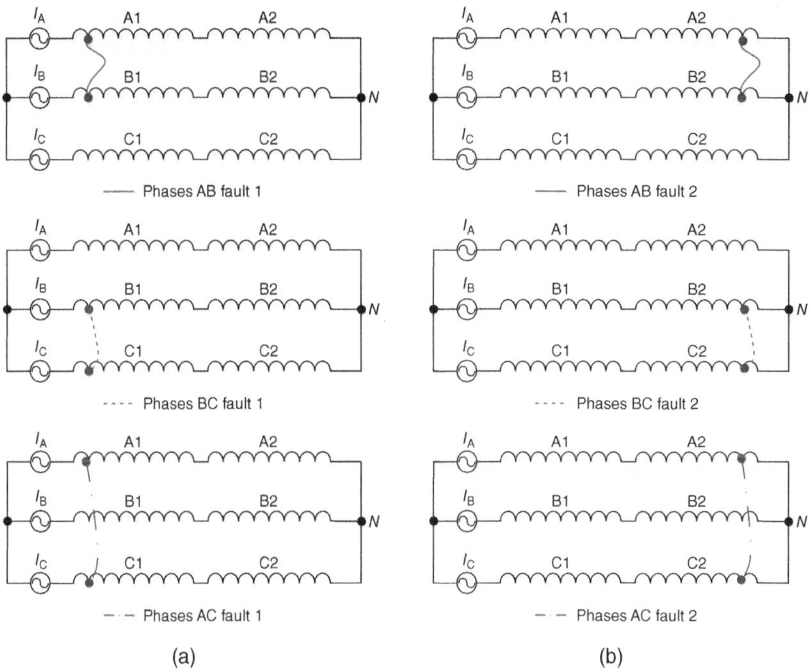

Figure 2.28 Illustration of interphase fault (a) at the terminal end and (b) at the neutral end.

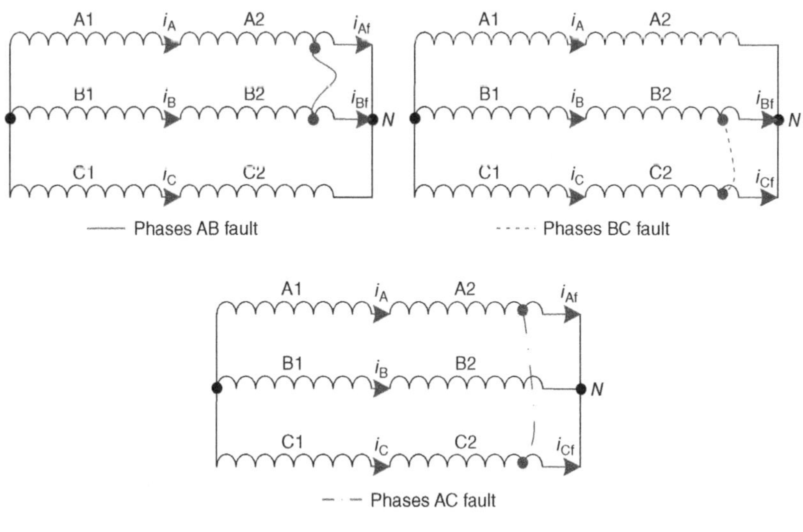

Figure 2.29 Interphase faults at the neutral end with TSC mitigation action.

The above three extreme interphase faults are simulated when the drive operates at 4000 rpm in the motoring mode and the two remaining 3-phase sets are fed with the rated currents. The resultant fault turn currents and phase currents in the faulted 3-phase sets are presented in Fig. 2.30. The phase currents are similar to those under the one 3-phase set short-circuit fault mode. However, the maximum RMS fault current is 99 A or 1.17 pu, which is much smaller than that of the single-turn intraphase fault shown in Fig. 2.22. The combined heating effect of the fault after the mitigation measure is much lower than the worst scenario of the intraphase turn fault. The lower fault current is attributed to two factors. First, the number of short-circuited turns involved in the worst-case interphase fault is two while that of the intraphase turn fault is one. This implies that loop inductance in the worst-case interphase turn fault is larger. Second, the bemfs in each turn of the two different phases have a phase shift. Hence, the net effect on each turn is reduced. The foregoing analysis shows that the interphase turn fault can be mitigated by TSC application, and the resultant fault current is lower than the worst intraphase turn fault. Further, the torque response during the postfault operation is very similar to that of the one 3-phase set TSC condition shown in Fig. 2.16 and hence not repeated.

2.3.6 Phase-to-Ground Fault

The insulation between phase conductors and stator lamination in electrical machines is typically provided by slot liners aided with epoxy resin. Such insulation component is also known as phase-to-ground insulation or ground-wall insulation. When this insulation fails, a short circuit between the phase conductors and the stator metallic housing takes place. The consequence of this fault and its mitigation measure are, however, dependent on the grounding system also known as the earthing system employed in an electrical drive.

There are three typical grounding systems, insulatum terra (IT), terra network (TN), and terra terra (TT), the details of which can be found in the International Standard IEC60364. TN and TT grounding systems are similar in that the live conductors of the power system are grounded to the earth and they only differ whether the earth connection is provided locally or via a protective earth (PE) conductor. Most electric drives used in stationary industrial environments employ either TN or TT grounding system. A typical example of the TN grounding in which the ground connection is provided by a separate PE conductor is illustrated in Fig. 2.31, where the star neural point formed by the live conductors of the 3-phase power supply is connected to the ground (earth).

In an IT grounding system, no live conductors are connected to earth for reference. Instead, voltage reference is provided by the metallic chassis of the electric system components connected together. Such systems are typical in mobile

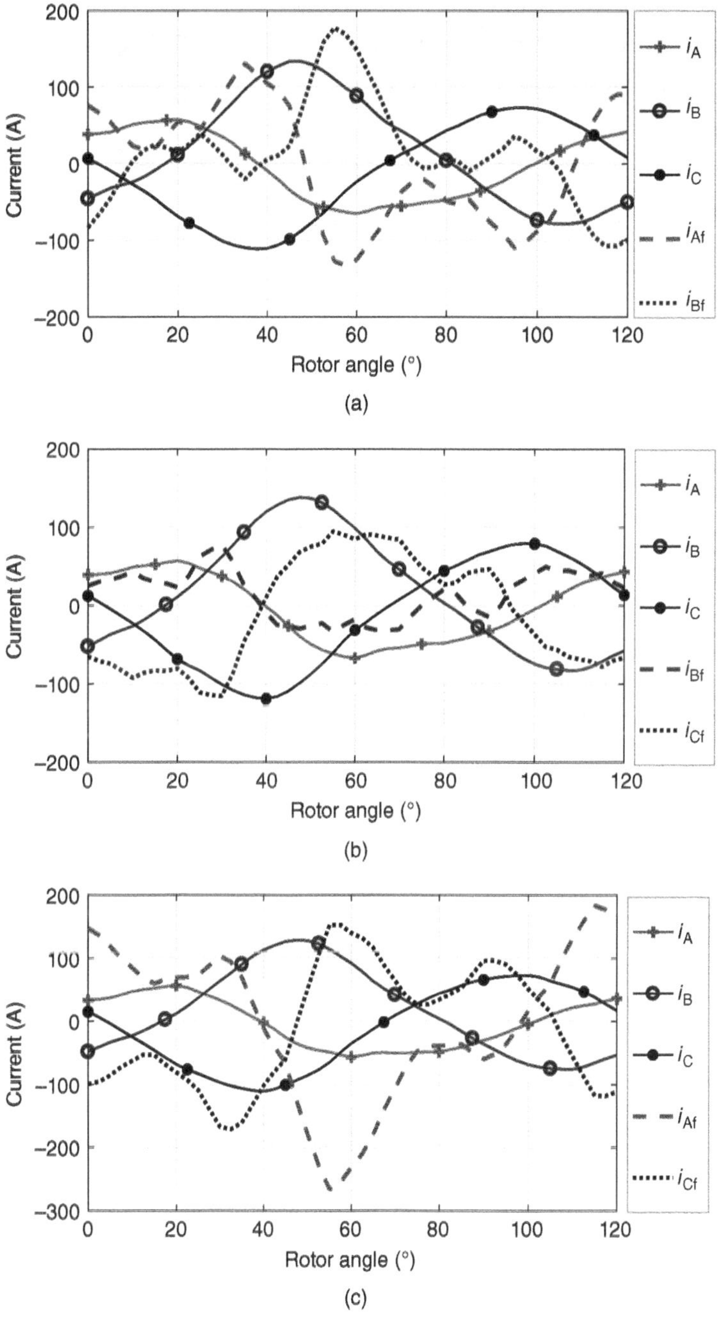

Figure 2.30 Short-circuit currents of the interphase turn faults (a) phases AB fault, (b) phases BC fault, and (c) phases AC fault.

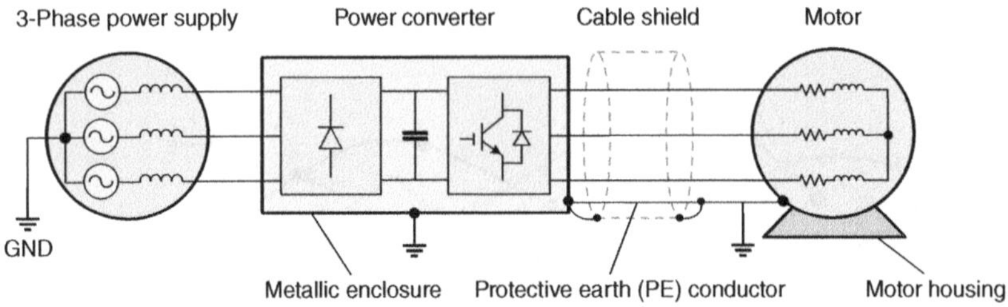

Figure 2.31 Schematic of TN grounding system of an electric drive.

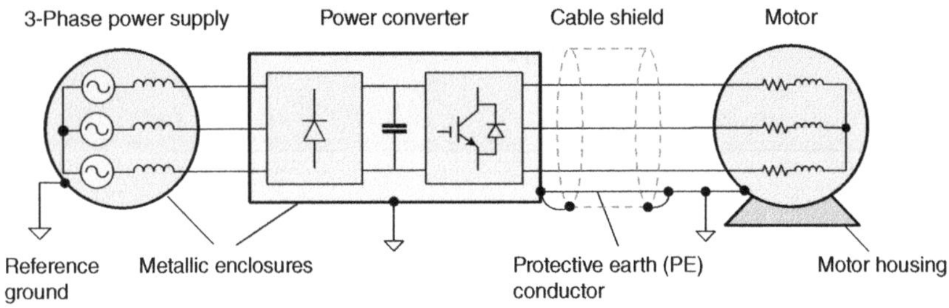

Figure 2.32 Schematic of IT grounding system of an electric drive.

vehicular applications, such as electric vehicle traction drives and electrical power systems in aircraft, where physical connection to earth is not possible or feasible. IT systems are also used in applications where better electromagnetic immunity is required. A typical IT-grounded drive system is shown in Fig. 2.32.

If any single point failure in the phase-to-ground insulation of the machine in an IT grounding system, no current return path exists since no active conductors are connected to the reference ground. Hence, no fault current will result and the fault consequence is benign. Such a fault will however change the relative phase voltages of the machine with respect to the ground reference, and this change can be used to detect such a fault. In general, the drive system will be able to continue its operation until the earliest possibility of repair to restore its insulation system. If this repair does not take place in time, catastrophic consequences may result if a second point failure in the phase-to-ground insulation occurs.

In contrast, if the phase-to-ground insulation fails at any single point in the electric machine in a TN or TT grounded system, as illustrated in Fig. 2.33, a circulating path exists for the fault current in the grounding system, via either earth connections or PE conductors. The fault current would cause serious deterioration of the drive performance and excessive thermal loading to the machine and converter and may eventually lead to a catastrophic failure.

The phase-to-ground insulation fault in a 3-phase drive with TN or TT grounding can be detected when a residual current device (RCD) is installed in the system.

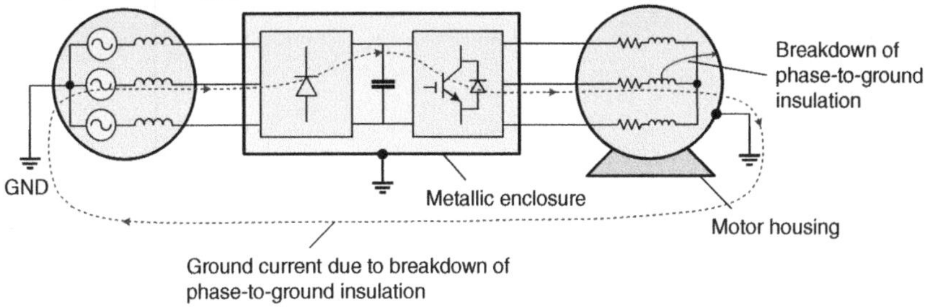

Figure 2.33 Ground fault current path caused by single point breakdown of phase-to-ground insulation in a motor drive in TN or TT grounding system (protective earth (PE) conductor and cable shields are omitted for clarity).

Upon detection of such a fault, the drive operation has to be terminated in order to avoid further damage to the system and other equipment. With the triple redundant 3×3-phase drive under study, the drive operation in the faulted 3-phase set can be stopped by deactivating the 3-phase inverter, and hence the fault current path is blocked by the inverter diodes. This leads to one 3-phase open-circuit fault as has been analyzed in Section 2.3.2, where the drive can continue its operations with the two remaining healthy 3-phase sets, albeit torque production is reduced by about one-third.

2.3.7 Uncontrolled Rectification Fault

When operating at high speed, the machine line bemf could exceed the voltage of the DC link, particularly in machine drives designed for a very wide constant power operation range (CPOR). This typically is not an issue if the inverter is functioning correctly, especially in field-weakening mode [13]. Nonetheless, if the inverter fails or applies self-protection due to device overheating or sensing errors, the gate drive signals may be inhibited. Consequently, the emf could significantly surpass the DC-link voltage, leading to uncontrolled rectifier operation via diodes. This situation might cause mechanical energy converted to electrical energy and fed back into the DC link, which could potentially harm the entire drive system if this regenerated power is not properly dissipated [14].

Imposing a limit on the maximum bemf in the design could negatively impact the dimensions of the machine and the inverters' VA capacity [15]. Therefore, there has been a compromise in design between the extent of CPOR and avoidance of uncontrolled generation operation. It has been established that machines with high rotor saliency are preferable because they can achieve broad CPOR while maintaining lower bemf levels [16].

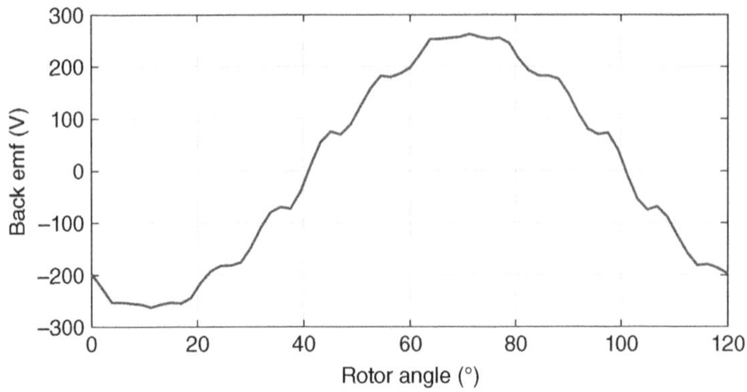

Figure 2.34 Line back-emf at max speed 19,200 rpm.

The PMA-SynRM, with high rotor saliency relying mainly on reluctance torque, is particularly suitable for broad CPOR applications while ensuring a lower bemf. This design minimizes the risk of uncontrolled generation operation if an inverter malfunction occurs.

Testing has shown that at the top speed of 19,200 rpm, the machine's line bemf peaks at 262 V as shown in Fig. 2.34, less than the DC-link voltage of 270 V. As a result, diodes will not be subjected to positive bias voltage and uncontrolled rectification is prevented, even in case of inverter failure, due to the low PM field inherent in PMA-SynRM design. However, it is crucial to note that not all PMA-SynRMs will have bemf lower than DC-link voltage, particularly in applications with wide CPOR. Therefore, the bemf should be evaluated for each specific design to avoid the possibility of uncontrolled rectification.

2.3.8 Demagnetization Fault

The PMA-SynRM's magnets are also at risk of demagnetization from high temperatures or strong armature reaction fields during heavy loads or fault conditions [17, 18]. Such demagnetization can reduce torque and increase currents within a closed-loop controlled drive, leading to a cycle that could further raise temperatures and exacerbate demagnetization.

A major concern may take place when the inverter falls out of synchronized control, resulting in voltages U that can be out of phase with the bemf E as shown in Fig. 2.35, causing substantial transient currents that might partially demagnetize the magnets. According to the PM demagnetization BH curve in Fig. 2.36, if the flux density of any magnet element is lower than the knee point during the fault process, it will operate on the recoil line with reduced remanence in subsequent steps.

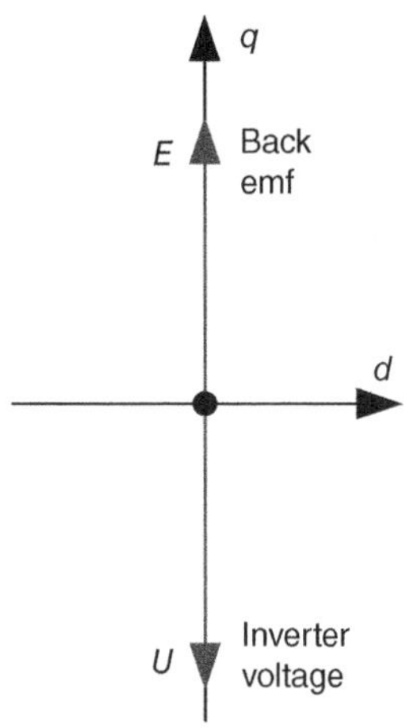

Figure 2.35 Illustration of loss of inverter synchronization.

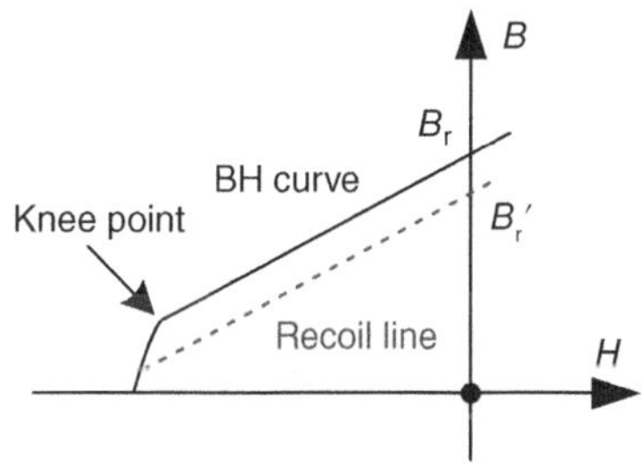

Figure 2.36 PM demagnetization BH curve.

Simulations have shown that in severe fault conditions, magnets may exhibit partial demagnetization [17, 18], reducing back EMF by roughly 31%, yet the torque declines by only about 9% as shown in Fig. 2.37. This is because PM torque constitutes only 30% of the total torque, indicating that the impact on torque capacity for the PMA-SynRM is relatively minor in the event of magnet demagnetization. Comparative studies have demonstrated that this configuration of PMA-SynRM has a stronger resistance to demagnetization relative to other types of PM machines, making it less susceptible to such faults [19].

2.3.9 DC Link Capacitor Fault

Within drive systems, the role of the DC link capacitor is to smooth out fluctuations in voltage across the DC link and provide a temporary energy buffer. This component is subjected to both voltage and thermal stresses during the operation of the inverter, contributing significantly to the rate of breakdowns in electrical drives [20].

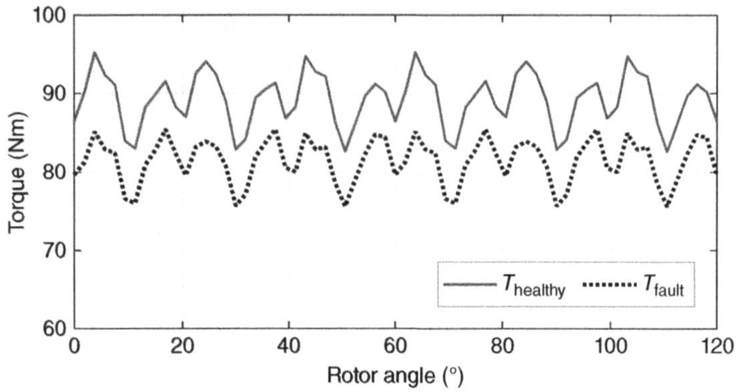

Figure 2.37 Output torque after demagnetization fault.

The predominant modes of capacitor degradation are the evaporation of the electrolyte and the occurrence of open circuits [21]. When an open-circuit fault develops, it is crucial to deactivate the relevant inverter because of the deficient filtering on the DC link, which can cause the drive system to become unstable [22]. As capacitors age, there is an increase in equivalent series resistance and a decrease in capacitance, both of which cause elevated current ripples and temperature rise, impairing the performance of the drive system. If significant changes in capacitor parameters are detected, the inverter should be shutdown to prevent a complete system failure. Similar to how the drive handles open or short-circuit faults in switches, a machine drive can withstand a DC link capacitor failure in one of the 3-phase inverters by maintaining approximately two-thirds of its operational torque, due to the triple redundancy. To tolerate any failure in the DC supply, each 3-phase inverter should have its own independent DC power source. In such a

Table 2.2 Summary of the faults in the machine drives.

Fault type	Output torque	Fault current
One set open circuit	56%	NA
One set short circuit	65%	96 A
Intraphase short circuit	~65%	340 A
Interphase short circuit	~65%	135 A
Phase-to-ground fault	56%	NA
Uncontrolled rectification fault	0	NA
Demagnetization fault	91%	NA
DC link capacitor failure	56%	NA

configuration, if one 3-phase inverter experiences a DC power loss, the system only suffers a one-third reduction in output torque.

The faults discussed above are summarized as shown in Table 2.2 in terms of postfault torque and resultant fault current. Therefore, it can be concluded that the intraphase single-turn fault in coil B2 under motoring mode is the worst fault case, which should be addressed in the machine design process.

2.4 Analysis of Fault Operation Behavior

Understanding of the machine drive fault behavior is of key importance to facilitate the fault modeling, diagnosis, control law design, and avoid further damage. Thus, the fault operation behavior of the triple redundant 3-phase PMA-SynRM is investigated. The mutual coupling mechanism of different 3-phase modules is explained by deriving MMF distribution. It is found that the 3-phase set module affects each other via an MMF offset component, causing second harmonics in the *dq* axis currents, voltages, and torque. The typical one 3-phase set open-circuit and short-circuit fault operation behaviors are investigated.

2.4.1 Analysis of Machine Winding MMF

First, the MMF associated with the machine drive's winding is calculated. The whole winding consists of three separate modules each spanning a mechanical arc of 360°, where one-third of this arc is allocated to one module. Each phase winding involves a pair of coils linked in series. According to the winding layout, the turn functions of all nine-phase coils are illustrated in Fig. 2.38. The turn functions for the phases C, F, and I appear inverted against phases A, B, D, E, G, and H, as these possess a reverse go-return polarity.

Subsequently, the average value is removed from the turn function to generate winding function as demonstrated in equation (2.2). Here, $n_x(\alpha)$ represents a specific coil's turn function, and $\langle n_x(\alpha)\rangle$ indicates the mean turn function value of the coils, as referenced in [23]. The winding function for coil A1, as an example, is depicted in Fig. 2.39.

$$N_x(\alpha) = n_x(\alpha) - \langle n_x(\alpha)\rangle \tag{2.2}$$

The winding MMF, which is calculated by multiplying each coil's winding functions with their respective currents, is described according to equation (2.3) [24].

$$\text{MMF}(\alpha) = \sum N_x(\alpha) i_x \tag{2.3}$$

To demonstrate the mutual coupling effect among the three 3-phase windings, the MMFs for ABC, DEF, and GHI sets are evaluated separately and depicted in

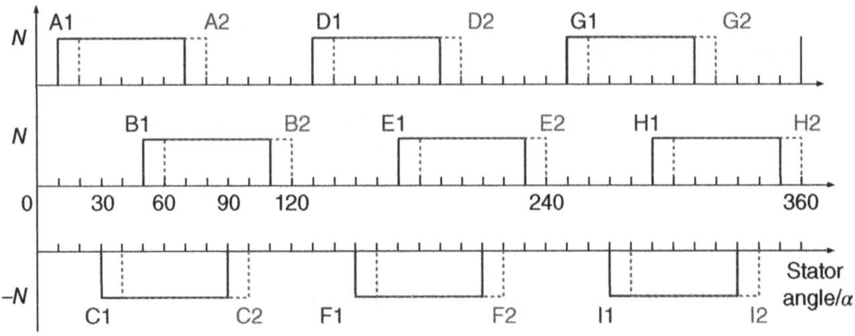

Figure 2.38 Turn functions of the nine-phase winding coils.

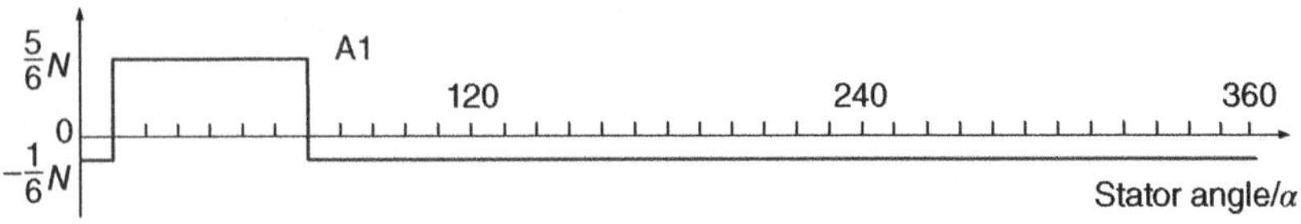

Figure 2.39 Winding function of coil A1.

Fig. 2.40. Here, a coil with nominal current represents one per-unit value (1 pu). The phase currents are defined by equation (2.4) for a given time instant.

$$
\begin{aligned}
i_A &= i_D = i_G = 1\ \text{pu} \\
i_B &= i_E = i_H = -0.5\ \text{pu} \\
i_C &= i_F = i_G = -0.5\ \text{pu}
\end{aligned}
\tag{2.4}
$$

Combining (2.3) and (2.4), the MMF distributions for the three 3-phase sets, ABC, DEF, and GHI can be obtained as shown in Fig. 2.40. It is seen that the three distributions are of the same shape with a phase displacement of 120 mechanical degrees against each other. It can also be observed, however, that the MMF distribution of one 3-phase set winding is not just confined within the space occupied by the windings, but also extends to the regions by the windings of the other two sets. This phenomenon suggests that the MMF from one set will produce flux linkages in the windings of the other two sets, meaning that the three 3-phase windings are magnetically coupled.

Since the MMF profiles from all three sets are identical, the features of the MMF distribution of the ABC set are further illustrated as an example in Fig. 2.41. The distribution can be decomposed into an AC component and a DC offset component, denoted as F^1_{ac} and F^{11}_{os}, respectively, over the space of the ABC windings, and two DC offset components, denoted as F^{21}_{os} over the space of the DEF set windings and F^{31}_{os} over the GHI set windings, respectively. These features

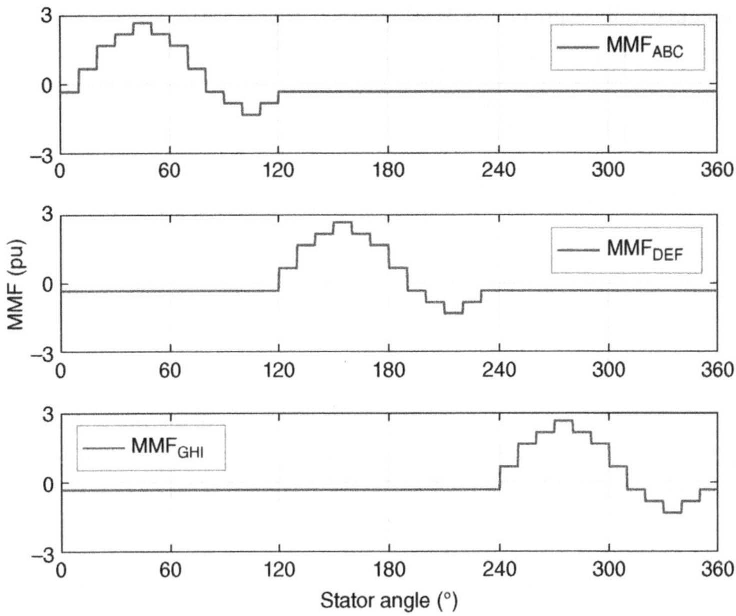

Figure 2.40 MMFs produced by each 3-phase set.

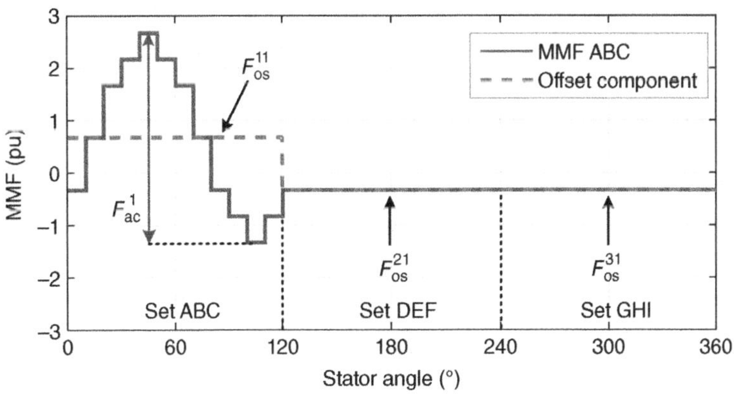

Figure 2.41 MMF produced by set ABC.

can be mathematically expressed as MMF_{ABC} in (2.5).

$$\text{MMF}_{\text{ABC}} = \begin{cases} F_{\text{ac}}^{1} + F_{\text{os}}^{11} & 0^\circ \leq \alpha < 120^\circ \\ F_{\text{os}}^{21} & 120^\circ \leq \alpha < 240^\circ \\ F_{\text{os}}^{31} & 240^\circ \leq \alpha < 360^\circ \end{cases}$$

$$F_{\text{ac}}^{1} = \sum_{n=1,3,5,\cdots} A_n \cos(3n\alpha + \delta_n)$$

$$F_{\text{os}}^{11} = \frac{2}{3}N(i_{\text{A}} + i_{\text{B}} - i_{\text{C}})$$

$$F_{\text{os}}^{21} = -\frac{1}{3}N(i_{\text{A}} + i_{\text{B}} - i_{\text{C}})$$

$$F_{\text{os}}^{31} = -\frac{1}{3}N(i_{\text{A}} + i_{\text{B}} - i_{\text{C}}) \tag{2.5}$$

A_n is the magnitude of the AC component associated with the nth space harmonic. The F_{ac}^{1} represents the AC component in healthy state when all the three sets of windings are loaded with the identical currents given in (2.4) [25]. The induced flux linkages can be described by the dq axis reference system in terms of inductance, currents, and rotor angle.

For notational convenience, set ABC, DEF, and GHI are denoted as set 1, 2, and 3, respectively. Hence, the offset components $F_{\text{os}}^{i,j}$ represent the MMF offset component of set i caused by set j, $i, j \in (1, 2, \textit{and } 3)$. The offset components can be calculated as in (2.5), and they vary with the phase currents.

During healthy operations, the current vectors or phasors are identical for all the three sets. The cumulative MMF over the space of a given set of windings is the sum of the MMFs of all three sets of windings. For instance, the MMF over the space of the ABC set, labeled as F^1, is given in equation (2.6).

$$F^1 = F_{\text{ac}}^{1} + F_{\text{os}}^{1}$$

$$F_{\text{os}}^{1} = F_{\text{os}}^{11} + F_{\text{os}}^{12} + F_{\text{os}}^{13}$$

$$= \frac{2}{3}N(i_{\text{A}} + i_{\text{B}} - i_{\text{C}}) - \frac{1}{3}N(i_{\text{D}} + i_{\text{E}} - i_{\text{F}}) - \frac{1}{3}N(i_{\text{G}} + i_{\text{H}} - i_{\text{I}}) \tag{2.6}$$

It should be noted that the magnitude of the MMF AC component over a 3-phase winding set is only dependent on the phase currents in the windings, while the offset component is on the currents of all three sets.

When the machine operates in healthy conditions, the currents in each of the three sets are identical, namely, $i_{\text{A}} = i_{\text{D}} = i_{\text{G}}$, $i_{\text{B}} = i_{\text{E}} = i_{\text{H}}$, and $i_{\text{C}} = i_{\text{F}} = i_{\text{I}}$, leading to cancellation of the offset components according to (2.6). Therefore, the machine operates symmetrically and can be controlled by classical PI current controller in the dq axis reference frame. Conversely, in the event of a fault, the offset component of the MMF deviates from zero, which interacts with the rotor magnetic field to produce a disturbance torque component to the machine drive. In summary, the different 3-phase sets experience mutual coupling through the MMF offset components, which become zero in healthy operations but a disturbance under a fault or during postfault operations. The above analysis serves as a knowledge base for the understanding of the machine fault operation under open-circuit and short-circuit conditions.

2.4.2 Analysis of Open-Circuit Fault

When an open-circuit fault in one phase is detected, all three phases in the winding set will be open circuited via inverter control as a remedy action. Consider the scenario where the open-circuit fault takes place in set ABC. Thus, the phase currents in set ABC are zero. For the purpose of general analysis, the currents in the remaining healthy sets are expressed by the currents in dq axis reference system as (i_d, i_q). The resulting currents in the healthy phases with time variation are given in (2.7), where ω represents the mechanical angular velocity of the machine, while p is the number of rotor pole pairs. The resulting MMF distribution across the stator air gap over 360 mechanical degrees is depicted in Fig. 2.42 for a given time instant.

$$
\begin{aligned}
i_{\mathrm{D}} &= i_{\mathrm{G}} = i_d \cos(p\omega t) - i_q \sin(p\omega t) \\
i_{\mathrm{E}} &= i_{\mathrm{H}} = i_d \cos(p\omega t - 120^\circ) - i_q \sin(p\omega t - 120^\circ) \\
i_{\mathrm{F}} &= i_{\mathrm{I}} = i_d \cos(p\omega t + 120^\circ) - i_q \sin(p\omega t + 120^\circ) \qquad (2.7)
\end{aligned}
$$

It can be noted that only an MMF offset component F_{os}^{1} exists over the region occupied by the open-circuited set ABC windings. The absence of phase currents in the fault set leads to a zero AC component in this area. Conversely, the regions over the healthy sets are characterized by an AC component and an offset MMF component. As previously stated, the magnitude of the AC component is dependent on the currents in the healthy sets, represented by (i_d, i_q). The offset MMF

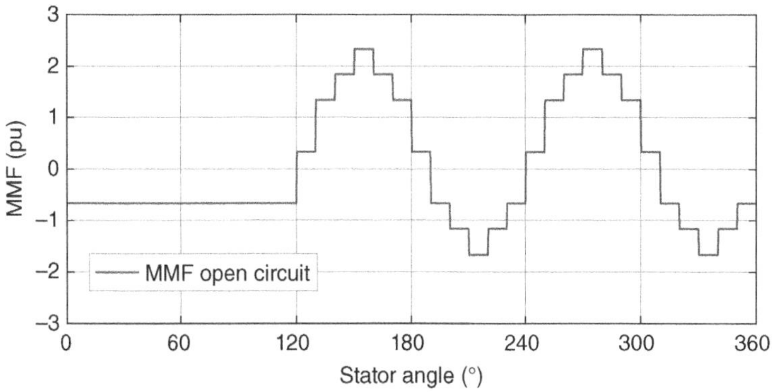

Figure 2.42 MMF in set ABC open-circuit condition.

components, F_{OS}^1, F_{OS}^2, F_{OS}^3, associated with the respective regions of the three sets of windings can be evaluated using equation (2.8) according to equation (2.6).

$$\begin{aligned} F_{\text{os}}^1 &= -\frac{1}{3}N(i_{\text{D}} + i_{\text{E}} - i_{\text{F}}) - \frac{1}{3}N(i_{\text{G}} + i_{\text{H}} - i_{\text{I}}) \\ F_{\text{os}}^2 &= \frac{2}{3}N(i_{\text{D}} + i_{\text{E}} - i_{\text{F}}) - \frac{1}{3}N(i_{\text{G}} + i_{\text{H}} - i_{\text{I}}) \\ F_{\text{os}}^3 &= -\frac{1}{3}N(i_{\text{D}} + i_{\text{E}} - i_{\text{F}}) + \frac{2}{3}N(i_{\text{G}} + i_{\text{H}} - i_{\text{I}}) \end{aligned} \tag{2.8}$$

By substituting equation (2.7) into (2.8), the following equations are derived:

$$\begin{aligned} F_{\text{os}}^1 &= -\frac{4}{3}N(i_d \cos(p\omega t - 60^\circ) - i_q \sin(p\omega t - 60^\circ)) \\ F_{\text{os}}^2 &= \frac{2}{3}N(i_d \cos(p\omega t - 60^\circ) - i_q \sin(p\omega t - 60^\circ)) \\ F_{\text{os}}^3 &= \frac{2}{3}N(i_d \cos(p\omega t - 60^\circ) - i_q \sin(p\omega t - 60^\circ)) \end{aligned} \tag{2.9}$$

The flux density in the air gap is dependent on the MMF distribution and hence determines the phase flux linkage. Since the MMFs of the set DEF and GHI are the same, the flux distribution of set DEF is further analyzed by way of example. While the MMF of the healthy set DEF consists of an AC component and an offset component, their resultant flux linkages are analyzed separately because of the difference in their distribution patterns. First, the dq axis flux linkages of set DEF produced by the MMF AC component are denoted by ψ_{ACd}^2, ψ_{ACq}^2 and can be expressed in terms of the dq axis inductances (L_d, L_q) and the dq axis currents (i_d, i_q) as shown in (2.10). This relationship is the same as that in healthy conditions.

$$\begin{aligned} \psi_{\text{ACd}}^2 &= L_d i_d \\ \psi_{\text{ACq}}^2 &= L_q i_q \end{aligned} \tag{2.10}$$

The phase flux linkages of the DEF set that arise from the offset MMF component are proportional to the magnitude of F_{OS}^2. The relationships are described in equation (2.11).

$$\begin{aligned} \psi_{\text{OSD}}^2 &= kF_{\text{os}}^2 \\ \psi_{\text{OSE}}^2 &= kF_{\text{os}}^2 \\ \psi_{\text{OSF}}^2 &= -kF_{\text{os}}^2 \end{aligned} \tag{2.11}$$

where k is a proportional constant associated with the permeance of the magnetic field distribution over the winding set. When magnetic saturation associated with the fundamental flux density components is not very high, this coefficient can be considered invariant with respect to the rotor angle due to the use of full-pitch distributed windings. Its value under a given operating condition is determined via FE computations [26, 27]. Notably, the flux linkage for phase F is reversely polarized, owing to the reverse turn function with respect to those of phases D and E.

Then, the phase flux linkages described in equation (2.11) are transformed to the $dq0$ axis components, (ψ^2_{OSd}, ψ^2_{OSq}, ψ^2_{OS0}), given in equation (2.12). It can be seen that the dq axis flux linkages contain a DC element alongside a second-order harmonic, whereas the 0-axis flux linkage is characterized solely by a component at the fundamental frequency. This is different from that in healthy operations, where only DC elements are present in the dq flux linkages and no 0-axis component exists. The presence of these harmonics is responsible for undesirable machine performance under the fault conditions, such as increased current distortion and torque ripple.

$$\begin{aligned}
\psi^2_{OSd} &= \frac{4k}{9}\left[i_d(1+\cos(2p\omega t-120°)) - i_q\sin(2p\omega t-120°)\right]\\
\psi^2_{OSq} &= \frac{4k}{9}\left[i_q(1-\cos(2p\omega t-120°)) - i_d\sin(2p\omega t-120°)\right]\\
\psi^2_{OS0} &= \frac{2k}{9}\left(i_d\cos(p\omega t-60°) - i_q\sin(p\omega t-60°)\right)
\end{aligned} \tag{2.12}$$

The impact of the open-circuit fault on machine performance has been assessed through FE simulations. When the machine is with the rated current of 120 A in the healthy sets, DEF and GHI at the base speed of 4000 rpm while set ABC experiencing an open circuit, the $dq0$ axis flux linkages of set DEF are depicted in Fig. 2.43. The presence of the second-order harmonics in the dq axes flux linkages and the component at the fundamental frequency in the 0-axis flux linkage are confirmed, validating the prior analysis. The flux linkages for the GHI set are similar, and thus, their details are omitted for brevity.

The interaction of the second harmonic d- and q-axis component of the DEF set with the constant rotor PM flux linkage produces the second-order

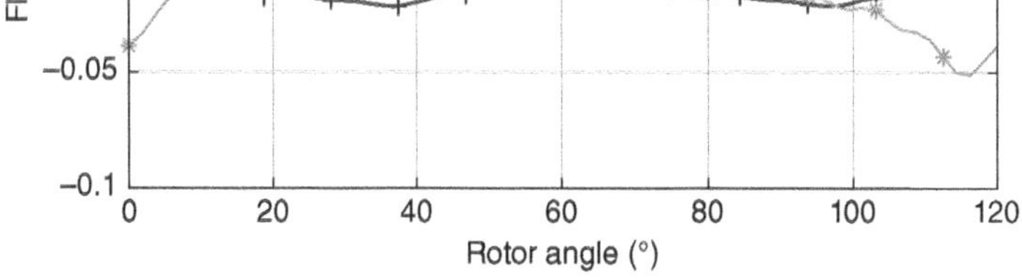

Figure 2.43 *dq0* axis flux linkages of set DEF with set ABC open circuited.

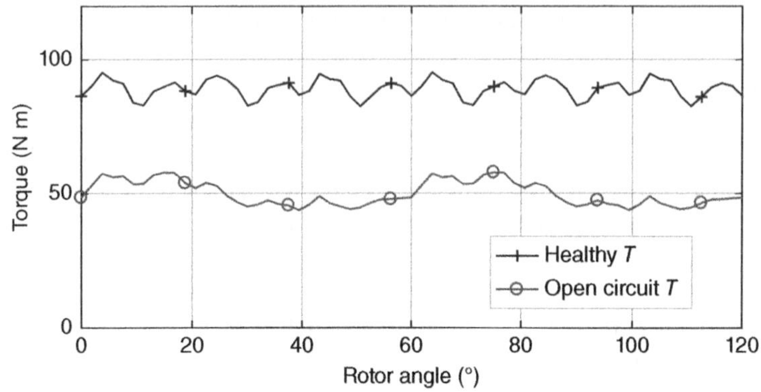

Figure 2.44 Torque waveform with set ABC open circuited.

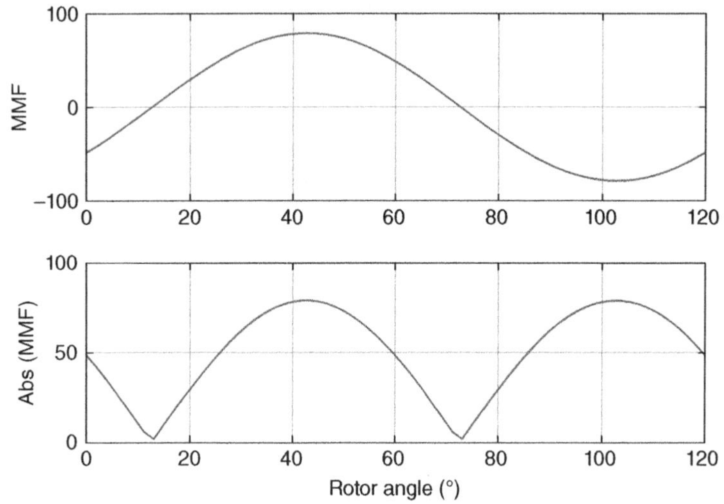

Figure 2.45 MMF and abs(MMF) offset component of set DEF when set ABC open circuited.

torque pulsation that is evident from the simulation results shown in Fig. 2.44, confirming the connection between the 2nd-order torque ripples and the MMF offset component. The MMF offset component's real-time value and its absolute value are represented in Fig. 2.45 as rotor angle varies. It is evident that the maximum value of open-circuit torque aligns with the zero MMF offset component. On the contrary, the minimum value coincides with the maximum value of MMF offset component.

It should be noted that the foregoing evaluations are performed with the assumption of ideal sinusoidal current inputs. During actual operations, each

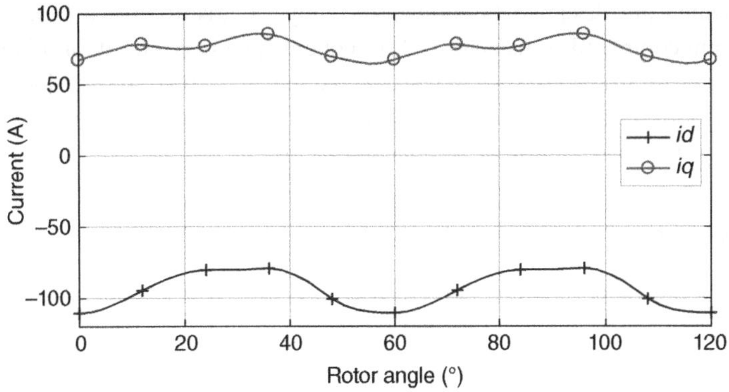

Figure 2.46 *dq* axis currents of set DEF with set ABC open circuited.

3-phase set of windings of the machine is typically driven by a 3-phase inverter under closed-loop proportional and integral (PI) current regulation. Thus, because of the limited bandwidth of the conventional PI controllers, the second harmonics in the *dq* axis flux linkages and the induced voltages under the open-circuit operation will interact with the PI current regulators, leading to further distortions in the currents and voltages. To appreciate this effect, further simulations are conducted considering the machine drive under a representative PI current controller. The resulting *dq* axis currents and voltages are displayed in Figs. 2.46 and 2.47. As anticipated, the disturbance caused by the MMF offset gives rise to the second-order harmonics in both the currents and voltages, which may undermine the drive stability during postfault operations. With the presence of the second-order harmonics in the *dq* axis flux linkages, the conventional PI

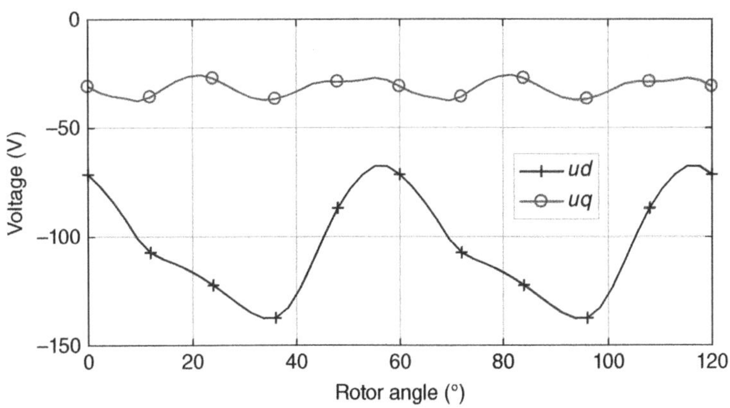

Figure 2.47 *dq* axis voltages of set DEF with set ABC open circuited.

controllers, which are designed to compensate only DC disturbances, cannot track the current commands satisfactorily. Consequently, there is a need for more sophisticated control strategies, such as negative sequence control or proportional resonant control, to enhance current tracking performance and ensure system stability during operations following an open fault.

2.4.3 Analysis of Short-Circuit Fault

In the event of a short-circuit fault, either in an inverter switch or in a phase winding of the machine, an effective mitigation strategy is to apply TSC on the fault set via inverter control. Meanwhile, the healthy sets continue their operation with appropriate currents required by the load. Consequently, the MMF generated by the currents in the healthy phases can be established in the same way as what has been described for an open-circuit fault. As discussed earlier, the currents in the healthy phases produce an MMF offset component over the fault set region, which can be evaluated by (2.9). The phase flux linkages ($\psi^1_{\text{OSA}}, \psi^1_{\text{OSB}}, \psi^1_{\text{OSC}}$) induced by the MMF offset component in the fault set ABC windings due to the currents in the healthy phases can be similarly deduced in (2.13) as in (2.11).

$$\begin{aligned}
\psi^1_{\text{OSA}} &= kF^1_{\text{os}} \\
\psi^1_{\text{OSB}} &= kF^1_{\text{os}} \\
\psi^1_{\text{OSC}} &= -kF^1_{\text{os}}
\end{aligned} \tag{2.13}$$

Converting the above flux linkages to $dq0$ axis, the following equation can be obtained:

$$\begin{aligned}
\psi^1_{\text{OS}d} &= \frac{8k}{9}\left[i_q \sin(2p\omega t - 120^\circ) - i_d(1 + \cos(2p\omega t - 120^\circ))\right] \\
\psi^1_{\text{OS}q} &= \frac{8k}{9}\left[i_d \sin(2p\omega t - 120^\circ) - i_q(1 - \cos(2p\omega t - 120^\circ))\right] \\
\psi^1_{\text{OS0}} &= -\frac{4k}{9}\left(i_d \cos(p\omega t - 60^\circ) - i_q \sin(p\omega t - 60^\circ)\right)
\end{aligned} \tag{2.14}$$

where ($\psi^1_{\text{OS}d}, \psi^1_{\text{OS}q}, \psi^1_{\text{OS0}}$) are the d, q, and 0 components of the flux linkages in (2.13). In addition, the magnetic field produced by the PM rotor will also link with the faulted 3-phase set of windings. The resultant flux linkages can be expressed in (2.15),

$$\begin{aligned}
\psi_{\text{mA}} &= \psi_{\text{m}} \cos(p\omega t) \\
\psi_{\text{mB}} &= \psi_{\text{m}} \cos(p\omega t - 120^\circ) \\
\psi_{\text{mC}} &= \psi_{\text{m}} \cos(p\omega t + 120^\circ)
\end{aligned} \tag{2.15}$$

where ψ_{m} is the magnitude of the PM flux linkage. ($\psi_{\text{mA}}, \psi_{\text{mB}}, \psi_{\text{mC}}$) denote the flux linkages in phases A, B, and C due to the rotor PM field, respectively. When they are expressed in the dq axis reference, only the d-axis flux linkage exists as the d-axis is aligned with PM field.

$$\psi_{\mathrm{md}}^{1} = \psi_{\mathrm{m}}, \psi_{\mathrm{m}q}^{1} = 0, \psi_{\mathrm{m0}}^{1} = 0 \tag{2.16}$$

where $(\psi_{\mathrm{md}}^{1}, \psi_{\mathrm{m}q}^{1}, \psi_{\mathrm{m0}}^{1})$ denotes the d-axis, q-axis, and 0-axis components of the flux linkage in phases A, B, and C due to the rotor PM field. Therefore, the total external flux linkages of the fault 3-phase set windings, due to the currents of the healthy sets and the rotor PM field, can be derived and given in (2.17), where $(\psi_{d}^{1}, \psi_{q}^{1}, \psi_{0}^{1})$ denotes the total d-axis, q-axis, and 0-axis flux linkages of the ABC set. It is well known that in the dq axis framework, the DC element corresponds to the positive sequence component of the flux linkages at the fundamental frequency while the second harmonics represent the negative sequence component at the fundamental frequency. This understanding provides a useful insight into the short-circuit currents, which are induced in the ABC set of windings after TSC.

$$\begin{aligned}
\psi_{d}^{1} &= \psi_{\mathrm{m}} + \frac{8k}{9}\left[i_q \sin(2p\omega t - 120^\circ) - i_d(1 + \cos(2p\omega t - 120^\circ))\right] \\
\psi_{q}^{1} &= \frac{8k}{9}\left[i_d \sin(2p\omega t - 120^\circ) - i_q(1 - \cos(2p\omega t - 120^\circ))\right] \\
\psi_{0}^{1} &= -\frac{4k}{9}\left(i_d \cos(p\omega t - 60^\circ) - i_q \sin(p\omega t - 60^\circ)\right)
\end{aligned} \tag{2.17}$$

According to Lenz's law, the induced short-circuit currents in the ABC set tend to reduce the external flux linkages given in (2.17). However, since the neutral point of the ABC windings is isolated, no 0-axis current will be present and the 0-axis flux linkage remains unchanged in the ABC phases. The resultant d-axis and q-axis currents, i_d^1, i_q^1, can be determined by (2.18). Owing to the presence of both the positive and negative sequence flux linkage components in the ABC windings, the resultant currents will contain similar sequence components [7]. Consequently, both the positive and negative sequence flux linkage components in the ABC phases will be much reduced.

$$\begin{aligned}
0 &= Ri_d^1 + \frac{d\psi_d^1}{dt} - p\omega\psi_q^1 \\
0 &= Ri_q^1 + \frac{d\psi_q^1}{dt} + p\omega\psi_d^1
\end{aligned} \tag{2.18}$$

The machine behavior under a representative post short-circuit fault operation is evaluated by FE simulation. The analysis is carried out with the rated load currents applied to the healthy DEF and GHI phases, while the ABC set undergoes a TSC at 4000 rpm. Figure 2.48 shows the resultant dq axis short-circuit currents of set ABC. These currents exhibit significant second harmonics, corresponding to the negative sequence components. For a more obvious illustration, the dq axis currents are transformed into the positive and negative sequence components shown in Fig. 2.49.

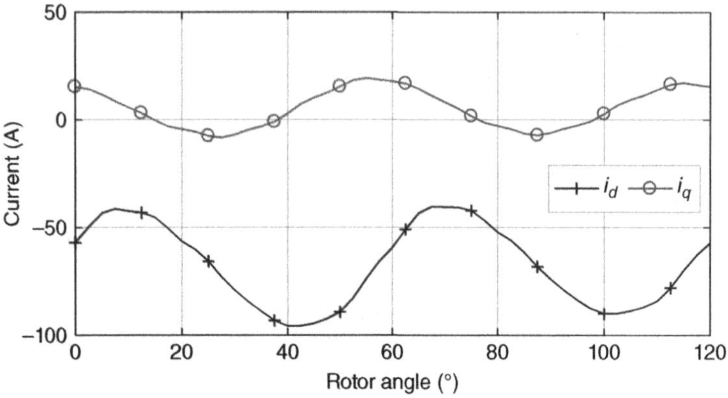

Figure 2.48 Short-circuit currents in set ABC.

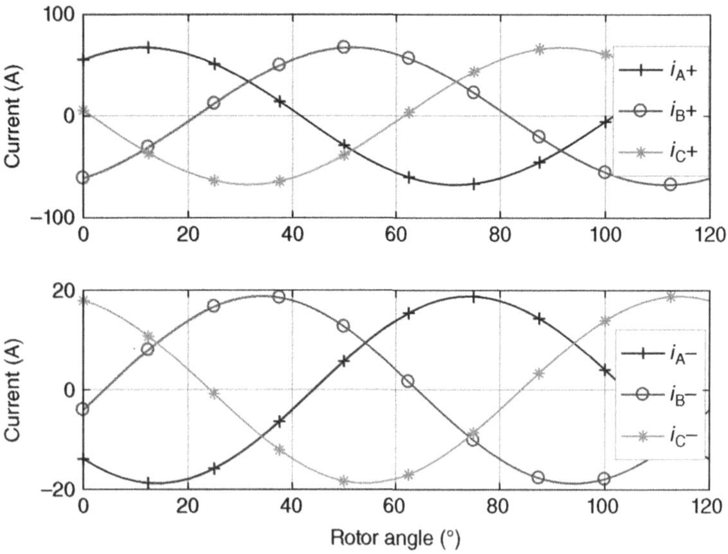

Figure 2.49 Positive and negative sequence components of the short-circuit currents in set ABC.

Figure 2.50 displays the residual flux linkages of the ABC set. It is evident that the flux linkages under the combined effect of the short-circuit currents and external flux linkages predominantly consist of zero-sequence components. This occurs as the positive and negative sequence components of the external flux linkage have been significantly reduced by their respective currents in Fig. 2.49, while the zero

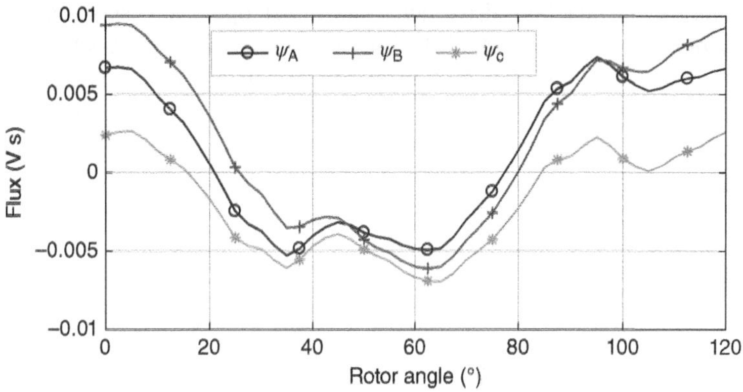

Figure 2.50 Phase flux linkages of set ABC when it is terminal short circuited.

sequence component remains owing to the absence of the 0-axis current path in the ABC windings.

It should be noted that the machine may also experience a turn fault [28] and phase-to-ground fault [29] if the turn-to-turn (or strands) insulation and the ground wall insulation degrade gradually to a sufficiently poor level. While these specific types of faults have not been evaluated in the foregoing analysis, the principles outlined offer useful insights into the potential behaviors of these fault types following the application of TSC as a mitigation measure.

2.5 Summary

In this chapter, a triple redundant 3×3-phase machine drive based on PMA SynRM with segregated windings has been described and its performances under healthy, fault, and postfault operation have been evaluated. The extensive performance evaluations show that the machine drive exhibits a very good balance between high performance in healthy operations and ability to tolerate various faults, including one set open circuit, one set short circuit, intraphase turn short circuit, interphase turn short circuit, single point phase-to-ground insulation fault, uncontrolled rectification, demagnetization, and DC capacitor failure. If any of these faults except for demagnetization occur to one 3-phase set drive, the output torque will just reduce by about one-third following the application of an appropriate mitigation measure. The reduction in torque under partial demagnetization is much lower. These desirable performances are mainly attributed to the combined torque mechanism of the machine topology with a high level of rotor saliency and low PM field, the segregated windings design, and the use of independent drives for each 3-phase winding set. In-depth physical understanding of the machine drive behaviors under open- and short-circuit fault operations has also been established by the mutual coupling mechanism through MMF distributions of each 3-phase winding. It has been shown that dominant second-order harmonics in the machine voltages, currents, and torque are caused by the MMF offset components through the mutual coupling mechanism.

References

1 Cao, W., Mecrow, B.C., Atkinson, G.J. et al. (2012). Overview of electric motor technologies used for more electric aircraft (MEA). *IEEE Transactions on Industrial Electronics* 59 (9): 3523–3531.

2 A. Boglietti, A. Cavagnino, A. Tenconi, and S. Vaschetto, "The safety critical electric machines and drives in the more electric aircraft: A survey," in *35th Annual Conference of IEEE Industrial Electronics, 2009. IECON '09*, 2009, pp. 2587–2594.

3 Yu-Seok, J., Seung-Ki, S., Schulz, S.E., and Patel, N.R. (2005). Fault detection and fault-tolerant control of interior permanent-magnet motor drive system for electric vehicle. *IEEE Transactions on Industry Applications* 41 (1): 46–51.

4 Ferrari, M., Bianchi, N., Doria, A., and Fornasiero, E. (2015). Design of synchronous reluctance motor for hybrid electric vehicles. *IEEE Transactions on Industry Applications* 51 (99): 3030–3040.

5 P. Guglielmi, N. G. Giraudo, G. M. Pellegrino, and A. Vagati, "P.M. assisted synchronous reluctance drive for minimal hybrid application," in *Industry Applications Conference, 2004. 39th IAS Annual Meeting. Conference Record of the 2004 IEEE*, 2004, vol. 1, pp. 1–306.

6 Q. Dinyu, L. Xiaogang, and T. A. Lipo, "Reluctance motor control for fault-tolerant capability," in *Electric Machines and Drives Conference Record, 1997. IEEE International*, 1997, pp. WA1/1.1–WA1/1.6.

7 Welchko, B.A., Jahns, T.M., Soong, W.L., and Nagashima, J.M. (2003). IPM synchronous machine drive response to symmetrical and asymmetrical short circuit faults. *IEEE Transactions on Energy Conversion* 18 (2): 291–298.

8 Welchko, B.A., Jahns, T.M., and Hiti, S. (2002). IPM synchronous machine drive response to a single-phase open circuit fault. *IEEE Transactions on Power Electronics* 17 (5): 764–771.

9 B. Wang, J. Wang, and A. Griffo, "A fault tolerant machine drive based on permanent magnet assisted synchronous reluctance machine" in *Energy Conversion Congress and Exposition (ECCE), 2016 IEEE*, Milwaukee, WI, 2016, pp. 1–8.

10 O. V. Thorsen and M. Dalva, "A survey of faults on induction motors in offshore oil industry, petrochemical industry, gas terminals and oil refineries," in *Petroleum and Chemical Industry Conference, 1994. Record of Conference Papers., Institute of Electrical and Electronics Engineers Incorporated Industry Applications Society 41st Annual*, 1994, pp. 1–9.

11 Chen, Q., Liu, G., Zhao, W. et al. (2014). Design and comparison of two fault-tolerant interior-permanent-magnet motors. *IEEE Transactions on Industrial Electronics* 61 (12): 6615–6623.

12 Patel, D.C. and Chandorkar, M.C. (2014). Modeling and analysis of stator interturn fault location effects on induction machines. *IEEE Transactions on Industrial Electronics* 61 (9): 4552–4564.

13 Jahns, T.M. and Caliskan, V. (1999). Uncontrolled generator operation of interior PM synchronous machines following high-speed inverter shutdown. *IEEE Transactions on Industry Applications* 35 (6): 1347–1357.

14 Chong-Zhi, L., Soong, W.L., Welchko, B.A., and Ertugrul, N. (2005). Uncontrolled generation in interior permanent-magnet machines. *IEEE Transactions on Industry Applications* 41 (4): 945–954.

15 S. H. Han, T. M. Jahns, M. Aydin, M. K. Guven, and W. L. Soong, "Impact of maximum back-EMF limits on the performance characteristics of interior permanent magnet synchronous machines," in *Conference Record of the 2006 IEEE Industry Applications Conference Forty-First IAS Annual Meeting*, 2006, vol. 4, pp. 1962–1969.

16 Pellegrino, G., Vagati, A., and Guglielmi, P. (2011). Design tradeoffs between constant power speed range, uncontrolled generator operation, and rated current of IPM motor drives. *IEEE Transactions on Industry Applications* 47 (5): 1995–2003.

17 Patel, V.I., Wang, J., and Nair, S.S. (2015). Demagnetization assessment of fractional-slot and distributed wound 6-phase permanent magnet machines. *IEEE Transactions on Magnetics* 51 (6): 1–11.

18 Nair, S.S., Patel, V.I., and Wang, J. (2016). Post-demagnetization performance assessment for interior permanent magnet AC machines. *IEEE Transactions on Magnetics* 52 (4): 1–10.

19 Fasolo, A., Alberti, L., and Bianchi, N. (2014). Performance comparison between switching-flux and IPM machines with rare-earth and ferrite PMs. *IEEE Transactions on Industry Applications* 50 (6): 3708–3716.

20 Sun, P., Gong, C., Du, X. et al. (2017). Online condition monitoring for both IGBT module and DC-link capacitor of power converter based on short-circuit current simultaneously. *IEEE Transactions on Industrial Electronics* 64 (5): 3662–3671.

21 T. Kamel, C. Diduch, Y. Bilestkiy, and L. Chang, “Fault diagnoses for the DC filters of power electronic converters,” in *2012 IEEE Energy Conversion Congress and Exposition (ECCE)*, 2012, pp. 2135–2141.

22 Wang, J. and Howe, D. (2008). A power shaping stabilizing control strategy for DC power systems with constant power loads. *IEEE Transactions on Power Electronics* 23 (6): 2982–2989.

23 Wang, B., Wang, J., Griffo, A., and Sen, B. (2018). A general modeling technique for a triple redundant 3×3-phase PMA SynRM. *IEEE Transactions on Industrial Electronics* 65 (11): 9068–9078.

24 Faiz, J., Ardekanei, I.T., and Toliyat, H.A. (2003). An evaluation of inductances of a squirrel-cage induction motor under mixed eccentric conditions. *IEEE Transactions on Energy Conversion* 18 (2): 252–258.

25 Chen, X., Wang, J., Patel, V.I., and Lazari, P. (2016). A nine-phase 18-slot 14-pole interior permanent magnet machine with low space harmonics for electric vehicle applications. *IEEE Transactions on Energy Conversion* 31 (3): 860–871.

26 B. Sen, J. Wang, and P. Lazari, “A detailed transient model of Interior Permanent Magnet motor accounting for saturation under stator turn fault,” in *2013 IEEE Energy Conversion Congress and Exposition (ECCE)*, 2013, pp. 3548–3555.

27 Wang, B., Wang, J., Griffo, A., and Hua, W. (2019). Effective turn fault mitigation by creating zero sequence current path for a triple redundant 3 × 3-phase PMA SynRM. *IEEE Transactions on Power Electronics* 34 (11): 11080–11089.

28 Ullah, Z. and Hur, J. (2020). Analysis of inter-turn-short fault in an FSCW IPM type brushless motor considering effect of control drive. *IEEE Transactions on Industry Applications* 56 (2): 1356–1367.

29 L. Wei and Z. Liu, “Identifying ground fault location in High Resistance Grounded systems for Adjustable Speed Drive at low speed,” in *2012 IEEE Energy Conversion Congress and Exposition (ECCE)*, 2012, pp. 3609–3616.

3

Design Optimization of Multiple 3-Phase Fault-Tolerant Machine

3.1 Introduction

This chapter describes a design optimization process for multiple 3-phase fault-tolerant machines based on the chosen permanent magnet-assisted synchronous reluctance machine (PMA-SynRM) topology with segregated windings for aerospace application as a case study. The design process of a fault-tolerant machine is more complex than that of a conventional machine since it is required to achieve desirable performance and enhanced fault tolerance while satisfying the electrical, thermal, mechanical, and volumetric constraints for all operation conditions. Especially for aerospace applications, the machine drives often operate in extremely harsh environment with limited space, high level of vibration, high temperature, and heavy load cycles, all of which make the design more challenging.

The design optimization of a fault-tolerant machine has attracted much attention in academia and industry. Various machine winding and redundancy configurations were analyzed in [1] for electrification in more electric aircraft to attain required reliability. In applications that require high power density and high efficiency, permanent magnet (PM) machines are usually the choice of design. By limiting the PM flux and choosing proper rotor saliency ratio, the undesirable braking torque and short-circuit current can be restricted at an acceptable level for applications such as electric power steering [2]. To limit short-circuit current to a thermally sustainable level, PM, or flux switching machines with fractional slot, single-layer concentrated windings may be employed and designed to have per unit phase inductance [3, 4]. However, excessive fault current may occur in the event of interturn short-circuit fault as a result of turn insulation failure, depending on the fault location in the slot. Further attempts have been made on the advanced conductor material, optimized conductor placement, and different slot pole combinations [5–7] to limit the turn fault current. While these measures have some effects on limiting the short-circuit currents, they also

Multiple 3-Phase Fault Tolerant Permanent Magnet Machine Drives: Design and Control, First Edition.
Bo Wang and Jiabin Wang.

increase AC losses in the machine windings and compromise performance in healthy operations. In general, the design of a fault-tolerant machine is more complex, requiring optimal trade-offs against performance, ability to tolerate various faults, and cost.

In this chapter, a comprehensive design methodology is presented and discussed for a fault-tolerant machine. The optimization process incorporates diverse design tools that evaluate electrical, magnetic, and thermal behaviors of candidate designs. Each design sample undergoes a dual assessment for performances under both healthy and fault-tolerant operations. The optimization aims to maximize the machine's efficiency at the rated operation point while ensuring sustainable postfault operation for the worst-fault case. The optimized machine is prototyped and tested under healthy and various fault modes of operation.

3.2 Design Specifications

Without loss of generality, a fault-tolerant machine required for aerospace application is considered and its specifications together with the cooling condition and volumetric constraints are listed in Table 3.1. The torque-speed envelope is shown in Fig. 3.1. This application requires high torque at low speed and a wide constant power operating range at high ambient temperature. The ratio of the base speed

Table 3.1 Machine specifications.

Specifications	Symbol	Value
Base speed	n_b	4000 rpm
Maximum speed	n_m	19,200 rpm
Peak power	P_p	50 kW
Peak torque below base speed	T_p	120 N m
Rated power at base speed	P_r	40 kW
Rated torque below base speed	T_r	95 N m
DC-link voltage	V_{dc}	270 V
Max phase current	I_m	200 A
Ambient temperature	T_{am}	100 °C
Cooling medium		Aeroshell oil
Coolant inlet temperature	T_{in}	110 °C
Max axial length		<400 mm
Max diameter		<220 mm
Redundancy		Triple or quadruple

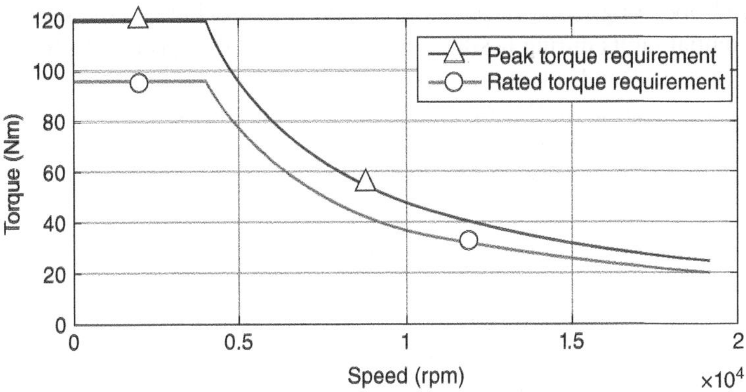

Figure 3.1 Required torque speed operation envelope.

to the maximum speed over which the machine operates in the constant power mode is 4.8.

To meet the high torque requirement at the high ambient temperature, it is necessary to employ oil cooling and class-R insulation for the windings. The anticipated lifespan for aerospace applications is around 300,000 hours, necessitating a maximum winding temperature limit of approximately 180 °C based on the chosen insulation class. Given the ambient temperature of 100 °C and a coolant inlet temperature of 110 °C, the allowable temperature rise is limited to 70 °C. This requires a carefully designed oil cooling system to prevent excessive temperature increase.

To facilitate fault-tolerant operations against all failure modes described in Chapter 1, a triple or quadruple redundancy is necessary for which the machine consists of three or four independent modules. Further, it is essential that all fault behaviors are thoroughly analyzed and appropriate measures are taken to prevent complete system failure. Consequently, the machine design must not only focus on maximizing the objective function, such as efficiency in healthy operations, but also meet the requirements for sustainable postfault operations against a range of potential faults, while satisfying electrical, mechanical, and thermal constraints in both normal and fault conditions. The need for this multifacet approach significantly adds to the complexity of the design process.

3.3 Design Optimization Process

Following a trade-off study for postfault performance against system complexity, a triple redundant fault-tolerant PMA-SynRM is selected and its design optimized through a comprehensive design optimization process. The main design objective is to maximize efficiency at the rated operation, while meeting the required torque-speed operating range in healthy conditions and being able to sustain various potential faults during postfault operation under the specific electrical, thermal, mechanical, and volumetric constraints.

The machine must be capable of withstanding all fault scenarios outlined in Chapter 1. As discussed in Chapter 2, accommodating an open circuit or terminal short circuit (TSC) fault is relatively straightforward because even under the short-circuit condition, the short-circuit currents and resultant thermal load are lower than the rated owing to the low back electromotive force (back-emf). The output torque and power will be reduced roughly by one-third, which is acceptable during postfault operations. As has been identified in Chapter 2, however, the severest fault is an interturn short circuit in a specific coil depending on modes of operation and direction of rotation, which occurs in the turn close to the slot opening, when the machine operates with the rated current. Although the probability of occurrence of the worst case is relatively low, it is prudent to analyze the behavior of the most critical fault and to ensure at the design stage that the machine can sustain such faults during postfault operations when TSC has been applied by the inverter as the mitigation measure. Additionally, the machine's back-emf must be limited against the available DC-link voltage to prevent unintended uncontrollable regeneration under inverter faults. The complex design process will be dealt with by employing an integrated design tool and methodology discussed subsequently.

3.3.1 Optimization Tools

To expedite the design process, the use of effective optimization techniques and computationally efficient tools becomes indispensable [8]. The integration of Cedrat's Flux electromagnetic analysis tool with the optimization tool GOT-It[1] creates a robust platform for electrical machine design optimization process. At the time of writing this book, Cedrat's Flux FE tool has been integrated into Altair's electromagnetic tools and design optimization platform,[2] which has become even more powerful and convenient. However, the principle and process are generic, and hence the optimization process described in this book is based on the combination of Cedrat FE tools and GOT-It optimization tool. GOT-It is a reliable and powerful optimization tool capable of addressing constrained or unconstrained optimization problems with one or multiple objectives. It offers a wide range of optimization algorithms, including genetic algorithm (GA) and sequential quadratic programming, allowing users to choose the most suitable method for their specific application.

First, a parametrized finite element (FE) model of the machine is built in Flux2D and it communicates with GOT-It during the optimization process. In GOT-It,

1 http://www.cedrat.com/en/software/got-it.html.
2 https://altair.com/flux/.

the parameters to be optimized, their ranges, objective functions, and constraints are flexibly defined by users in Python scripts. Each time the optimizer requests a run of the FE model, GOT-It feeds the geometry parameters to the model, initiating the solving process automatically. The results, in terms of objectives and constraints, are then returned as functions of the geometry design parameters. Subsequently, an optimization algorithm assesses these results to determine the global optimum point for the FE-based optimization problem involving multiple geometry parameters [9].

A sequential surrogate optimization (SSO) process is adopted for the design. The SSO analyzes the objectives and constraints of each design derived from the FE model by constructing two response surfaces – one for the constraints and one for the objectives. These surfaces capture the estimated behaviors of the objectives and constraints and are then combined to create a surrogate. As this surrogate is an analytical function, the SSO can utilize GA or sequential quadratic program to refine the parameter search region, bringing it closer to the global optimum design. This iterative process continues, typically converging to the optimum design within a defined level of tolerance after about three generations of iterations. The entire optimization process is depicted in the flow chart shown in Fig. 3.2.

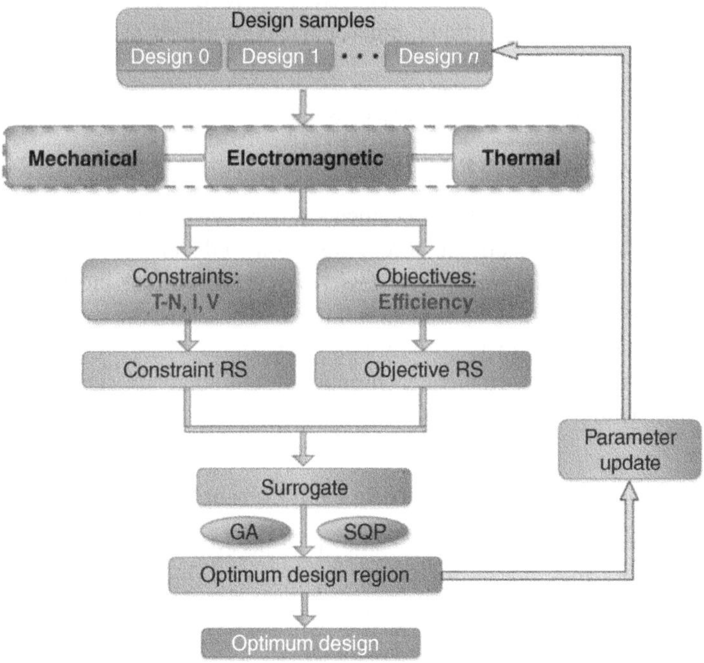

Figure 3.2 Flow chart of SSO optimization process.

3.3.2 Flux Linkage Machine Model Based on FE Computation

Typically, an FE machine model takes currents as the input and produces torque and induced voltages, etc., as the outputs. Direct FE solutions of such a model are not convenient if the voltage constraints during constant power operations are satisfied by iterative FE solutions with varying current vectors. Further, in the design process of interior permanent magnet (IPM) machines such as the PMA-SynRM, it is necessary to determine the optimal current vector for a given torque and speed command. To satisfy these requirements in a computationally efficient manner, a flux linkage machine model is established first through FE analysis and used to determine the optimal current vector for a specified torque-speed input, enabling performance evaluation of the design sample at any specific torque and speed combinations [8, 9], including those of constant power operation.

To build the model, static electromagnetic FE computations are performed to obtain the d-axis and q-axis flux linkage maps against the d-axis and q-axis currents (i_d, i_q) as shown in Fig. 3.3. For the studied machine design, an array of (6 by 7) is selected for the (i_d, i_q) currents. Thus, the dq-axis flux linkage variations as functions of the dq-axis currents are established. Subsequently, the flux linkages at any operating point of (i_d, i_q) can be determined through the flux linkage map. By employing equation (3.1), it becomes possible to derive the torque and voltage as functions of the current vector.

$$\begin{aligned} T(i_d, i_q) &= 1.5p[\psi_d(i_d, i_q) \times i_q - \psi_q(i_d, i_q) \times i_d] \\ V(i_d, i_q) &= \omega\sqrt{\psi_d(i_d, i_q)^2 + \psi_q(i_d, i_q)^2} \end{aligned} \tag{3.1}$$

To determine an optimal (i_d, i_q) for a specified torque-speed command, a search algorithm is devised, based on the torque function and voltage constraint. This algorithm aims to minimize power loss while ensuring that the voltage remains at or below the DC-link voltage limit if the speed exceeds the base speed. Notably, this search algorithm is entirely analytical and does not rely on FE computations, thus being very efficient. Consequently, this approach significantly accelerates the machine design process.

3.3.3 Full Model Integration for Interturn Fault Current Prediction

For the design of the fault-tolerant machine, it is necessary to evaluate performances in the healthy and fault-tolerant operations of each design sample during the optimization. As discussed previously, by imposing a limit on the

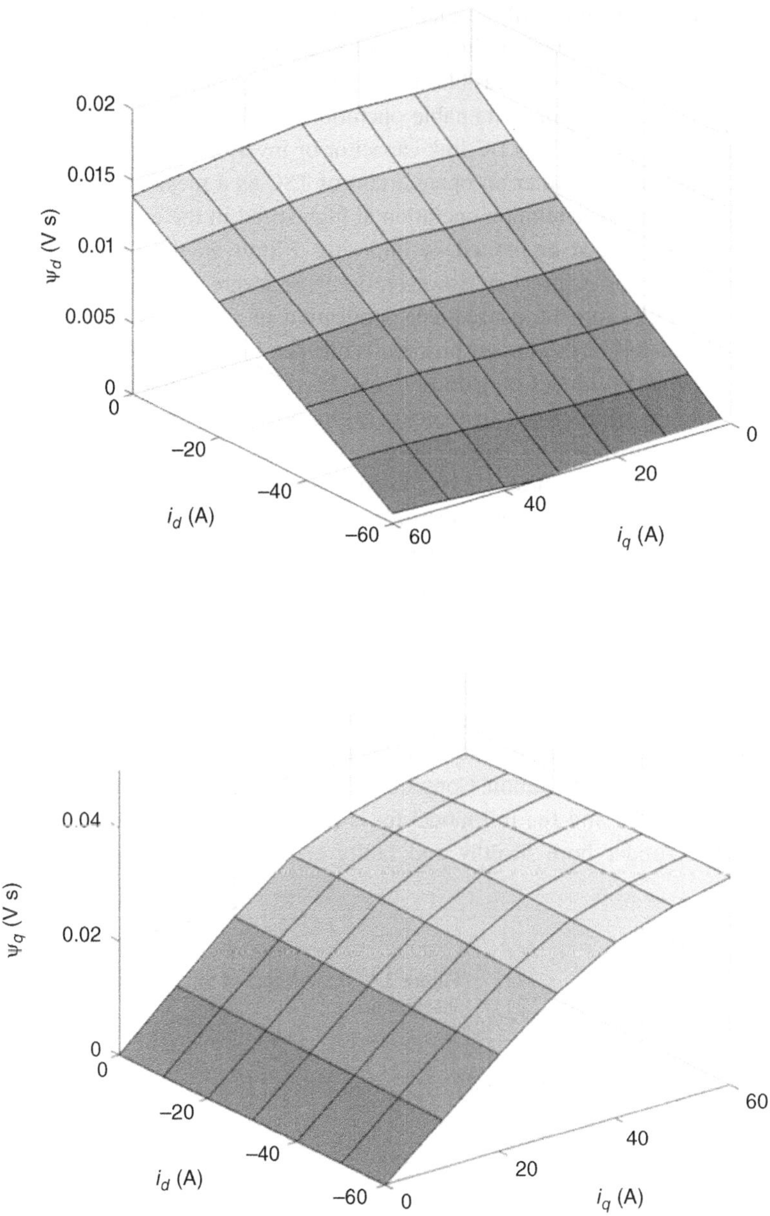

Figure 3.3 Flux maps of ψ_d and ψ_q versus i_d and i_q.

TSC current, the machine will be capable of sustained operations with the two remaining 3-phase healthy sets in the event of an open-circuit or TSC in one 3-phase set. Hence, when a feasible design has been produced, no additional evaluation is necessary for sustainable operations in these two failure modes. Other types of faults, such as DC-link capacitor or inverter fault in one 3-phase set, would also lead to either an open circuit or TSC as a means of mitigation. The uncontrollable regeneration operation at high speed in the event of inverter failures can be avoided or limited by imposing a limit on the back-emf as a constraint in the design process. However, the most severe fault is a single-turn short-circuit fault, and the consequent application of the mitigation measure has to be evaluated in the design process. While the location of the worst-case scenario depends on modes of operation and directions of rotation, a representative single-turn short circuit that occurs in the turn close to the slot opening in coil B2 during motoring mode can be assumed. The resultant turn fault current would be the highest, leading to the most thermally demanding case. To obtain a feasible design, it is imperative to ensure that the turn fault current and currents in the other part and phases when TSC as the mitigation measure has been applied are within the defined thermal limit. Hence, as part of the optimization process, the worst-turn fault current in B2 coil after application of TSC on the ABC phases should be evaluated for each candidate design. To reduce computation time, the machine performance under healthy conditions is evaluated using the 1/6 model by leveraging its symmetry. However, the prediction of turn fault current necessitates the use of the full machine model due to the asymmetric condition. Consequently, the optimization process utilizes both the 1/6 model and the full model for each design, enabling evaluation of performances under both healthy and faulty operation modes, as illustrated in Fig. 3.4.

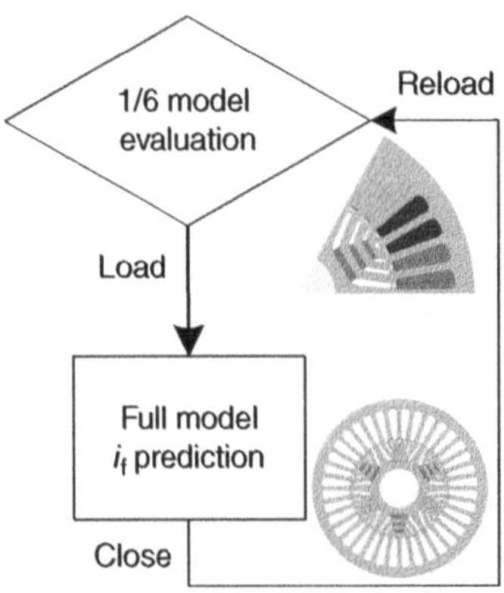

Figure 3.4 Flow chart of full machine model integration.

3.3.4 Cooling Design

For the selected thermal class of insulation and required service lifetime of 300,000 hours, it is necessary to limit the maximum winding temperature below 180 °C under continuous healthy operation. Given the demanding operating conditions and the high ambient temperature, effective heat dissipation via oil cooling is required. Spiral cooling jackets, housing ducts, and slot ducts are three common cooling solutions [10]. Among them, the spiral cooling jacket in the stator housing is commonly used for liquid cooling and has been chosen for this machine design.

The spiral cooling jacket can be integrated on the housing frame as shown in Fig. 3.5. The grooves are cut out on the cooling chamber to carry the coolant oil all around the stator housing surface and fins. The cooling sleeve fits outside of the cooling chamber, making it a sealed environment for the coolant fluid. Coolant oil is flowing in the path determined by the grooves to cool the machine. An external pump and cooling system cool the coolant oil and provide the circulating pressure. Since the diameter for the machine including housing is limited to 220 mm, the maximum available diameter for the stator lamination is set as 180 mm, leaving 40 mm for the spiral cooling jacket space. The stator is inserted into the housing frame by thermal fitting. To have a better cooling effect, the windings are potted with a high thermal conductivity material (Stycast 2676FT, 1.3 (W m)/K, service temperature, 185–260 °C) to enhance the heat dissipation capability. The grooves in the jacket are appropriately designed to maximize the cooling effect while ensuring structural integrity in Motor-CAD.[3]

The cooling effect of the spiral cooling jacket is evaluated in Motor-CAD. The leading parameters and losses of a representative design are listed in Table 3.2. The losses are evaluated with the rated torque at the base speed, which is the worst case for the cooling system in continuous operations over the

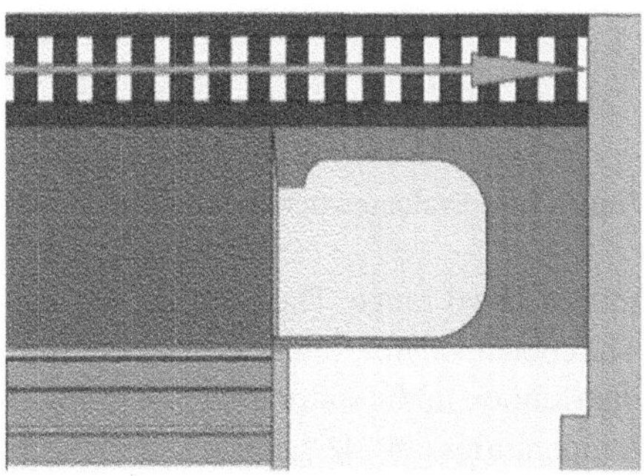

Figure 3.5 Illustration of spiral cooling jacket.

3 https://www.ansys.com/products/electronics/ansys-motor-cad.

Table 3.2 Leading parameters and losses of the initial sample.

Parameter	**Spiral housing jacket**
Stator outer diameter	180 mm
Housing diameter	220 mm
Air gap	0.75 mm
Copper loss (at 160 °C)	1234 W
Iron loss	324 W

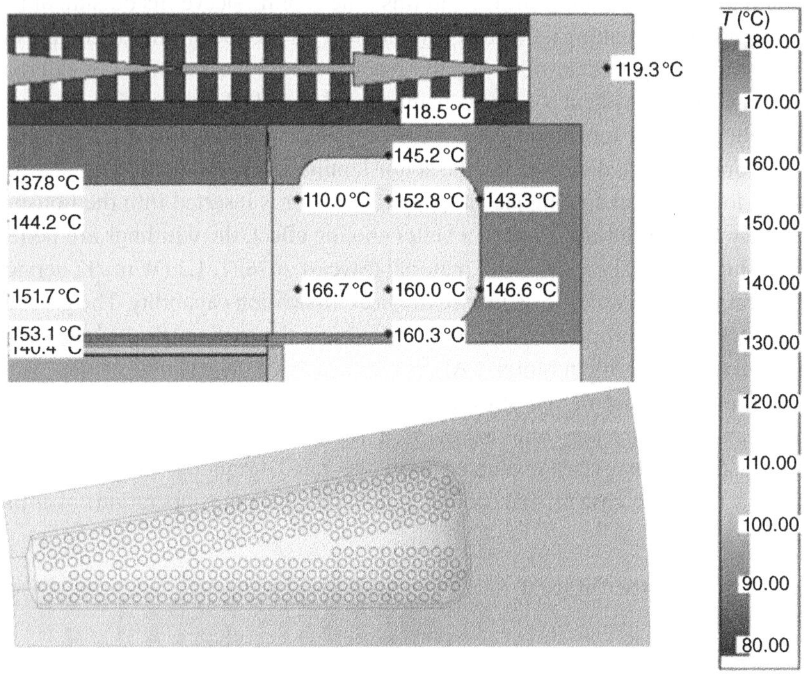

Figure 3.6 Evaluated temperature rise in Motor-CAD.

torque speed range. The losses are loaded to the thermal model in Motor-CAD. The coolant oil flow rate is set to a minimum of 6 L/min. The resultant temperature distribution in the stator is shown in Fig. 3.6. The average and maximum winding temperatures are 148.2 and 167 °C, respectively.

It is seen that the maximum temperature, which occurs in the winding overhang region close to the inner bore of the lamination side, is under the defined thermal limit (180 °C) with ~13 °C headroom. This demonstrates the required continuous output torque and power with the spiral cooling jacket design can be achieved under the volumetric limit.

3.3.5 Lumped Parameter Thermal Model

The spiral housing jacket has been designed as the cooling solution for this machine in Motor-CAD environment. However, the Motor-CAD model cannot be incorporated into the automated optimization process for each design sample. To circumvent this problem, a lumped parameter thermal model is developed for fast temperature prediction as shown in Fig. 3.7 [11].

The thermal dissipation in the machine is assumed to be predominant in the radial direction, and hence the thermal dissipation in the axial direction is neglected. The thermal network models only half a tooth pitch due to symmetry. In the thermal network, the thermal resistances of each part can be derived using the governing principle of the heat conduction [12]. The thermal resistance of the stator housing shown in Fig. 3.7 is evaluated in Motor_CAD, which captures the cooling capability of the spiral cooling jacket. Therefore, the average slot winding temperature of the machine can be predicted by this lumped parameter thermal model during design optimization in the form of Python script. As seen in Fig. 3.6, the maximum temperature is located in the winding overhang region rather than in the middle slot. Thus, 15° is added to the estimated average slot winding temperature to account for the maximum temperature rise in the windings. The result is sent to the GOT-IT software to determine the next step optimization.

The single-turn fault close to slot opening in B2 coil has been identified as the worst-fault scenario and the fault current as well as the currents in other

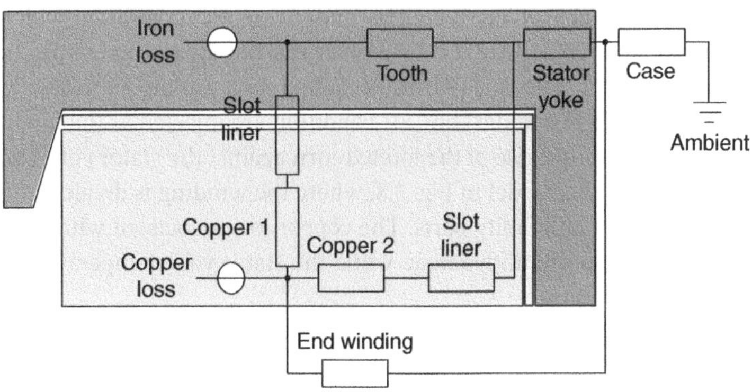

Figure 3.7 Lumped parameter thermal model of the machine.

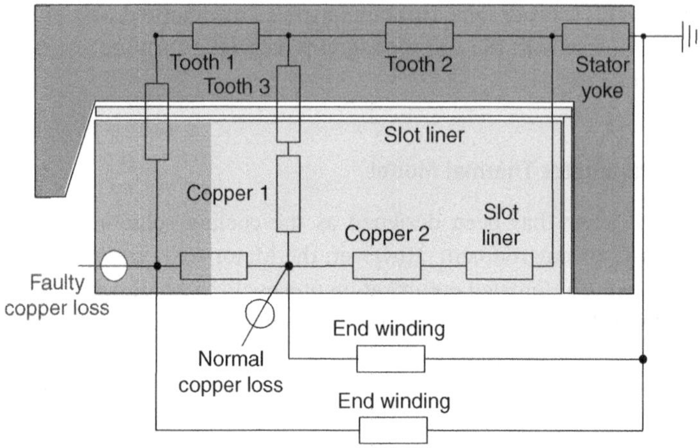

Figure 3.8 Lumped parameter thermal model for turn fault.

parts of the windings during postfault operations are evaluated with the full machine model during optimization. The resultant hotspot temperature under this extreme fault condition should also be estimated to assess the feasibility of a candidate design. For this purpose, a lumped parameter thermal model representative of this fault condition is developed as shown in Fig. 3.8, where an additional node and two thermal resistances are used to model the heat generation and dissipation in the short-circuited turn. The thermal analysis can therefore be performed against this condition. It should be noted that TSC has been applied under this turn fault condition, only the current of the faulted turn is noticeably higher than the rated but it has minor thermal influence on the other regions because the high heat intensity is confined with the one turn volume. Thus, the temperature of the stator yoke T_y is mainly determined by the phase currents in the healthy parts of the windings under the TSC condition, which are close to the characteristic currents. To predict the hot spot temperature in the faulted turn, the stator yoke T_y is first determined with the model shown in Fig. 3.7, where the total losses under the TSC condition are injected as the input. Subsequently, the temperature rise of the faulted turn against the stator yoke can be calculated by the adapted model in Fig. 3.8, where the winding is divided into two parts, i.e. the healthy and faulty parts. The copper loss associated with each part is injected into its representative node while the stator yoke temperature is set as the reference temperature.

Finally, the temperature rise of the faulty turn, ΔT_f, with respect to the stator yoke is estimated by the thermal model for postfault operations under the turn fault with TSC application. Thus, the actual temperature of the faulty copper winding is obtained by (3.2).

$$T_{\mathrm{f}} = \Delta T_{\mathrm{f}} + T_{\mathrm{y}} \tag{3.2}$$

The temperature rises in both healthy and the worst-fault conditions will be fed to the GOT-IT optimizer as constraints. If any of the temperatures have exceeded the maximum permissible limit, the design sample will be discarded. It should be noted that sustainable postfault operation is usually required for a short period of time only. Hence, the maximum permissible temperature in the faulted turn can be slightly higher than the maximum continuous temperature of the selected thermal class of insulation.

3.3.6 Optimization Objective, Constraints, and Parameters

In the optimization tool GOT-IT, the optimization objective is set to maximize the efficiency at the rated operating point. The efficiency is evaluated for each design sample when the copper loss and iron loss are quantified by the FE tools. From the torque-speed envelope, the peak torque at the base speed point is 120 (N m)/4000 rpm and the peak torque at the maximum speed is 25 (N m)/19,200 rpm. These two points determine the bas shape of the envelope under the current and voltage constraints shown in Table 3.3. The maximum current density shown in the table is used just as initial gauge and indeed the maximum winding temperature is imposed as a strict constraint during the design process. The maximum current limit is imposed to facilitate the use of a predetermined power module.

To increase power density and reduce iron loss, 0.2 mm thick Cobalt iron laminations are used. The stator lamination is optimized for high saturation flux density while the rotor lamination is optimized for good mechanical strength. To ensure

Table 3.3 Design constraints of the machine.

Name	Value
Max current	200 A
DC-link voltage	270 V
Maximum current density	12.5 A/mm^2
Max temperature in the windings	180 °C
Von Mises stress	450 MPa

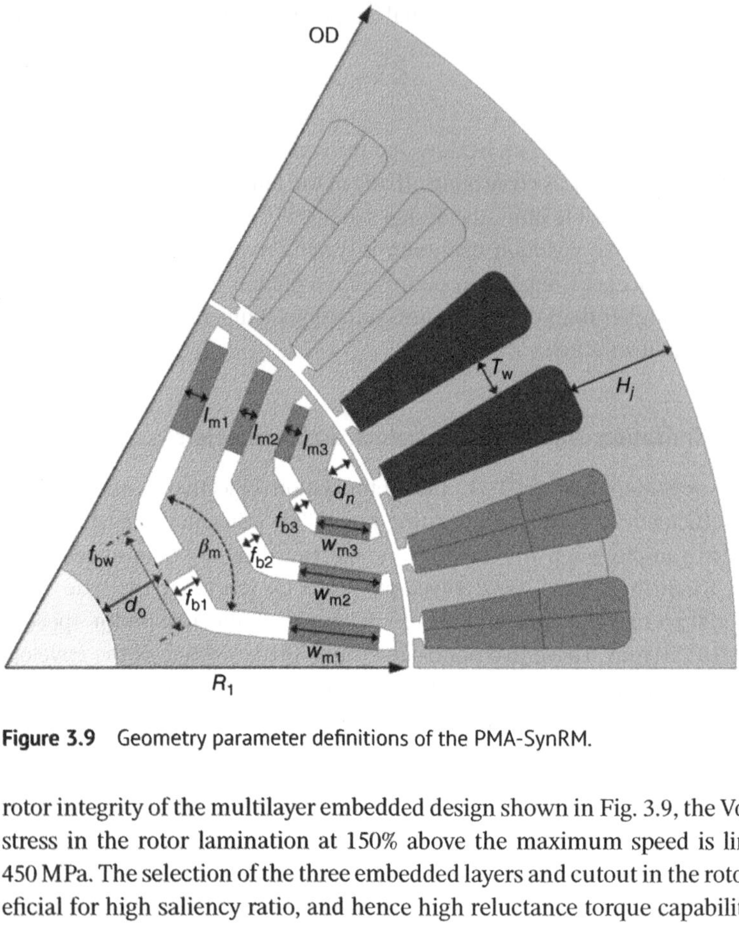

Figure 3.9 Geometry parameter definitions of the PMA-SynRM.

rotor integrity of the multilayer embedded design shown in Fig. 3.9, the Von Mises stress in the rotor lamination at 150% above the maximum speed is limited to 450 MPa. The selection of the three embedded layers and cutout in the rotor is beneficial for high saliency ratio, and hence high reluctance torque capability, while ensuring good structural integrity and manufacturability. In addition, as has been discussed previously, the maximum back-emf is also limited below the available DC-link voltage, while the short-circuit current after TSC application is limited below its rated current for a feasible design.

After taking into account the space for stator housing and spiral cooling ducts, the maximum available stator outer diameter of 180 mm is selected for the purpose of fully exploiting the reluctance torque capability, while facilitating a low current density in the slots. The airgap is fixed to 0.75 mm for a good balance between high rotor saliency and the ability to tolerate mechanical unbalance and bearing eccentricity. The axial length should be preliminarily scanned within its maximum available range to produce a reference design with a minimum length that satisfies the performance requirements, thermal limit, and other design constraints.

Table 3.4 Parameters to be optimized.

Name	Symbol	Initial value	Search range
Back iron thickness	H_j	10 mm	8–14
Tooth width	T_w	6 mm	4.5–7
Rotor radius	R_1	42 mm	45–55
Magnet thickness	l_{m1}, l_{m2}, l_{m3}	3, 2.5, 2 mm	*(1.2–0.8)
Magnet width	w_{m1}, w_{m2}, w_{m3}	18, 12, 8 mm	*(1.2–0.8)
Middle flux barrier thickness	f_{b1}, f_{b2}, f_{b3}	4, 3, 2 mm	*(1.2–0.8)
Middle flux barrier width	f_{bw}	10 mm	7–12
Magnet layer's angular span	β_m	160°	140–175
Magnet layer's depth	d_0	8 mm	6–15
Turn number	TN	8	6–10

*The geometry parameters are varied between (0.8–1.2) times of the initial value.

The leading design parameters to be optimized for the PMA-SynRM are illustrated in Fig. 3.9, and they are also listed in Table 3.4 together with their initial values and search ranges. The initial values are obtained from the reference design of the preliminary scan. These parameters are selected as they have a significant influence on the performance and the ability of the machine to tolerate various faults. The stator tooth width T_w and the yoke thickness H_j affect both the electric and magnetic loadings of the stator, and hence an optimal balance between the two loadings would maximize its efficiency or torque capability. Likewise, the rotor radius R_1 determines the split ratio between the rotor and stator diameters and consequently provides a trade-off for the rotor magnetic loading and the stator electric loading. Best designs are often achieved when an optimal balance of the two loadings is reached. The remaining parameters mainly influence the rotor saliency and the torque production mechanism (i.e. the ratio of PM torque to the reluctance torque). They also control the magnetic saturation level in the rotor bridges and back iron and affect the mechanical stress of the bridge, hence the rotor mechanical integrity. An optimal combination of these parameters would yield the best designs while satisfying all performance requirements and electric, thermal, and mechanical constraints, including the ability to tolerate various faults.

The techniques discussed in Sections 3.3.2–3.3.5, namely, the machine flux-linkage model, the turn fault current prediction, and lumped parameter thermal models are all integrated to speed up computations in the global optimization platform through Python scripts. The objective is to maximize the efficiency at the rated operating point, while satisfying the required torque-speed operating range and the design constraints listed in Table 3.3 as well as those required to tolerate various faults, such as the maximum back-emf limit and the limit on TSC current. The optimization process starts from the initial design parameters and searches in the design parameter space through a few generations of iteration controlled by the GA within Got-It until the objective function converges to the global maximum within a specified tolerance range.

3.4 Selected Design Alternatives and Performance Comparison

The efficiency evolution of 405 candidate designs after three generations of GA iterations during the optimization is shown in Fig. 3.10. It is seen that after each generation of iterations, the design samples move toward a higher efficiency range. The efficiency of the final design increases from the initial 95.10% to 96.23%.

It can be observed from Fig. 3.10 that the differences in efficiency of the designs in the third round of GA iterations are very small. This implies that there are many designs that yield similar efficiency. After examination of these design samples, two designs, namely, sample A and sample B as shown in Fig. 3.11 appear to be most promising. Sample A exhibits ~0.5% higher efficiency than sample B. The key features of the two samples are compared in Table 3.5. Sample A utilizes 18% more magnets than sample B while sample B has a larger stator

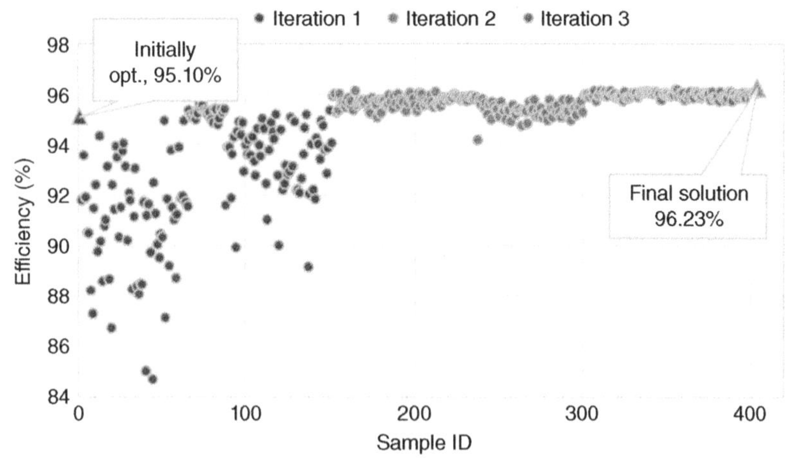

Figure 3.10 Efficiencies of the design samples during design iteration.

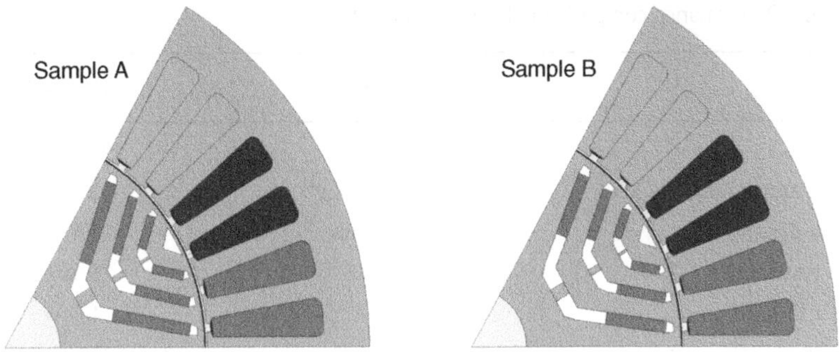

Figure 3.11 Cross sections of samples A and B.

Table 3.5 Key features of samples A and B.

	Sample A	Sample B
No. of turns/coil	8	8
Magnets usage	1.02 kg	0.86 kg
Rotor radius	49.3 mm	51.75 mm
Peak line back-emf at 19,200 rpm	356 V	257 V
Inductance difference $(L_q - L_d)$	0.85 mH	0.93 mH
Characteristic current	108.6 A	75.5 A

inner bore. The different geometries lead to more PM torque production in sample A with higher back-emf while more reluctance torque is exploited in sample B due to the higher saliency ratio. Due to higher magnets usage, the back-emf of sample A at the maximum speed is greater than the DC-link voltage 270 V; therefore, uncontrolled rectification may occur if the inverter fails. This would violate the back-emf constraint while it is used to illustrate the trade-off in the designs. Further, the high back-emf will incur a higher fault current in case of a turn fault. In contrast, the back-emf of sample B is 30% lower, which eliminates the possibility of uncontrolled regeneration in an event of the inverter failure. Its fault current in case of the worst-case turn-to-turn short circuit is also much lower.

The electrical and thermal performances of the two candidates at the rated operation point are further compared in Table 3.6, where an iron loss built factor of 1.3 is considered. Sample A requires less current with the help of a stronger PM field while sample B exhibits a larger gamma angle which is beneficial for reluctance torque. The current density and the copper loss of sample A are 15% and 25.8%

Table 3.6 Performance comparison of samples A and B.

	Sample A	**Sample B**
Efficiency	96.23%	95.65%
Current and gamma angle	113 A/48.9°	120 A/54.5°
Current density	8.8 A/mm^2	10 A/mm^2
Phase resistance	21.3 mΩ	22.6 mΩ
Copper loss	1223 W	1464 W
Iron loss (1.3×)	266 W	275 W
Copper temperature	166 °C	179 °C
Turn fault current (RMS)	320 A	266 A
Fault turn temperature	295 °C	232 °C

less than those of sample B, respectively. As a result, the efficiency of sample A is ~0.5% higher. The predicted maximum winding temperature is 166 and 179 °C for samples A and B, respectively.

Since the machine is aimed for fault-tolerant application, the performances during postfault operation against the worst-case fault scenario are also examined. Assuming the worst-fault scenario of single-turn short circuit occurs in B2 coil, the RMS currents in the faulted turn after TSC application are 320 and 266 A for samples A and B, respectively. They are significantly higher than the rated value. The predicted hotspot temperatures in the faulted turn are 295 and 232 °C for samples A and B, respectively. Clearly, the hotspot temperature in sample A is 75 °C higher than the maximum continuous operation temperature 220 °C of class-R insulation. This high temperature would degrade the winding insulation quickly and lead to further damage over a short period of time. As for sample B, it is only 13 °C higher and the insulation can sustain a much longer time, at least a few hundred hours. As has been pointed out previously, sustainable postfault operation is usually required for a short period of time only. Postfault repair or replacement will be carried out such that the system recovers to its full capacity.

Based on the foregoing discussions, the fault-tolerant capability of sample A is not satisfactory. Hence, sample B is selected as the final design whose leading design parameters are listed in Table 3.7. The trade-off of the two candidates is quite common in fault-tolerant machine design. The ability to tolerate fault usually requires some form of compromise on performances during healthy operations. The goal of the foregoing fault-tolerant machine design optimization is to find the best design with the highest efficiency and acceptable levels of fault tolerance.

Table 3.7 Design parameters of samples A and B.

Name	Symbol	Sample A	Sample B
Back iron thickness	H_j	11.7 mm	10.25 mm
Tooth width	T_w	4.55 mm	5.1 mm
Rotor radius	R_1	49.3 mm	51.75 mm
Magnet thickness	l_{m1}, l_{m2}, l_{m3}	2.8, 2.3, 2 mm	2.8, 2.3, 2 mm
Magnet width	w_{m1}, w_{m2}, w_{m3}	19.4, 13.6, 7.8 mm	14.9, 12, 6.9 mm
Middle flux barrier thickness	f_{b1}, f_{b2}, f_{b3}	4.4, 3.3, 2.6 mm	3.75, 2.8, 2.25 mm
Middle flux barrier width	f_{bw}	7.26 mm	10.64 mm
Magnet layer's angular span	β_m	160.6°	157.6°
Magnet layer's depth	d_0	7.05 mm	9.83 mm
Turn number	TN	8	8

3.5 Test Setup of Fault-Tolerant Machine Drives

The above-optimized machine is manufactured as shown in Fig. 3.12 for the stator lamination stack and rotor stack with the magnets assembled. The stator stack is skewed with one stator slot to reduce the torque ripple and voltage harmonic, as a result, the output torque is also lower compared with the designed value in previous sections. One key feature of the fault-tolerant machine is the segregated windings for three electrically independent 3-phase systems as illustrated in Fig. 3.13(a). It can be seen that the windings of the different 3-phase sets have been

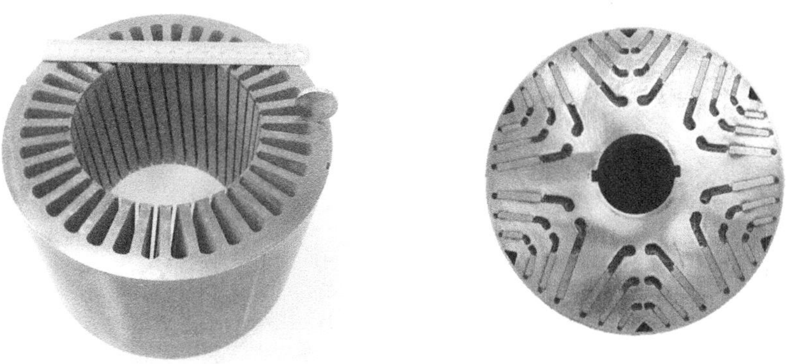

(a) (b)

Figure 3.12 Stator and rotor stack: (a) stator stack and (b) rotor stack. Wang et al. [17]/With permission of IEEE.

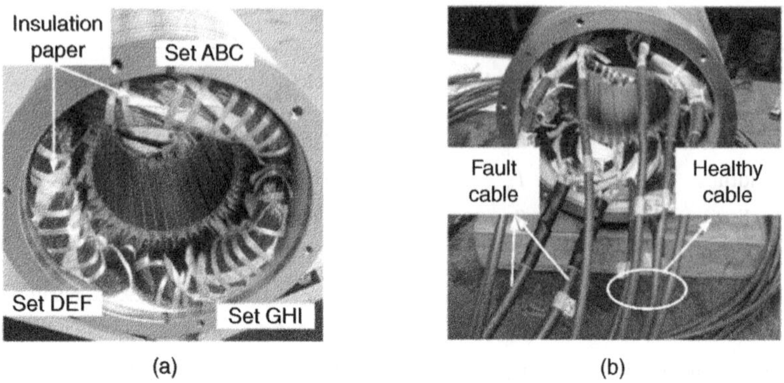

Figure 3.13 Machine windings and terminal leads: (a) segregated windings and (b) terminal leads. Wang et al. [17]/With permission of IEEE.

separated in the winding configuration with no overlapping. K-type thermocouples have been buried in the slots and end windings to monitor the winding temperatures during test operations. The cable leads are shown in Fig. 3.13(b), where the thinner cables are the leads of the 9-phase windings, which carry the phase currents. To emulate the worst-interturn fault case, a single-turn tap is brought out in coil B2 of 3-phase set ABC. The fault emulation tap and the neutral point have been connected to two thick cables to minimize the additional impedance in the emulated short-circuit loop. The thick cable leads are connected to a high

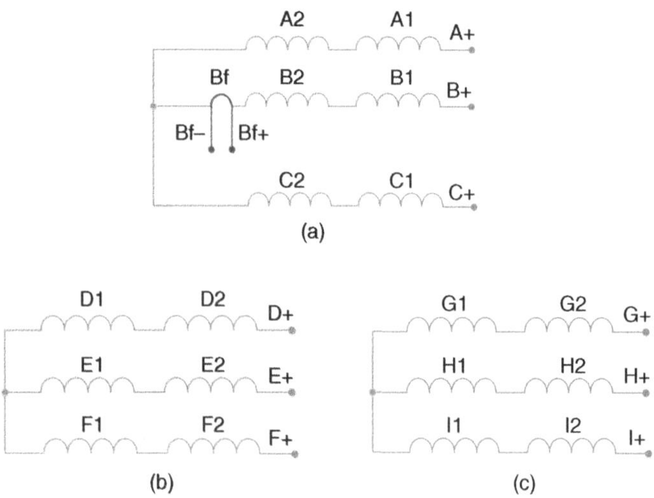

Figure 3.14 Terminal connection of the three 3-phase sets: (a) set ABC, (b) set DEF, and (c) set GHI.

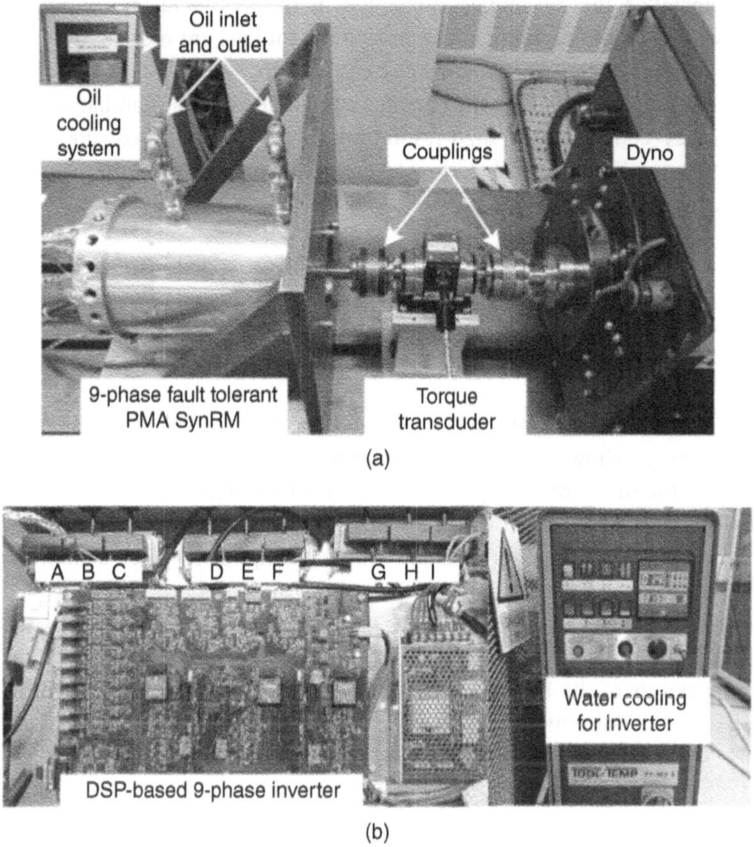

Figure 3.15 Real test bench: (a) 9-phase PMA-SynRM with dyno and (b) 9-phase inverter. Wang et al. [17]/With permission of IEEE.

current relay to emulate the interturn fault in a controlled manner. The terminal connections of the 9-phase PMA-SynRM windings are plotted in Fig. 3.14 for a more clear illustration. It is worth noting that the fault turn located in coil B2 represents the worst-case scenario in motoring operations when the rotor rotates anticlockwise. However, as discussed in Chapter 2, the result is equivalent to the worst-fault location in coil A1 if the rotor rotates clockwise. Hence, tests with the emulated turn fault cover the two worst cases.

The manufactured prototype is mounted on the testbed and connected to the AVL dynamometer via couplings with a torque transducer as shown in Fig. 3.15. The dynamometer is operated in speed control mode at a given speed command. The machine is fed by a DSP-based 9-phase inverter, which is configured as three 3-phase inverters in torque control mode. Machine temperatures are recorded by the preinstalled thermocouples embedded in the slots and end windings. During tests, the phase currents in the machine windings are measured by the current sensors in the inverters and the current data are stored in the DSP RAM. More accurate measurements of the currents can be taken with high-bandwidth current probes connected to high-resolution, high-bandwidth oscilloscopes. However, the number of available current probes is insufficient to measure all 9-phase currents simultaneously.

3.6 Test Under Healthy Conditions

3.6.1 No-Load Test

First, the phase back-emfs of ABC phases are measured and compared with the predictions, as shown in Fig. 3.16, when the rotor speed is 4000 rpm. It is observed that the measured and predicted waveforms match very well albeit the measured magnitude is ~2% lower than the prediction. This is mainly caused by deviations in material properties and manufacturing tolerances. Further, the line back-emfs of the other two 3-phase sets DEF and GHI are also measured and compared with the predictions in Fig. 3.17. In the machine under study, the phase back-emfs of the three 3-phase sets are in phase with each other. Since the star neutral points of the DEF and GHI sets are not connected to any external terminal, direct measurements of their phase back-emfs are not possible. Thus,

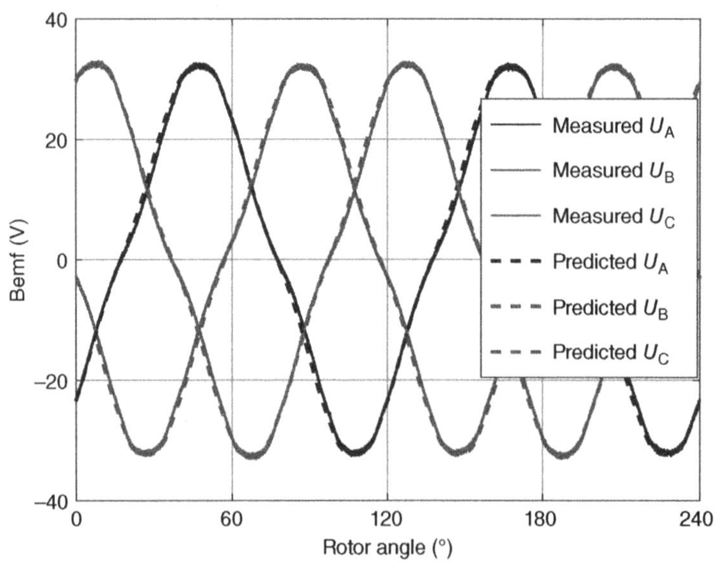

Figure 3.16 Comparison of phase back-emf waveforms at 4000 rpm.

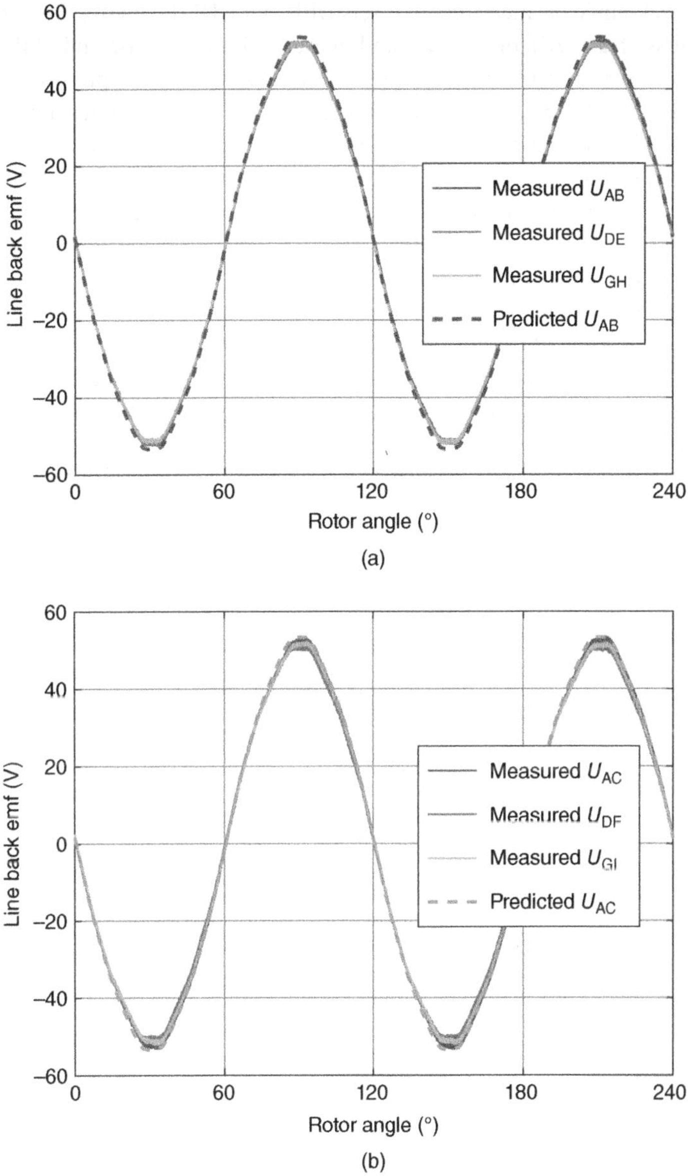

Figure 3.17 Comparison of line back-emf waveforms at 4000 rpm: (a) line back-emf U_{AB} and (b) line back-emf $U_{AC.}$

only the line back-emfs are measurable in sets DEF and GHI. Likewise, the line back-emfs across phases AB, phases DE, and phases GH are in phase and of the same magnitude. Similar relations hold for the line back-emfs across phases BC, phases EF, and phases HI, etc., as is evident in Fig. 3.17. The nearly identical phase and line back-emfs observed in the three 3-phase sets confirm good symmetry in the three 3-phase systems.

Based on the measurements shown in Fig. 3.17, the line back-emf at the maximum speed 19,200 rpm can be derived as 246.5 V, which is lower than the DC-link voltage. Therefore, uncontrolled rectifier operation will not occur even if the inverter fails. This fail-safe feature is desirable for the PM machine drives, which are required to operate at high speed under field weakening control.

The characteristic current I_{ch} represents the induced short-circuit currents when all 3-phase windings are directly short circuited at their terminals. It can be inferred by (3.3), where ψ_m is the phase PM flux linkage and L_d is the d-axis inductance for each 3-phase set. For the triple 3-phase machine under study, it can also be measured by short circuiting all three 3-phase sets.

$$I_{ch} = \psi_m / L_d \tag{3.3}$$

Figure 3.18 presents the characteristic currents measured at 4000 rpm. It can be seen that the peak current is 53.6 A, which is slightly lower than the predicted value. This difference could be attributed to two factors, slightly lower measured

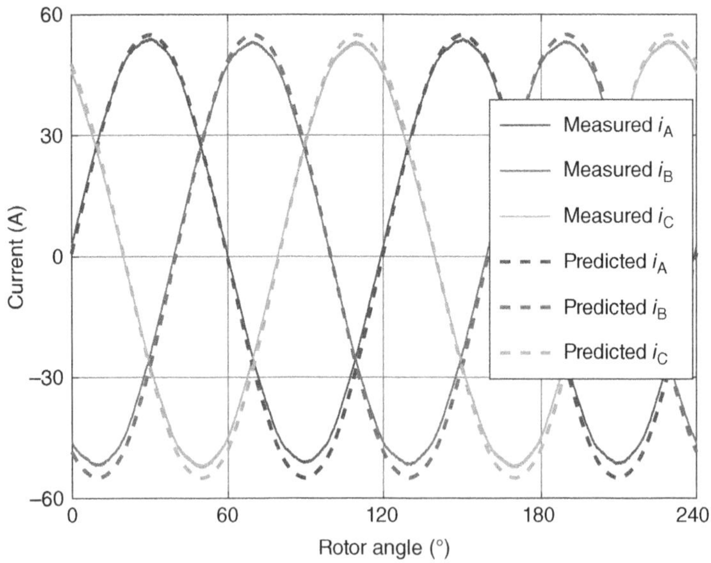

Figure 3.18 Phase currents of set ABC at 4000 rpm with three 3-phase sets short circuited.

back-emfs and the additional impedance introduced by the terminal cables. The low characteristic current is primarily a result of the low PM field, an important feature of the selected machine topology and design. A lower characteristic current than the rated current is crucial for achieving the required fault tolerance since the mitigation measure against all modes of turn-to-turn short circuit, including intraphase and interphase short circuit, and inverter short circuit is to apply TSC. The resultant short-circuit currents in the healthy phases pose no thermal risks to the machine if they are lower than the rated.

3.6.2 On-Load Test

Load tests on the prototype machine are performed when it is fed by three independent 3-phase inverters. As shown previously, the electric and magnetic behaviors of the three 3-phase winding sets in healthy conditions are virtually identical and also similar to that of a conventional 3-phase machine. Hence, conventional vector control, also known as field-oriented control, is employed to regulate the currents in the three 3-phase inverter drives. The current references for the three-inverter drive modules are taken from the speed controller if present, but each inverter module tracks its references independently using its individual current controller. As a result, if a failure occurs in one 3-phase set, it does not significantly impact the operation of the other sets. Each drive can independently respond to accommodate the fault.

Initially, the test is performed at 4000 rpm when each 3-phase drive operates with the rated current. By way of example, the measured phase currents of set ABC with high-bandwidth current probes under this condition are depicted in Fig. 3.19. Since the available number of such high-bandwidth probes is not enough to measure the 9-phase currents simultaneously, the phase currents measured by the inverter current sensors in each 3-phase drive and stored in the DSP memory are shown in Fig. 3.20. It is seen that the motor currents are effectively controlled in all three phase drives while exhibiting minimal distortion near their peaks. Switching harmonics are observable in the waveforms measured by the high-bandwidth probes, but they are not visible in those measured by the current sensors in each 3-phase drive. This is because the current measurement in the drive is performed at 10 kHz switching frequency, and as such the measurement is not able to capture any transient within a switching interval. In addition, anti-aliasing filters with a corner frequency of 5 kHz are also deployed in the current measurement channels of the inverters. Thus, high-frequency components are also attenuated by these filters. Furthermore, Fig. 3.20 shows that the phase currents of each 3-phase drive essentially overlap, demonstrating good symmetry of the machine operation. Overall, the load tests have validated performance of the triple redundant machine drive in healthy operations.

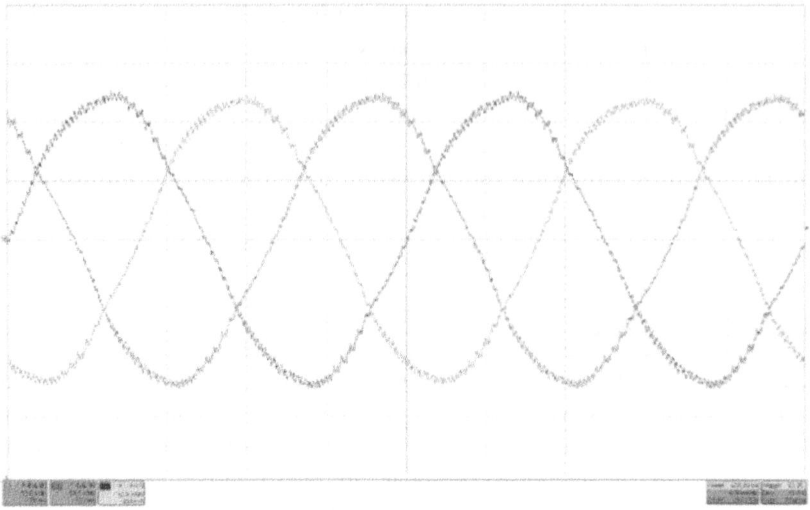

Figure 3.19 Phase currents of set ABC at 4000 rpm in healthy condition, x-axis, 1 ms/div, y-axis, 50 A/div.

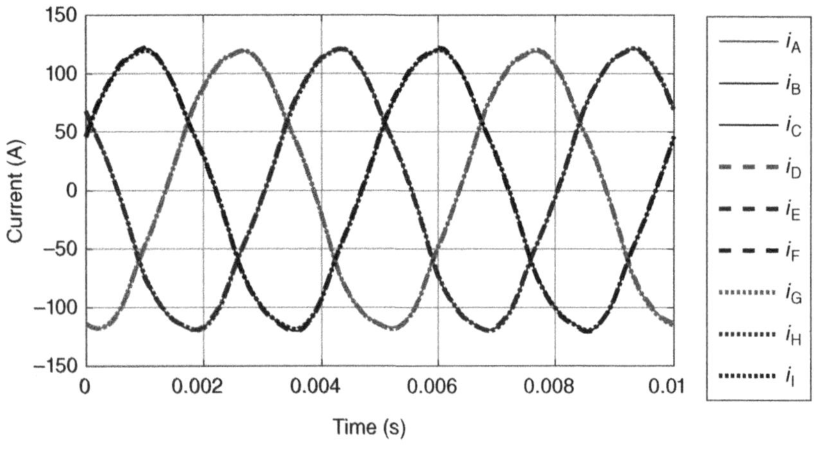

Figure 3.20 9-phase currents at 4000 rpm in healthy condition.

For PM machines with a high saliency ratio, such as IPM and PMA-SynRM under this study, it is important to operate at the maximum torque per Ampere (MTPA) condition. Hence, load tests are performed to determine the current vector angle with respect to the q-axis for MTPA operation. For a given current magnitude, the MTPA point is determined by varying the current vector angle in the drive, denoted as gamma angle γ, with respect to the q-axis as shown in Fig. 3.21

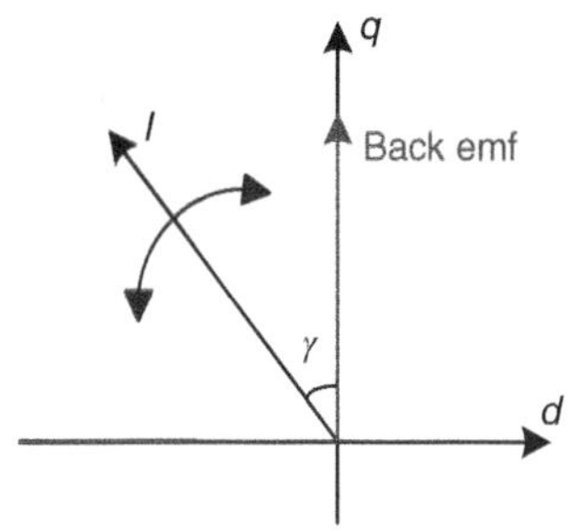

Figure 3.21 Illustration of current vector and gamma angle γ.

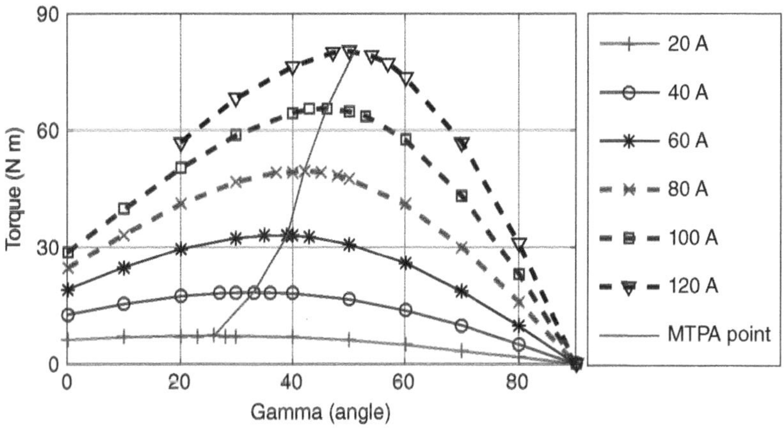

Figure 3.22 Torque variations with current magnitude and gamma angle.

at 500 rpm. The tests are repeated over a range of current magnitudes from 20 to 120 A, and the measured torque versus gamma angle loci is shown in Fig. 3.22. As can be seen, the optimum gamma angle, which corresponds to the MTPA condition for a given current magnitude, increases with the current magnitude (or load torque).

Further tests under MTPA operations with varying current magnitude are performed. Figure 3.23 compares the measured and predicted torque variations with load current under the MTPA conditions. The measured torque is approximately 5% lower than the predicted values. The prediction error is mainly due to the discrepancies in material properties (such as BH curves, remanence of magnets, etc.) and manufacturing tolerances. The prediction error is also evident in the comparison of the measured and predicted back-emfs shown in Fig. 3.17.

While it is impossible to separate the PM torque and reluctance torque in the torque measurements, based on the measured optimal gamma angles, the ratio of reluctance torque to total output can be determined using the frozen permeability method [13], and its variation with load current is illustrated in Fig. 3.24.

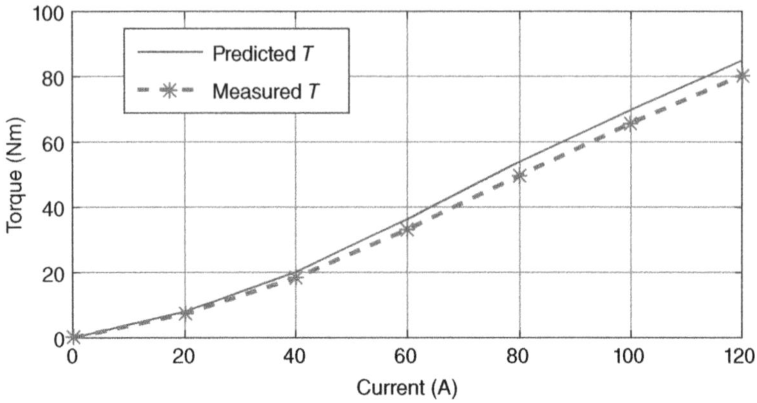

Figure 3.23 Torque comparison in healthy condition.

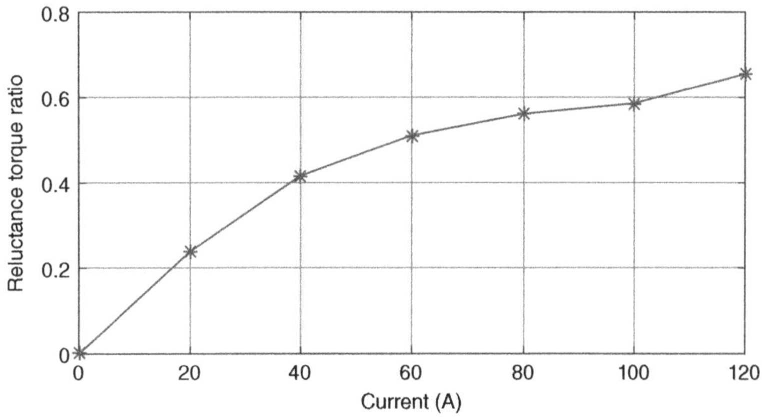

Figure 3.24 Reluctance torque ratio against the current magnitude.

It is observed that when the current magnitude exceeds 50% of the rated value, the reluctance torque becomes dominant. This indicates that even in the unlikely event of partial demagnetization, the machine has the capability to output substantial torque by leveraging the reluctance torque.

The efficiency of the motor drive is measured by a Yokogawa power analyzer. This is possible because the voltages and currents in each 3-phase drive are essentially identical. Hence, by measurement of the input power to one 3-phase drive and multiplying the result by three, the input power to the machine can be measured by the power analyzer. For the output power, the speed and torque measured in the torque transducer are fed to the power analyzer as usual. The efficiency map of the machine up to 12,000 rpm has been obtained and is illustrated in Fig. 3.25.

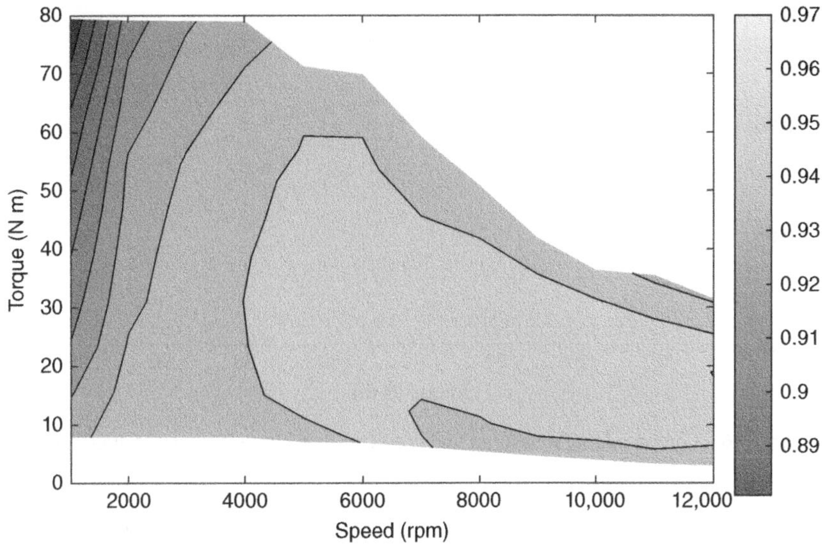

Figure 3.25 Measured machine efficiency map up to 12,000 rpm.

Under the base speed, the machine operates in accordance with the previously determined MTPA condition, while in the high-speed region, field weakening control is employed to reduce phase voltages within the voltage limit. Observing the efficiency map, it is notable that the machine achieves an efficiency exceeding 95% across most torque ranges. Furthermore, Fig. 3.26 presents a comparison of the measured and predicted efficiency at 4000 rpm when the load torque varies up to 80 N m. The measured efficiency exhibits similar trends to the predicted values, albeit slightly lower. The lower measured efficiency can be attributed to the lower measured torque capability than the prediction shown in Fig. 3.23. In addition, imperfect winding formulation and insertion as well as connection leads result in an increase in the phase resistance and hence in the copper loss.

3.6.3 Thermal Test in Healthy Condition

The thermal test is performed when the machine drive operates at the rated conditions for 2 hours, with oil cooling provided at a flow rate of 7 L/min. The temperature variations at the oil inlet, oil outlet, and windings are recorded using the embedded thermocouples, as depicted in Fig. 3.27. Due to the limited capacity of the heat exchanger, the temperatures at the oil inlet and outlet experience an increase from 24 to 44 °C and from 25 to 57 °C, respectively. It is observed that the temperatures in the slots, specifically in B1 slot, E2 slot, and B2 slot (where a tap for emulating a single-turn short circuit is placed), are relatively

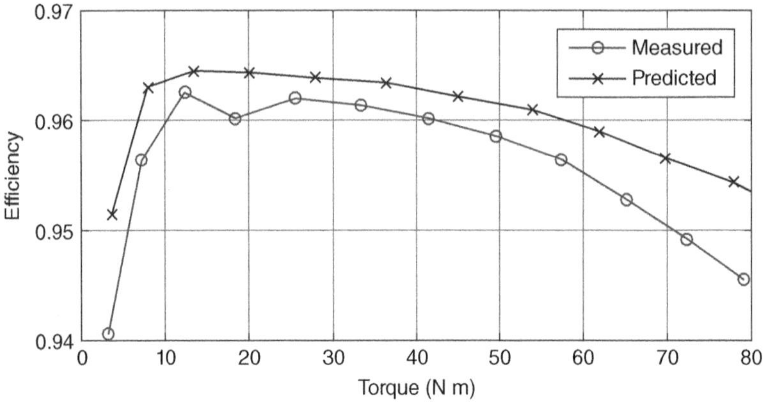

Figure 3.26 Comparison of measured and predicted machine efficiency at 4000 rpm.

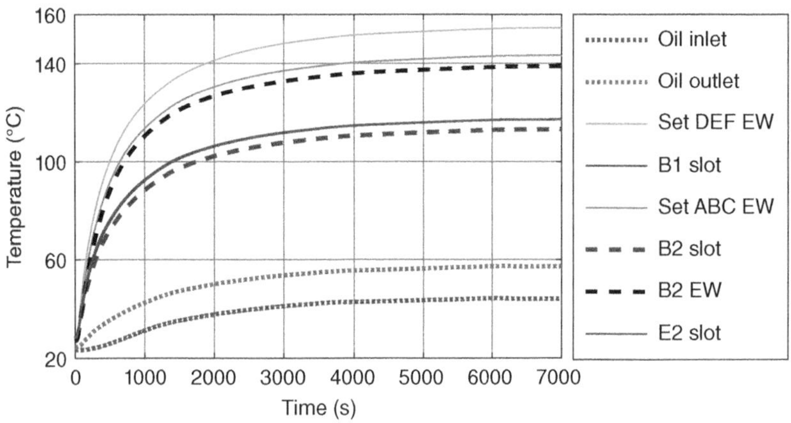

Figure 3.27 Thermal test results with 120 A at 4000 rpm in healthy condition.

close, with differences of less than 4°. The measured temperature of 142 °C in the end winding region of the ABC set is about 13° higher than those in the slots. However, the temperature in the end winding of the DEF set is 11° higher than that in the ABC set. This disparity is likely attributed to the loss caused by the neutral connection. For ABC set, the neutral is connected outside the machine for easy access, while the neutral of the DEF set is terminated within the end winding region. The connection of the neutral is achieved through soldering the 3-phase terminals, resulting in additional resistance and loss. Consequently, the thermocouple located close to the neutral point of the DEF set indicates a higher temperature rise in the end winding region. Upon updating the Motor_CAD thermal model with the measured copper loss, the predicted maximum temperature is determined to be 138 °C, which closely aligns with the measurement of the ABC set. This emphasizes the significance of accurately incorporating copper loss data into thermal models for precise temperature predictions.

It should be noted that the ambient temperature of the testing is different from that of the design. Hence, the temperature rise would be more appropriate for the comparison with the prediction. The measured temperature rise against the ambient (cooling oil) in the windings embedded in slot regions is about 58 °C, quite close to the prediction. However, the measured temperature rise in the end winding regions, the DEF set in particular, is ~12 °C higher than the prediction. Therefore, if the machine drive operates in the ambient temperature specified in the design, the maximum temperature is likely to reach 190 °C. While this temperature is much lower than the maximum permissible continuous operating temperature of class R insulation, the estimated lifetime of the machine will be reduced to 150,000 hours. The higher measured temperature than the prediction can be attributed to a number of factors. First, the measured efficiency is lower than the prediction, which implies the losses are greater, and hence higher temperature. Further, additional leads and resistance to form the star connection result in additional losses in the end winding region, where thermal dissipation is less effective. As the prototype machine is manually wound, the measured phase resistance is 10% larger than the FE prediction, which leads to an increase in copper loss. The foregoing assessments and discussions show that a small discrepancy between predictions and measurements may have a large impact on thermal life of the machine. For machine drives that require a long lifetime, it is essential that all these discrepancies are accounted for through designs and tests.

3.7 Test Under Fault Conditions

In this section, the machine fault behaviors and its torque capability during post-fault operations are investigated experimentally for open-circuit, short-circuit, and interturn short-circuit faults over a wide operation range.

3.7.1 Test Under One 3-Phase Set in Open Circuit

First, the prototype drive system is tested with one 3-phase set in an open-circuit configuration. During the test, set ABC was open circuited by deactivating all the IGBTs of the inverter, while the two remaining 3-phase sets, DEF and GHI, were fed by their inverters and controlled independently via the current controllers in each inverter. The reference currents maintained the same as in normal operation. The waveforms of the measured currents at the base speed are depicted in Fig. 3.28. Notably, these waveforms reveal more distortions than

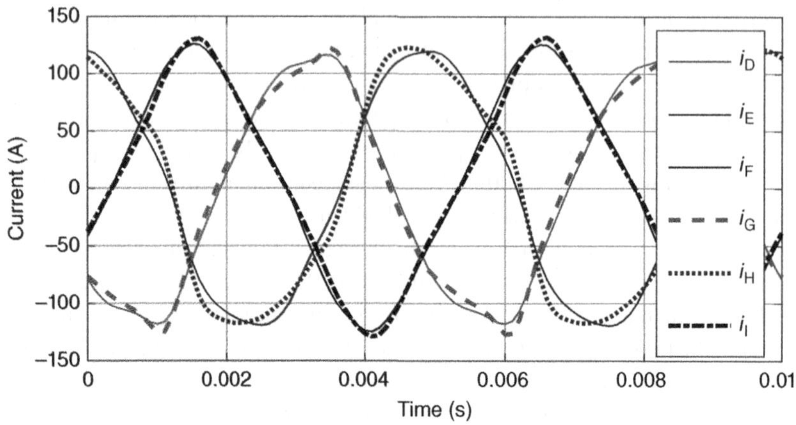

Figure 3.28 Healthy phase currents at 4000 rpm with set ABC open circuited.

those observed in healthy conditions caused by imbalances stemming from magnetic coupling between the open-circuited set and the healthy sets and the limited current control bandwidth. Furthermore, the measured torque variation with current magnitude under the open circuit configuration is compared in Fig. 3.29 with that of the healthy condition. It demonstrates the drive is capable of continuous operation and able to deliver 58% of healthy torque when one 3-phase set is open circuited. Given that the total thermal loading for a given current magnitude is reduced under the open circuit condition, it is possible to increase the current magnitude in the healthy 3-phase sets so that the torque reduction is smaller.

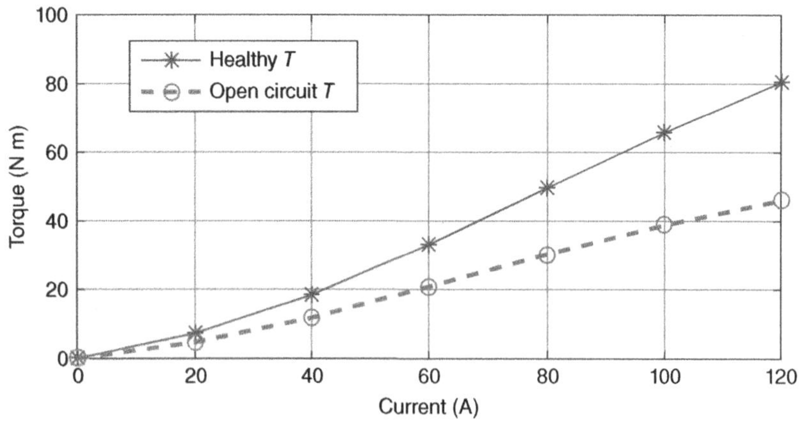

Figure 3.29 Torque comparison with set ABC open circuited.

3.7.2 Test Under One 3-Phase Set in Short Circuit

In the event of a short circuit occurring in either the inverter switches or the windings of one 3-phase set, it is important to apply TSC to prevent further damage. To test postfault operation under this failure mode, a short circuit is created in set ABC via its inverter switches, as shown in Fig. 3.30, while the current excitations in the healthy sets DEF and GHI are varied from 20 to 120 A at the rated speed of 4000 rpm. The resulting torque variation is illustrated in Fig. 3.31, alongside those measured under healthy conditions. Comparatively, the torque capability under this failure mode is also about two-thirds of the rated, slightly higher than that of the open circuit mode. It is worth noting that the machine exhibits a small braking torque when the currents in the two healthy sets are zero because of the losses incurred in the short-circuited phases.

Figure 3.32 shows the induced short-circuit currents in set ABC when the two healthy sets are excited with a current of 120 A. The currents in phases A, B, and C display a high degree of imbalance due to the magnetic coupling in the three

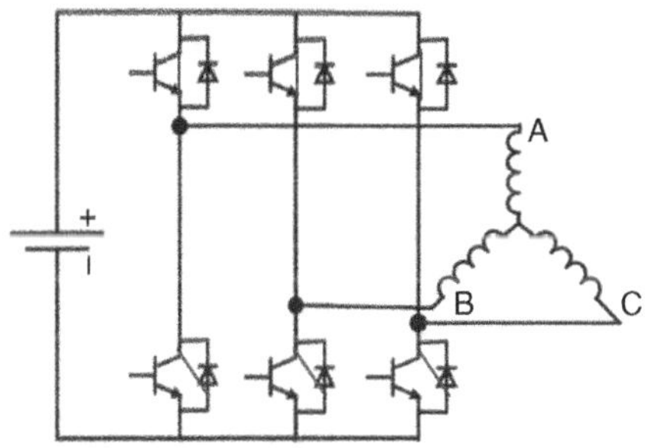

Figure 3.30 Illustration of TSC in the inverter side.

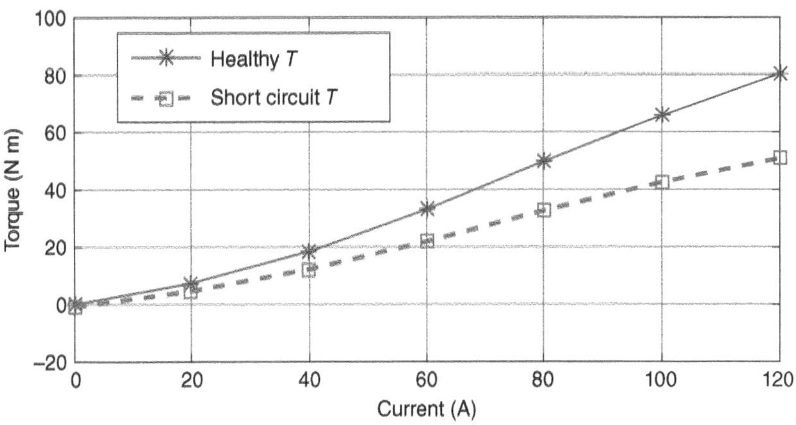

Figure 3.31 Torque comparison with set ABC short circuited.

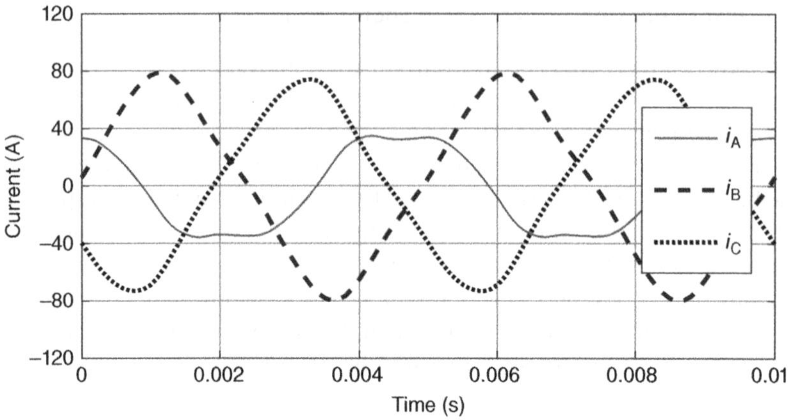

Figure 3.32 Short-circuit phase currents at 4000 rpm.

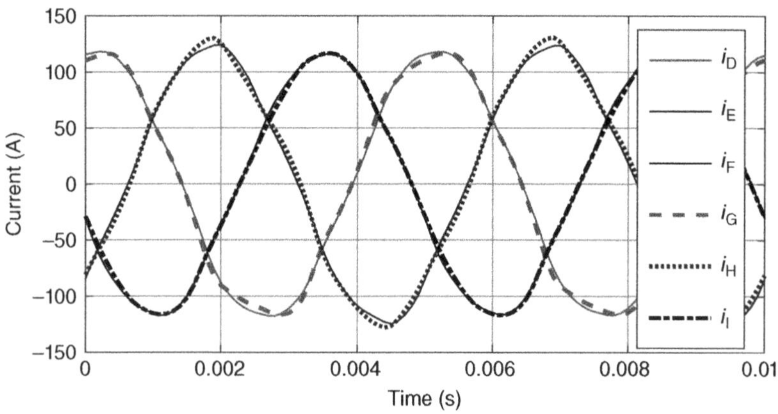

Figure 3.33 Healthy phase currents at 4000 rpm with set ABC short circuited.

3-phase sets. Notably, the RMS current in phase A is significantly lower than that in phases B and C. However, due to the low PM field, they are all below the rated value. As a result, the machine drive can withstand the short-circuit fault without any thermal risks. Furthermore, Fig. 3.33 displays the phase currents in the healthy 3-phase sets, where slight distortion caused by the imbalance can also be observed.

3.7.3 Interturn Short-Circuit Test

An interturn short-circuit fault may occur in any of the six coils of one 3-phase winding set as shown in Chapter 2. For the machine under tests, a single-turn

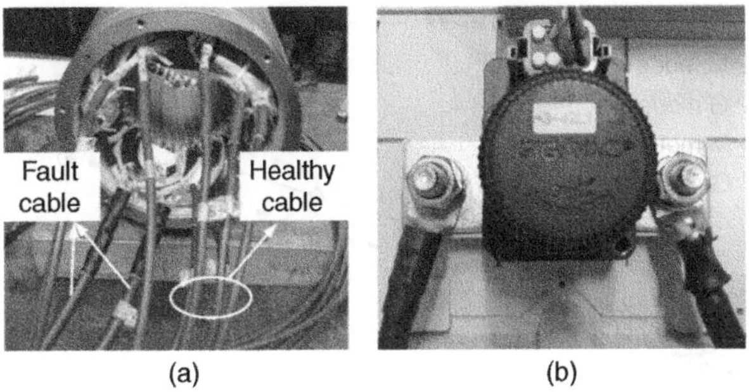

Figure 3.34 Test setup of interturn short-circuit fault: (a) cable leads and (b) control relay. Wang et al. [17]/With permission of IEEE.

short circuit in coil B2 can be performed with the additional tap and connection leads as shown in Figs. 3.13(b) and 3.14(a). The selection of the fault emulation is based on the foregoing theoretical analysis, which has shown that a representative worst-case scenario, leading to the largest short-circuit current, occurs when a single turn located in coil B2 is short circuited during motoring operation in anticlockwise direction of rotation. First, a test is performed without TSC application when the machine rotates anticlockwise. To avoid the damage due to excessive fault current, the turn fault is tested at 1000 rpm when all 3-phase sets are excited with 80 A load current and the fault is injected by closing the relay connected to the fault emulation cables, as shown in Fig. 3.34 for 0.2 second. A current probe is employed to measure the fault current during the test, and the result is shown in Fig. 3.35. It is seen that the turn fault current reaches 440 A. Clearly, the fault current is ~4 times greater than the rated, and without a mitigation measure the insulation in

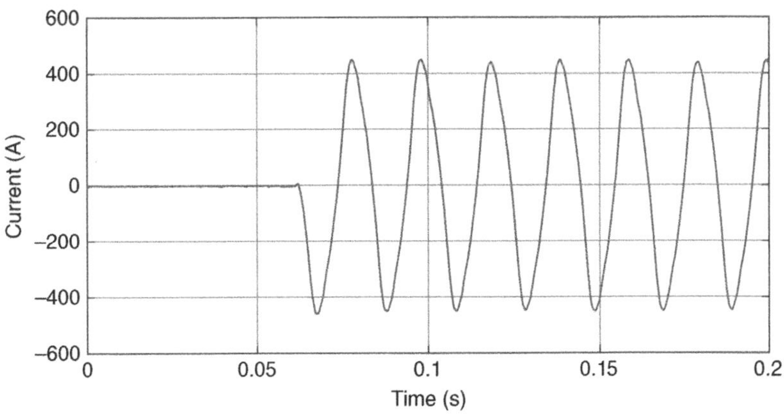

Figure 3.35 Turn fault current with 80 A load current at 1000 rpm.

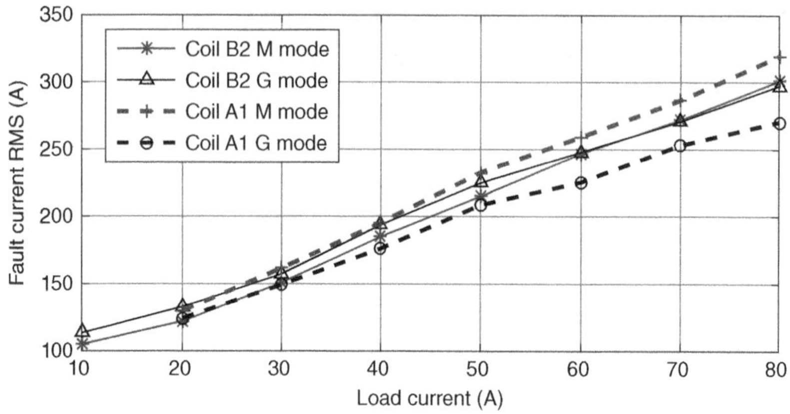

Figure 3.36 Turn fault current variations in coil B2 and A1 at 1000 rpm in motoring and generating modes.

the surrounding region can be quickly damaged due to a rapid increase in temperature. The tests are repeated with a variation of load currents from 10 to 80 A at 1000 rpm. By rotating anticlockwise in both motoring and generation modes, the emulated turn-to-turn short circuit corresponds to one-turn short-circuit fault in coil B2. Conversely, by rotating clockwise in motoring and generation modes, it represents one-turn short-circuit fault in coil A1. The measured variation of the RMS fault currents with the load currents is plotted in Fig. 3.36 for the four fault modes. As can be seen, the RMS fault currents under the four modes are similar and increase with the load currents.

As predicted, small differences in the fault currents between the motoring and generating modes are present because of the differences in the interaction of the armature reaction field with the PM field in the two operating modes. It is seen that even if with a very light load of 20 A, the short-circuit current is notably greater than the rate value. The tests show that such types of faults are impossible to be tolerated without a mitigation action. It is also essential to emphasize that the fault current would escalate even further if either the load current or speed increases. Consequently, application of mitigation measures becomes crucial in order to reduce the fault current under such fault conditions.

Subsequently, the reduction of the turn fault current with the application of TSC to the ABC set via its inverter switches is tested at 4000 rpm when the healthy 3-phase sets are excited with the rated current of 120 A. The fault current in the short-circuited turn and the currents in the rest of the 3-phase windings are shown in Fig. 3.37. It is seen that the peak fault current has been reduced to 220 A and its RMS value to 130 A while the phase currents in the rest of the faulted 3-phase windings are very close to those observed under the TSC of the three phases

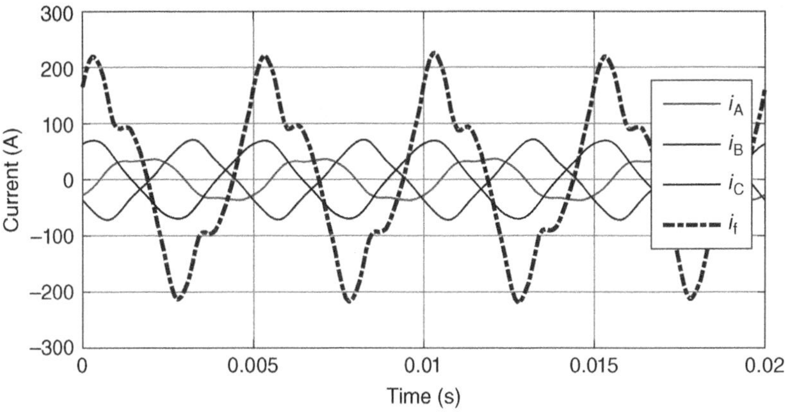

Figure 3.37 Turn fault current in coil B2 and phase currents of set ABC with 120 A load current at 4000 rpm in motoring mode after TSC.

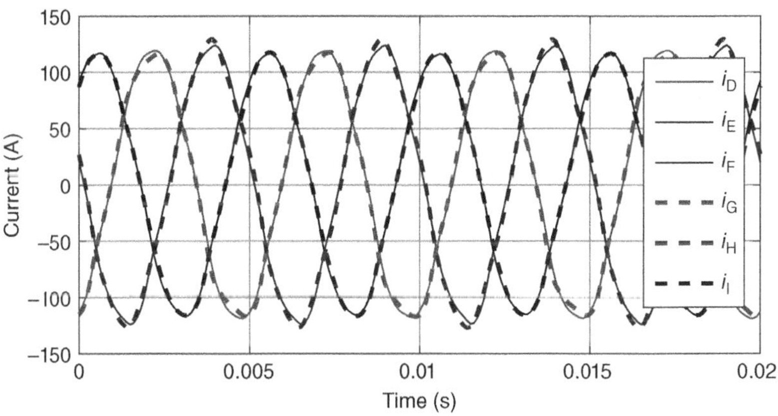

Figure 3.38 Healthy phase currents with turn fault in coil B2 when load current is set to 120 A in motoring mode at 4000 rpm after TSC.

shown in Fig. 3.32. During the test, the currents in the healthy sets remain well controlled, as shown in Fig. 3.38. This represents postfault operation after the TSC mitigation measure is applied against the interturn short-circuit fault and demonstrates the ability of the drive to tolerate such faults with reduced output torque and power.

To investigate the mitigation effect of TSC application against the turn faults, further tests with the emulated turn fault when the machine operates at 4000 rpm are conducted by varying the healthy sets currents from 20 to 120 A, for the faults in coil B2 and coil A1 in both motoring and generating modes. The variations

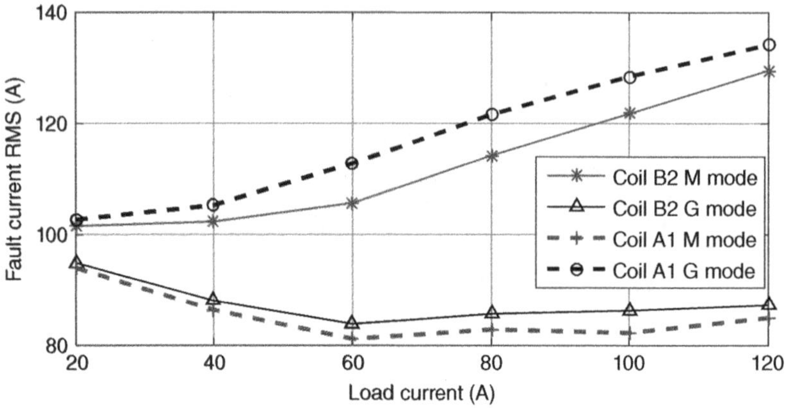

Figure 3.39 Turn fault current variations in coil B2 and A1 at 4000 rpm in motoring and generating modes after TSC.

of the measured RMS currents in the faulted turns with the load current in the healthy phases are depicted in Fig. 3.39. It is evident that in motoring mode, the fault current in coil B2 is much greater than that in coil A1, whereas the opposite is observed in generation mode. These findings are consistent with the results discussed in Chapter 2. The worst-case scenario in motoring mode occurs with a single-turn fault in coil B2 when the other healthy 3-phase sets are excited with the rated current. A similar conclusion can be drawn for the worst case in coil A1 in the generating mode.

The drive output torques during the postfault operations under the emulated turn fault after TSC applications have also been measured and assessed in the tests, and the results are very similar to those shown in Fig. 3.31 under the TSC of one 3-phase set. That is, the drive is capable of delivering approximately 2/3 of the torque compared to healthy operations.

3.7.4 Thermal Test Under Turn Fault Condition with TSC

To assess the sustainability of the drive to tolerate the most severe turn fault, the temperature rises in various parts of the windings are measured with the embedded thermal couplers. A 2-hour test is performed with a single-turn fault injected in coil B2 and TSC applied while the remaining two healthy 3-phase sets carry the rated current of 120 A. The recorded temperatures in different parts of the machine are presented in Fig. 3.40. It is seen that the measured temperatures in the slot and end winding sections, where the fault turn is located, are lower than those in the healthy DEF windings, despite the current in the faulted turn being 1.5 pu. In addition, the temperatures of coil B1 and the ABC

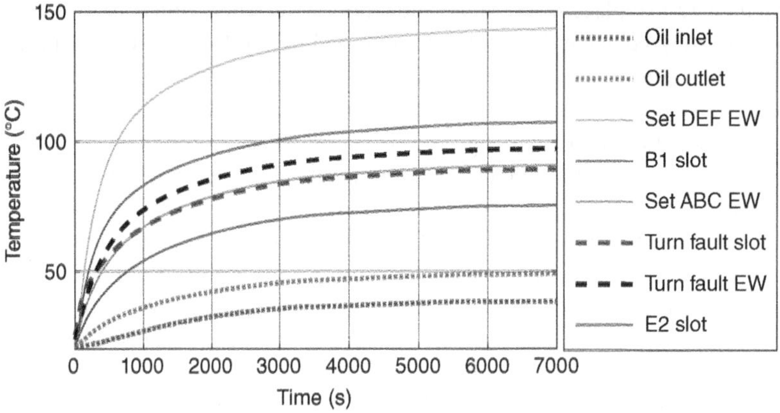

Figure 3.40 Thermal test results under single-turn fault in coil B2 with 120 A at 4000 rpm in motoring mode after TSC.

end winding are also lower than those of the DEF windings. By comparing Figs. 3.27 and 3.40, it can be observed that the temperatures of the DEF windings are even lower than those in the healthy condition. This is attributable to the fact that after TSC application, the RMS currents in the healthy parts of the ABC windings are considerably lower than the rated as is evident in Fig. 3.37. Therefore, despite the higher current in the faulted turn, the total thermal loading of the ABC windings and the machine is reduced, leading to the reduction in temperature rise in comparison to the healthy condition. The thermal tests further confirm that the machine drive can sustain the most severe turn fault with TSC application without any thermal risk, albeit the output torque will be reduced by approximately 1/3.

Compared with the predicted maximum temperature rise of 122 °C in the faulted turn against the ambient temperature of 110 °C for the worst-case scenario as discussed in Section 3.4, the measured temperature rise of ~52 °C is considerably lower. This is mainly due to the factor that the measured fault current in the short-circuit turn is much lower than the prediction for the reasons discussed previously. Further, there is not guarantee that the measurement captured the maximum temperature in the faulted turn. Under the fault condition, indeed, the temperature distribution in the machine no longer exhibits per slot-pitch symmetry, and hence more sophisticated modeling techniques are required. Further information on this can be found in [11, 14].

It should be noted that other faults, such as mechanical failure, demagnetization fault, interphase turn fault, and DC capacitor failure, may also occur in the machine drive which are not tested. Mechanical failures, including bearing failure and broken rotor, are usually avoided by vibration monitoring or by regular maintenance and replacement. Demagnetization under worst-fault conditions has been fully analyzed using the techniques reported in [15, 16]. It has been shown that no demagnetization will take place under any interturn short circuit and TSC conditions. This is because although interturn fault current can be many times of the rated, it flows in a few turns, and its overall effect on the magnets which are partly protected by the rotor core is insignificant. The repeated tests with one turn short circuit partly support the above findings as no partial demagnetization is seen, although the measured fault current is much lower than the prediction due to the reasons discussed previously. As pointed out earlier, the PMA-SynRM mainly relies on the reluctance torque. Hence, the machine drive can still output considerable torque even if an unlikely demagnetization fault occurs. The interphase turn fault is not tested since it is less severe than the single-turn intraphase fault. The DC capacitor failure is also not tested but the resultant behavior is similar to open circuit in one 3-phase set. Obviously, it can be sustained by the triple redundant machine drive with appropriate condition monitoring techniques in place.

3.8 Summary

In this chapter, a 40 kW triple redundant 3-phase PMA-SynRM has been designed for aerospace application. Various computationally efficient prediction techniques and tools are integrated into an optimization platform for the machine design to maximize objective functions while satisfying all the electrical, thermal, and mechanical constraints. The machine is optimized to exhibit high efficiency and good fault tolerance against different faults. A machine prototype is built and comprehensive experimental tests have been performed on an established test bench under healthy and various fault modes. The measured results confirm the optimized machine drive exhibits good performance and excellent fault tolerance, including the worst-fault case single-turn short circuit.

References

1 J. W. Bennett, G. J. Atkinson, B. C. Mecrow, and D. J. Atkinson, "Fault-tolerant design considerations and control strategies for aerospace drives," *IEEE Transactions on Industrial Electronics*, vol. 59, no. 5, pp. 2049–2058, 2012.

2 N. Bianchi and S. Bolognani, "Design of a fault-tolerant IPM motor for electric power steering," in *IEEE 36th Power Electronics Specialists Conference, 2005. PESC '05*, 2005, p. 2873.

3 B. C. Mecrow *et al.*, "Design and testing of a four-phase fault-tolerant permanent-magnet machine for an engine fuel pump," *IEEE Transactions on Energy Conversion*, vol. 19, no. 4, pp. 671–678, 2004.

4 T. Raminosoa, C. Gerada, and M. Galea, "Design considerations for a fault-tolerant flux-switching permanent-magnet machine," *IEEE Transactions on Industrial Electronics*, vol. 58, no. 7, pp. 2818–2825, 2011.

5 P. Arumugam, "Design optimization on conductor placement in the slot of permanent magnet machines to restrict turn-turn short-circuit fault current," *IEEE Transactions on Magnetics*, vol. 52, no. 99, pp. 1–8, 2016.

6 J. Dusek, P. Arumugam, C. Brunson, E. K. Amankwah, T. Hamiti, and C. Gerada, "Impact of slot/pole combination on inter-turn short-circuit current in fault-tolerant permanent magnet machines," *IEEE Transactions on Magnetics*, vol. 52, no. 4, pp. 1–9, 2016.

7 P. Arumugam, T. Hamiti, C. Brunson, and C. Gerada, "Analysis of vertical strip wound fault-tolerant permanent magnet synchronous machines," *IEEE Transactions on Industrial Electronics*, vol. 61, no. 3, pp. 1158–1168, 2014.

8 L. Chen, J. Wang, P. Lazari, and X. Chen, "Optimizations of a permanent magnet machine targeting different driving cycles for electric vehicles," in *Electric Machines & Drives Conference (IEMDC), 2013 IEEE International*, 2013, pp. 855–862.

9 L. Chen, J. Wang, P. Lombard, P. Lazari, and V. Leconte, "Design optimisation of permanent magnet assisted synchronous reluctance machines for electric vehicle applications," *in 2012 XXth International Conference on Electrical Machines (ICEM)*, 2012, pp. 2647–2653.

10 V. I. Patel, J. Wang, W. Wang, and X. Chen, "Analysis and design of 6-phase fractional slot per pole per phase permanent magnet machines with low space harmonics," in *2013 IEEE International Electric Machines & Drives Conference (IEMDC)*, 2013, pp. 386–393.

11 P. H. Mellor, D. Roberts, and D. R. Turner, "Lumped parameter thermal model for electrical machines of TEFC design," *IEE Proceedings B, Electric Power Applications*, vol. 138, no. 5, pp. 205–218, 1991.

12 J. Wang, W. Wang, K. Atallah, and D. Howe, "Design of a linear permanent magnet motor for active vehicle suspension," in *2009 IEEE International Electric Machines and Drives Conference*, 2009, pp. 585–591.

13 Q. Chen, G. Liu, W. Zhao, L. Sun, M. Shao, and Z. Liu, "Design and comparison of two fault-tolerant interior-permanent-magnet motors," *IEEE Transactions on Industrial Electronics*, vol. 61, no. 12, pp. 6615–6623, 2014.

14 Y. Shi, J. Wang, and B. Wang, "Electromagnetic-thermal coupled simulation under various fault conditions of a triple redundant 9-phase PMASynRM," *IEEE Transactions on Industry Applications*, vol. 56, no. 1, pp. 128–137, 2020.

15 S. S. Nair, V. I. Patel, and J. Wang, "Post-demagnetization performance assessment for interior permanent magnet AC machines," *IEEE Transactions on Magnetics*, vol. 52, no. 4, pp. 1–10, 2016.

16 V. I. Patel, J. Wang, and S. S. Nair, "Demagnetization assessment of fractional-slot and distributed wound 6-phase permanent magnet machines," *IEEE Transactions on Magnetics*, vol. 51, no. 6, pp. 1–11, 2015.

17 B. Wang, J. Wang, A. Griffo, and B. Sen, "Experimental Assessments of a Triple Redundant Nine-Phase Fault-Tolerant PMA SynRM Drive," *IEEE Transactions on Industrial Electronics*, vol. 66, no. 1, pp. 772–783, 2017.

4

General Modeling Technique for Multiple 3-Phase Machine Drive

4.1 Introduction

Electrical machine drive behaviors under a fault condition are notably different from those in healthy operations due to physical changes arising from the fault and the influence of fault current in case of any type of short-circuit faults. Therefore, an effective fault modeling technique is essential for developing a fault-tolerant machine drive to assess its behaviors and fault impacts in a computationally efficient way and to facilitate the development of suitable fault detection and postfault control techniques.

However, since occurrence of a fault in a machine drive will break the symmetrical operating conditions as well as periodic boundary conditions, conventional modeling techniques for electrical machines, such as lumped parameter circuit and the techniques based on the synchronous rotating reference frame, are not applicable. Numerical techniques, such as finite element (FE) analysis, are often used to assess machine performance under a fault condition. However, since the periodic boundary condition is no longer true in a faulted machine, a FE model of the full machine geometry is required for the performance predictions, which leads to considerable computational resources and times. Further, since there are not only various failure modes (or fault types) but also many influential factors in each fault type, numerical analysis-based techniques would consume vast amounts of resources and effort to cover all possible fault scenarios, and hence become ineffective for developing fault detection techniques and postfault control strategies. In addition, such techniques do not provide much insight into the underlying physical behavior of a machine under fault conditions. Effective fault modeling is, therefore, a key challenge for the development of a fault-tolerant machine drive.

For the triple 3-phase permanent magnet-assisted synchronous reluctance machine (PMA-SynRM) drive under study, although the different 3-phase modules are physically, thermally, and electrically isolated, they are not magnetically

Multiple 3-Phase Fault Tolerant Permanent Magnet Machine Drives: Design and Control, First Edition.
Bo Wang and Jiabin Wang.

isolated. Indeed, the three 3-phase sets are mutually coupled and cannot be regarded as independent when the machine operates in a fault mode. In addition, the winding inductance cannot be considered constant due to the nonlinear magnetic circuit of the complex rotor geometry. The machine drive could operate in healthy, open-circuit, and short-circuit conditions. More generally, the magnitude and phase angle of the currents in the three winding sets may be all different and this condition is referred to as unbalanced current operation mode. For the triple 3-phase drive, it has six independent current variables in *dq* axis. Together with the rotor position, the machine flux linkages are nonlinear functions of seven variables. The current state-of-the-art modeling techniques are not sufficient to describe this complex machine drive in a computationally efficient manner. Hence, it is valuable to develop a modeling technique that can represent the machine behavior and capture its main characteristics.

More specifically for an interturn short-circuit fault which only involves a few turns, very large current is induced in the short-circuit path [1]. It leads to uneven currents in the faulty turns, which further increases the difficulty level of modeling [2, 3]. The turn fault current is not only affected by the currents in the fault set but also influenced by the currents in the healthy sets. In addition, it is also dependent on the number of faulted turns, their position in a slot, and the location of the coil in which the fault occurs as shown in Fig. 4.1 since the flux linkage of the fault turns change accordingly. After the remedial terminal short-circuit (TSC) action, the resultant current in the faulted turns is critical and needs to be predicted accurately when the currents in each 3-phase set are all different. However, it is computationally expensive to evaluate all the possible fault scenarios to identify the worst case by repeated FE simulations. While accurate fault modeling is a great challenge for the triple 3-phase machine drive, a fast-modeling technique

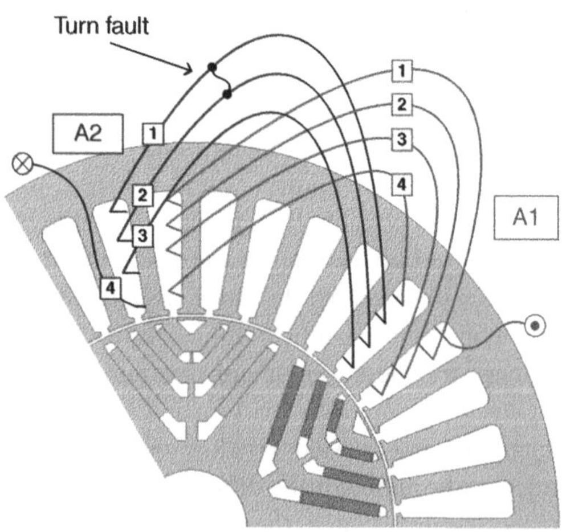

Figure 4.1 Illustration of turn fault location in a slot and phase winding.

that accounts all factors affecting the fault behavior, including the currents in each 3-phase set, the number of faulted turns, and the associated coil location and slot position, will be of great value.

To this end, a general modeling technique is established for the triple 3-phase PMA-SynRM considering various fault scenarios, including the healthy, open-circuit, short-circuit, and unbalanced current operation. The model can be adapted for interturn fault accounting the factors described earlier. In this way, the machine performance and behavior under healthy and fault conditions can be predicted in a flexible and computationally efficient way.

4.2 General Modeling Technique for 3-Phase Winding Sets

4.2.1 MMF Analysis for 3-Phase Winding Sets

According to the magneto-motive force (MMF) analysis in Chapter 2, the MMF across the air gap produced by the currents in set ABC windings when the stator and rotor laminations are assumed to be infinitely permeable is replotted in Fig. 4.2. The currents in set ABC produce the MMF not only in its own spatial region but also in the regions of the other two sets of windings. It has been shown that the MMF in set ABC region consists of an AC component and an offset component while only an offset component exists in the other regions. Likewise, the MMFs produced by the currents in the other two winding sets can be obtained with appropriate phase shifts as shown in Fig. 4.3.

It is seen that the currents in each 3-phase winding set produce an AC component only in the region occupied by its own winding while they contribute to the

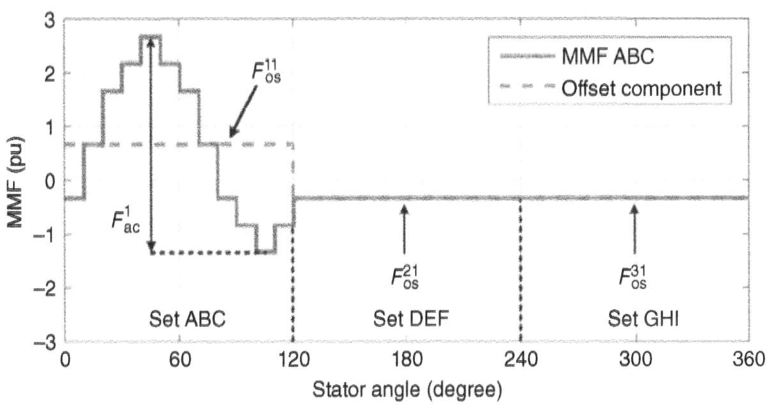

Figure 4.2 MMF produced by set ABC.

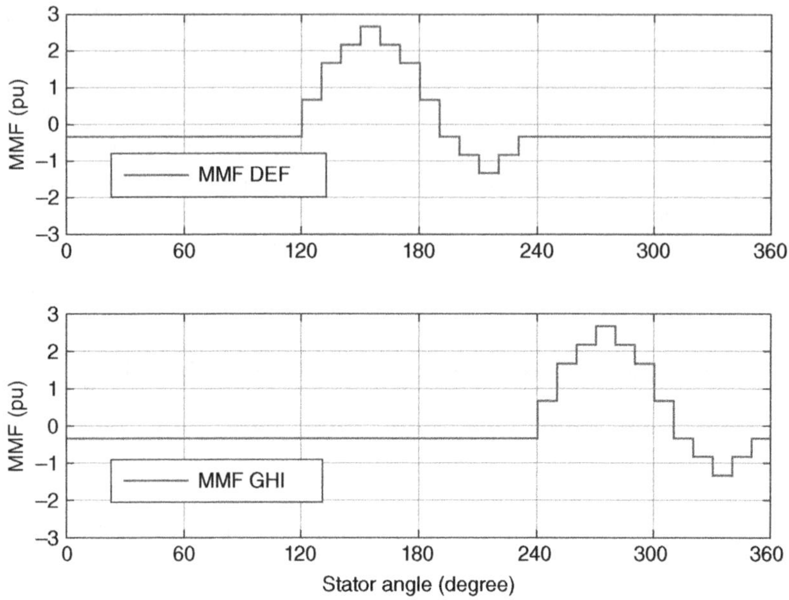

Figure 4.3 MMFs produced by sets DEF and GHI.

offset component in the regions occupied by the other windings. The total MMF in one 3-phase set region can be obtained by adding the MMF contributions of all three sets. Thus, the total MMF over set ABC, denoted as F^1, can be expressed as

$$\begin{aligned} F^1 &= F^1_{\mathrm{ac}} + F^1_{\mathrm{os}} \\ F^1_{\mathrm{os}} &= F^{11}_{\mathrm{os}} + F^{12}_{\mathrm{os}} + F^{13}_{\mathrm{os}} \\ &= \frac{2N}{3}(i_{\mathrm{A}} + i_{\mathrm{B}} - i_{\mathrm{C}}) - \frac{N}{3}(i_{\mathrm{D}} + i_{\mathrm{E}} - i_{\mathrm{F}}) - \frac{N}{3}(i_{\mathrm{G}} + i_{\mathrm{H}} - i_{\mathrm{I}}) \end{aligned} \tag{4.1}$$

where F^1_{os} denotes the net MMF offset component in the region occupied by the ABC windings and N is the number of series connected turns per coil. According to (4.1), the total MMF in one 3-phase set is also composed of an AC component F^1_{ac} and an offset component F^1_{os}. The AC component is purely determined by the currents in its own set and rotor angular position while the offset component is the combined effect of the currents in all three sets. Due to nonexistence of any zero-sequence current in a 3-phase winding set, the phase currents i_{A}, i_{B}, i_{C} can be represented as i_{d1}, i_{q1} in the dq-axis reference frame. Together with the offset component F^1_{os}, the MMF over set ABC F^1 can be expressed as a function of $(\theta, i_{d1}, i_{q1}, F^1_{\mathrm{os}})$:

$$F^1 = F^1_{\mathrm{ac}} + F^1_{\mathrm{os}} = f\left(\theta, i_{d1}, i_{q1}, F^1_{\mathrm{os}}\right) \tag{4.2}$$

where θ is the rotor angle. Similarly, the total MMFs over sets DEF and GHI can be expressed as

$$\begin{aligned}
F^2 &= F^2_{ac} + F^2_{os} = f\left(\theta, i_{d2}, i_{q2}, F^2_{os}\right)\\
F^3 &= F^2_{ac} + F^2_{os} = f\left(\theta, i_{d3}, i_{q3}, F^3_{os}\right)\\
F^2_{os} &= \frac{2N}{3}(i_D + i_E - i_F) - \frac{N}{3}(i_A + i_B - i_C) - \frac{N}{3}(i_G + i_H - i_I)\\
F^3_{os} &= \frac{2N}{3}(i_G + i_H - i_I) - \frac{N}{3}(i_A + i_B - i_C) - \frac{N}{3}(i_D + i_E - i_F)
\end{aligned} \tag{4.3}$$

The variables are similarly defined for the second and third winding sets, DEF and GHI respectively. It can be shown that under healthy conditions the dq-axis currents of each 3-phase set are identical while the offset components are cancelled according to (4.1). Consequently, all three sets will produce the same AC component as in a conventional 3-phase machine with three pole pairs. In case of one set open-circuit condition, the AC component is zero for the fault set due to zero phase currents but the offset component still exists due to the currents in the other two 3-phase sets, whilst both an AC and offset component will exist in the other two sets. If one set is short circuited, short-circuit currents will be induced in the fault set, which are governed by the circuit laws. The resultant MMF can be deduced accordingly as a combination of the AC component and offset component for each set. In the extreme condition where the currents in each set are all different, the MMF of each set can still be expressed by (4.2) and (4.3) with independent dq-axis currents in the three sets. It is important to point out that the above descriptions of the MMFs associated with each 3-phase winding set are generic and applicable to any current waveform.

It follows that in any operation conditions the MMF of the whole machine can be partitioned in three regions. The MMF in each region associated with a 3-phase set can be expressed as a function $f(\theta, i_d, i_q, F_{os})$ with four variables, where i_d, i_q are the currents of the winding set in the region while F_{os} is the combined offset component due to the currents in all three sets of windings.

The MMF analysis confirms that different 3-phase sets are not magnetically isolated and their mutual coupling effect can be represented by the MMF offset component. And only in healthy conditions, the offset component is cancelled by each other due to the same current excitation and hence the machine behaves as a conventional 3-phase machine. This condition is known as balanced or symmetric operating condition. In contrast, in any fault conditions the residual MMF offset component is non-zero and it will lead to unbalanced or asymmetrical operation conditions. Such understanding provides an important physical insight into the cause of changes in machine behavior under a fault or unbalanced operation condition.

4.2.2 General Modeling Approach

As discussed earlier, the MMF in one 3-phase set region can be described by the function of four variables (θ, i_d, i_q, F_{os}). Physically as shown in Fig. 4.4, the flux density in the air gap is determined by the MMF described by (i_d, i_q, F_{os}) and the permanent magnet (PM) field, whose combined effect produces flux linkages and torque. Therefore, the flux linkages and torque of each 3-phase set can be expressed by a 4D function $\mathcal{F}(\theta, i_d, i_q, F_{os})$. This function can be obtained by FE simulations of the full machine model to establish the flux linkages and torque as functions of the four variables in the form of 4D lookup tables, as described in Section 4.2.3. As a result, the performance of each 3-phase set can be predicted by interpolating the 4D tables via the four variables (θ, i_d, i_q, F_{os}) together with any circuit equations and physical governing laws for the drive operation.

Thus, a general modeling method is established as shown in Fig. 4.5. First, the (i_d, i_q) currents of each 3-phase set together with θ are fed into the MMF offset calculation block to derive F_{os} for each set according to (4.1) and (4.3). Then, the resultant four variables (θ, i_d, i_q, F_{os}) for each 3-phase set form the inputs to the 4D tables, which determine the resultant flux linkages and torque of the 3-phase winding set. It should be noted that the flux linkages and torque of each set are interpolated separately based on the same lookup tables since the electrical offset angle between any two adjacent winding sets is 360°. The total torque is obtained by adding the torque contributions of all three sets.

According to the modeling scheme, the flux linkages and torque component of each 3-phase set can be determined as long as the currents of all three winding sets are available, no matter the machine is operating in healthy, open-circuit, short-circuit, or unbalanced current conditions. Thus, the proposed model can predict the machine behavior under healthy, open-circuit, short-circuit, and unbalanced current operating conditions without any modification, and in doing so, significant efforts are saved to predict the machine performance under

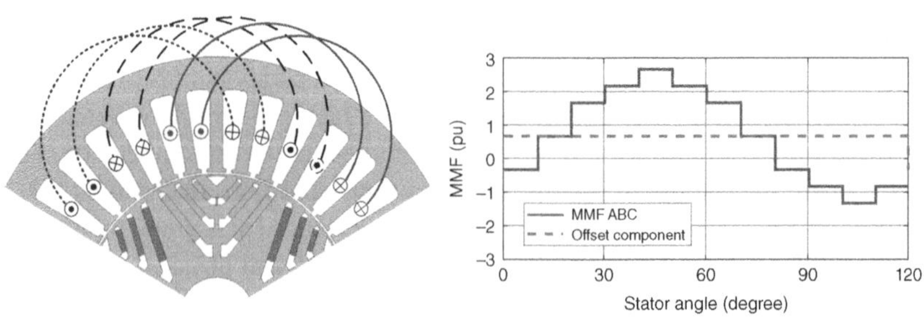

Figure 4.4 One 3-phase module and its MMF in the air gap.

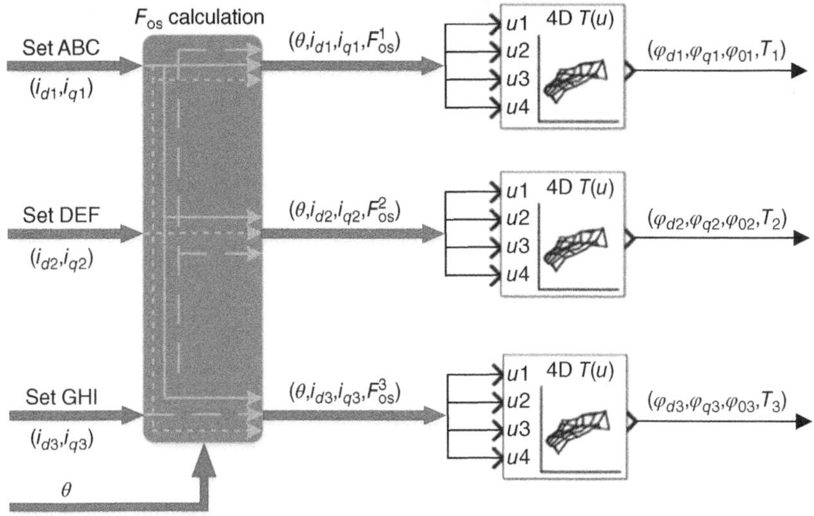

Figure 4.5 General modeling diagram.

different conditions. It should be noted that the modeling technique can only deal with open-circuit, short-circuit faults and unbalanced current operations in the 3-phase winding set level at this stage. It will be adapted for modeling with interturn short-circuit fault in Section 4.4.

4.2.3 4D Table Construction

Now the key task is to construct the 4D lookup tables to determine the flux linkages and torque of one 3-phase set as functions of the four variables $(\theta, i_d, i_q, F_{os})$, which capture the combined effects of the winding MMFs and PM field over that 3-phase set. Since the PMA-SynRM is subject to magnetic saturation, FE analysis is performed to extract the data for the 4D tables. It is worth noting that the geometry of the three 3-phase sets is identical as shown in Fig. 4.4 because the electrical phase shift between each set is 360°. Therefore, the 4D tables only need to be constructed for one 3-phase set and the other two sets can use the same table. Hence, FE simulations are performed to extract the flux linkages and torque data of set ABC region in the full machine model, shown in Fig. 4.6, for a given MMF distribution.

The first step of table construction is to determine the currents in the nine phases to generate the MMF over set ABC defined by $(\theta, i_d, i_q, F_{os})$. As discussed previously, the MMF AC component is determined by the currents of its own set winding. Thus, for given (i_d, i_q) and the rotor electrical angle θ, the phase currents

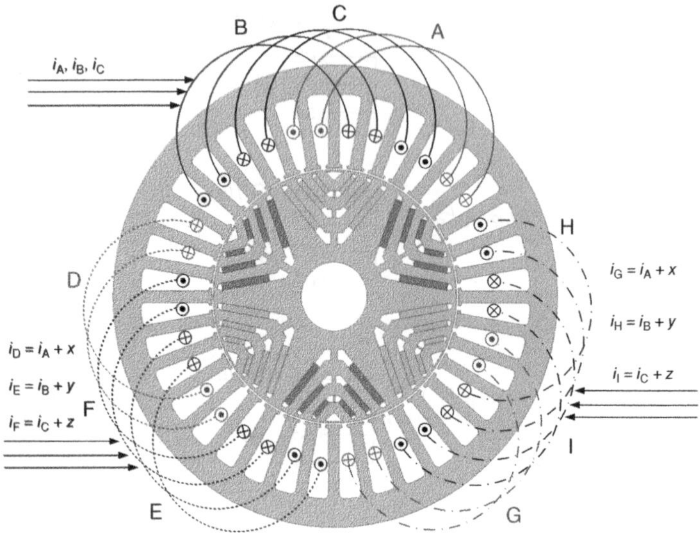

Figure 4.6 Full machine model for 4D table generation.

of set ABC can be derived by the *dq* transform as

$$\begin{aligned} i_{\mathrm{A}} &= i_d \cos\theta - i_q \sin\theta \\ i_{\mathrm{B}} &= i_d \cos(\theta - 120^\circ) - i_q \sin(\theta - 120^\circ) \\ i_{\mathrm{C}} &= i_d \cos(\theta + 120^\circ) - i_q \sin(\theta + 120^\circ) \end{aligned} \tag{4.4}$$

However, the MMF offset component F_{os} is the combined effect of currents in all three 3-phase sets. To represent this offset component, specific currents should be injected to the DEF and GHI sets of windings to generate the required F_{os}. For this purpose, the currents in sets DEF and GHI that differ from those in set ABC are specified as (x, y, z) in (4.5), which are to be determined to produce the required F_{os}.

$$\begin{aligned} i_{\mathrm{D}} &= i_{\mathrm{G}} = i_{\mathrm{A}} + x \\ i_{\mathrm{E}} &= i_{\mathrm{H}} = i_{\mathrm{B}} + y \\ i_{\mathrm{F}} &= i_{\mathrm{I}} = i_{\mathrm{C}} + z \end{aligned} \tag{4.5}$$

As a result, the F_{os} over set ABC is obtained by substituting (4.4) and (4.5) into (4.1) as shown in (4.6).

$$F_{\mathrm{os}} = -\frac{2}{3}N(x + y - z) \tag{4.6}$$

Since zero sequence current cannot circulate in a 3-phase system with no neutral connection, the 3-phase currents are confined by (4.7).

$$x + y + z = 0 \tag{4.7}$$

Based on (4.6) and (4.7), it can be deduced that

$$\begin{aligned} x + y &= -\frac{3}{4N} F_{\text{os}} \\ z &= \frac{3}{4N} F_{\text{os}} \end{aligned} \tag{4.8}$$

To obtain a unique solution for x, y, their values are chosen as to minimize the peak currents over one electrical period for phases D, E, F and G, H, I. This reduces the localized saturation effect in the stator teeth where these windings are accommodated. Such a condition is closer to the assumption that the stator and rotor laminations are infinitely permeable for the derivation of (4.1). Thus, x, y are determined in (4.9).

$$x = y = -\frac{3}{8N} F_{\text{os}} \tag{4.9}$$

Therefore, for a given set MMF described by (θ, i_d, i_q, F_{os}), their physical effect is replicated over set ABC by injecting the currents given in (4.4) and (4.5) in the full machine FE model as shown in Fig. 4.6. Then, the resultant flux linkages and torque are calculated for the ABC set region. The flux linkages are directly obtained by FE tools while the torque is calculated by Maxwell stress tensor in the air gap under set ABC region. As a result, the 4D lookup table data are extracted by scanning the four variables (θ, i_d, i_q, F_{os}) over their respective ranges. It is important to point out that in fault or unbalanced operations, MMF offset components will be present in each 3-phase set, which leads to a zero-sequence component of the flux linkages. Hence, the outputs of the 4D table for each set include d-axis, q-axis, and zero-sequence flux linkages and torque.

It is worth noting that the MMF offset value calculated by (4.1) assumes infinitely permeable stator and rotor cores. Thus, the reconstructed MMF offset component may not exactly represent the actual value in the air gap. However, the saturation effect of the MMF AC component, slotting effect, and space harmonics are accounted for by the FE model directly. Therefore, it would only cause a very small error to the prediction. The accuracy of the model will be examined by FE simulations and experimental tests.

4.2.4 Model Integration with Voltage Equations

As shown in Fig. 4.5, the inputs of the general model are the dq-axis currents of the three 3-phase winding sets. However, in most simulation studies and real-time operations, the machine is fed by voltage-source inverters. Thus, the inputs to the model are typically voltages and the machine currents as the outputs which are then fed to the general model. For this purpose, the governing *dq*-axis voltage equations for each 3-phase winding set can be utilized to predict the *dq*-axis currents using (4.10), where u_{di}, u_{qi} are the *dq*-axis voltages applied to the *i*th set of the 3-phase windings.

$$\begin{aligned} i_{di} &= \left(u_{di} + \omega\psi_{qi} - \frac{d\psi_{di}}{dt}\right)/R_s \\ i_{qi} &= \left(u_{qi} - \omega\psi_{di} - \frac{d\psi_{qi}}{dt}\right)/R_s \end{aligned} \tag{4.10}$$

Further, any control law, such as proportional and integral (PI) control, can also be included in the simulation model as shown in Fig. 4.7 for assessments of the drive system behaviors. As can be seen, each 3-phase module is controlled by an independent current loop. If the machine operates in healthy conditions, all three winding sets will exhibit the same behavior and no MMF offset components will be produced. In case of an open-circuit fault in one 3-phase set, the currents of the fault set are forced to zero while the other remaining two sets continue their operation with the control units. The influence of the open circuit in one 3-phase set on the other two remaining 3-phase sets will be accounted for by the MMF offset components evaluated in the model shown in Fig. 4.5. If one 3-phase set is terminally short circuited, the voltages of the fault set are forced to zero and the short-circuit currents will be calculated according to (4.10). As a result, by appropriate configuration of the voltage equation blocks, the model shown in Fig. 4.17 can be used to simulate the machine drive behaviors in both healthy and fault conditions as

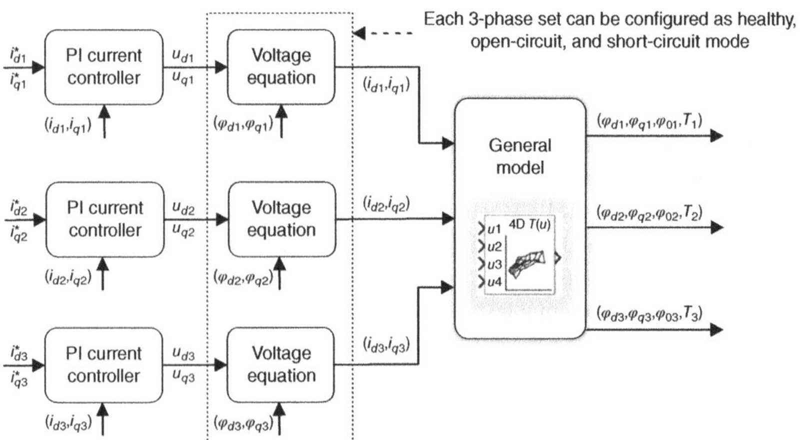

Figure 4.7 General model integrated with voltage equations and current controllers.

well as for the developments of fault detection techniques and postfault control algorithms.

4.3 Study on Model Accuracy and Computational Efficiency

This section evaluates the accuracy and computational efficiency of the modeling method by comparing its predictions to the results obtained from direct FE analysis. The comparisons are made against different scenarios, including healthy operation, open circuit, and short circuit in one 3-phase set and unbalanced current conditions. For comparison, the general model shown in Fig. 4.5 is built in Matlab/Simulink with 4D lookup tables, and the *dq*-axis currents and rotor angle as the inputs. The same inputs are fed into the two-dimensional (2D) FE model of the machine, and the outputs predicted from the two models are compared for each operation condition described below. Further, the times taken for the predictions by each model on the same computing platform are also compared.

4.3.1 Healthy Operation

Under normal (or healthy) operating conditions, the currents in each 3-phase set are assumed to be balanced with equal magnitude, resulting in a zero F_{03} for all three 3-phase sets. As a result, the behavior of the machine is solely determined by the variables (θ, i_d, i_q). Simulations with the 4D table-based model and by direct FE analysis were conducted at the base speed 4000 rpm and rated current of 120 A in magnitude. The predicted torque and flux linkages with the two methods are compared in Figs. 4.8 and 4.9, respectively. The figures reveal that the predictions with

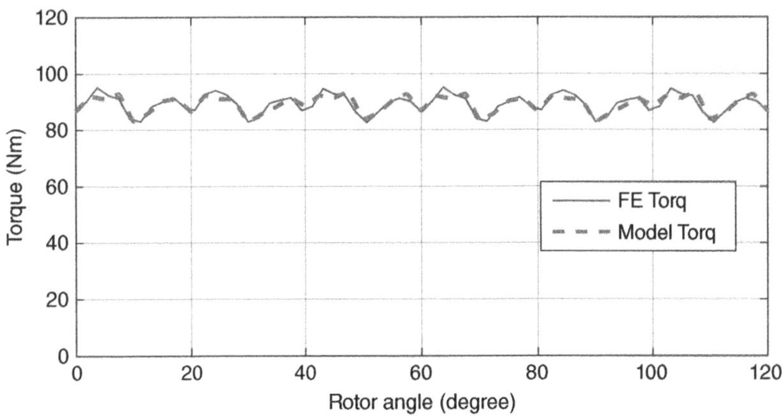

Figure 4.8 Comparison of predicted torque by two models in healthy condition.

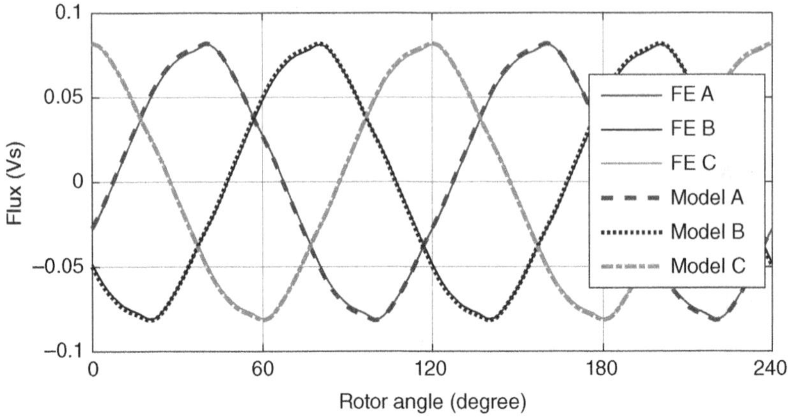

Figure 4.9 Comparison of predicted phases ABC flux linkages by two models in healthy condition.

the 4D table model closely match with those obtained from the FE analysis [4]. The relative absolute average differences in the torque and flux linkages predicted by the two models are 0.06% and 0.88%, respectively, where the torque is normalized to the FE-predicted average and flux linkages to the FE-predicted fundamental magnitude. It should be noted that in healthy operations when the currents in each 3-phase set form balanced sinusoids with equal magnitude, the FE model can be reduced to one sixth of the full geometry and the simulation over one electric period takes ~5 minutes. In contrast, the prediction with the 4D table-based model over the same period is much faster and takes ~0.2 second. However, the computational efficiency of the 4D table model is far more significant in predictions of fault and unbalanced current operations as will be discussed in the subsequent sections.

4.3.2 Open Circuit

Assuming an open-circuit fault in set ABC, the currents in phases ABC are zero, while the currents in sets DEF and GHI remain the same as those at the rated condition in the healthy mode. As depicted in the 4D table-based model shown in Fig. 4.5, now the currents in set ABC (i_{d1}, i_{q1}) are zero while the dq-axis currents in the other two healthy 3-phase sets remain constant at the rated. The MMF offset components in the three 3-phase sets should be formed according to (4.1) and (4.3) in the 4D table-based model. Hence, the four variables (θ, i_d, i_q, F_{os}) for each 3-phase set are derived and inputted into the 4D lookup tables to predict the flux linkages and torque.

Figures 4.10 and 4.11 compare the flux linkages and torques predicted by the 4D model and FE analysis. As can be seen, the predicted torque matches closely

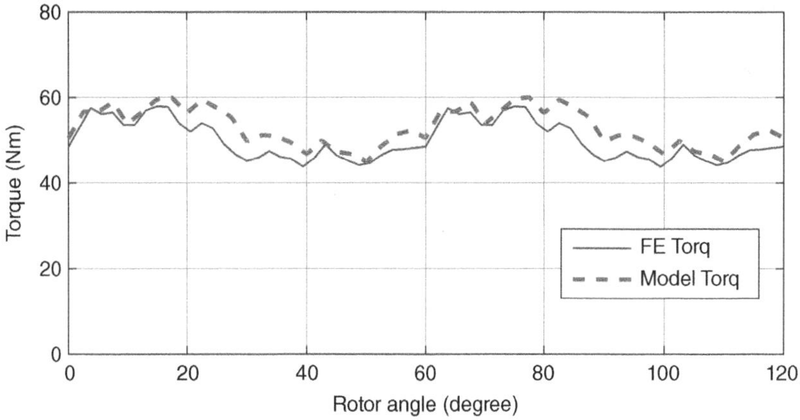

Figure 4.10 Comparison of predicted torques by two models in open-circuit condition.

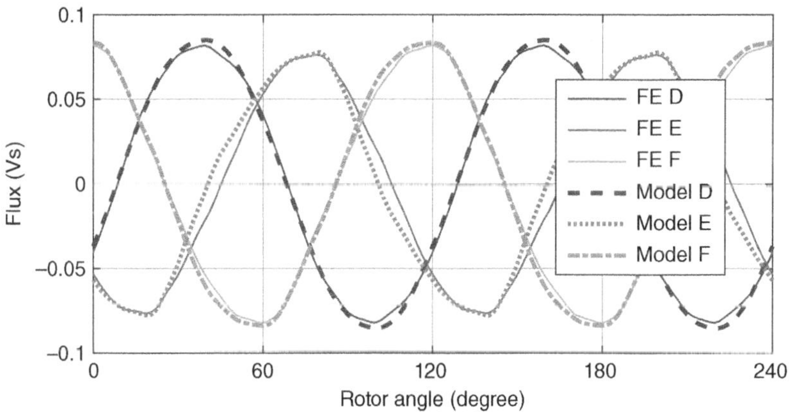

Figure 4.11 Comparison of predicted flux linkages of healthy DEF set by two models.

with the trend observed in the FE results, albeit slightly higher. The 2nd harmonic in the torque waveform is well captured by the 4D table-based prediction. Since the electromagnetic behaviors of sets DEF and GHI are similar under the fault condition, only the flux linkages of set DEF are illustrated in Fig. 4.11, where it is seen that the 4D table-based predictions also closely follow the FE results, displaying minor differences. It is evident that the flux linkages are no longer symmetric under the fault condition because of the mutual magnetic coupling between the different 3-phase sets via the non-zero MMF offset component which is plotted in Fig. 4.12. It is a pulsating component that gives rise to torque ripples but does not contribute to net electromagnetic energy conversion.

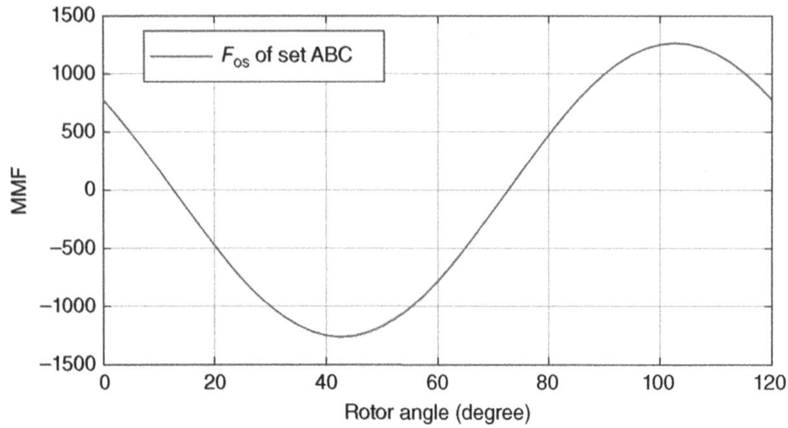

Figure 4.12 MMF offset component of set ABC under open-circuit condition.

It is also evident that, compared with the predictions in the healthy operation, the differences in the 4D table and FE predictions are slightly larger. The relative absolute averaging errors for the torque and flux linkages are 5% and 3.3%, being quite small and acceptable. The increase in the 4D table-based prediction errors can be attributed to the increase in the MMF offset magnitudes in the 4D table model, which are derived by assuming infinitely permeable laminations.

It should be noted that under the open-circuit fault, the FE simulation requires the full model of the 2D machine geometry and takes ~30 minutes to complete for one electrical period. In contrast, the simulation with the 4D table-based model over the same period still only takes ~0.2 second, representing an increase in computational efficiency by about 9000 times.

4.3.3 Short Circuit

In case of a switch or winding short circuit, TSC is applied to the faulted 3-phase set as a mitigating measure. Since the zero-sequence current is zero in the 3-phase windings with no neutral connection, the induced short-circuit currents are determined by the flux linkages of the faulty set according to (4.11).

$$i_d = \left(\omega\psi_q - \frac{d\psi_d}{dt}\right)/R_s$$
$$i_q = -\left(\omega\psi_d + \frac{d\psi_q}{dt}\right)/R_s \tag{4.11}$$

The derived dq-axis currents, using the model-predicted dq-axis flux linkages, are fed to the inputs associated with the ABC set in the 4D table-based model while the dq-axis currents in other two healthy sets are kept to their values at the rated operation. The MMF offset components are processed accordingly. The same operation under the short circuit is also simulated by the full FE models. Figure 4.13 demonstrates the close match between the 4D model-predicted torque and the FE result. Good agreements are also observed in the predicted short-circuit currents and flux linkages by the two models as depicted in Figs. 4.14–4.16. It is evident that the short-circuit currents in each phase of set ABC are asymmetrical and distorted. It is also seen that the flux linkages in the faulty 3-phase set are nearly an order of magnitude lower than those in the two healthy 3-phase sets as they have effectively been neutralized by the short-circuit currents except for the residual zero-sequence component due to the absence of zero-sequence current in the

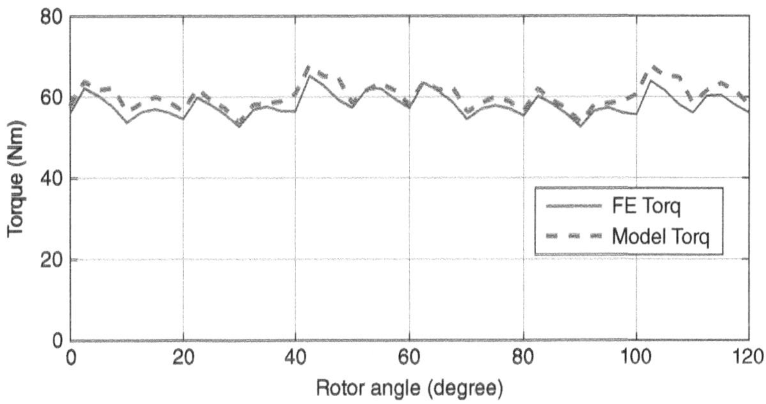

Figure 4.13 Comparison of predicted torques by two models in short-circuit condition.

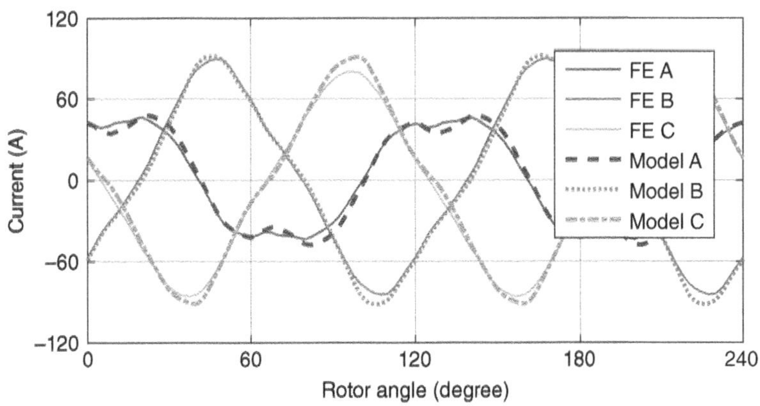

Figure 4.14 Comparison of predicted short-circuit currents by two models.

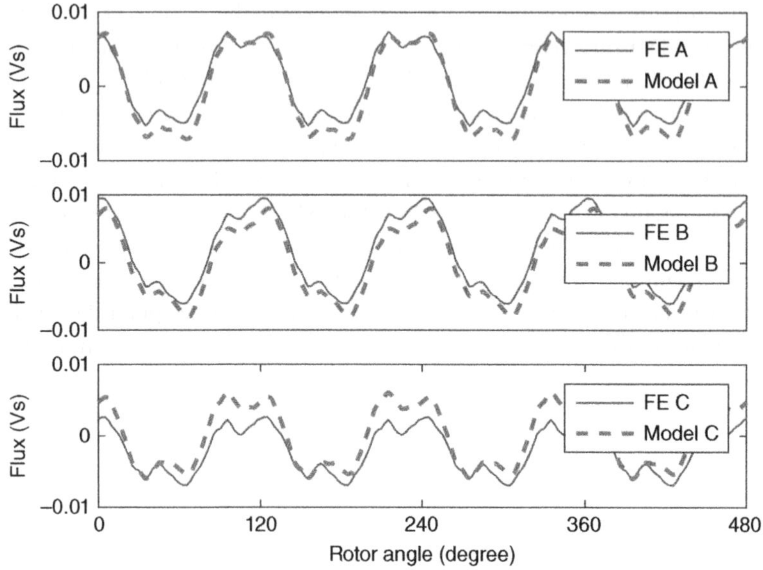

Figure 4.15 Comparison of predicted flux linkages in the short-circuit set by the two models.

ABC set. Meanwhile, the predicted flux linkages of the healthy DEF set by the two models are also well matched as shown in Fig. 4.16.

Like the open-circuit fault, the FE prediction for the short-circuit operation over one electrical period also requires the full model and takes 30 minutes to complete. In addition, the short-circuit currents require at least 10 electrical cycles to reach the steady state. Thus, in total, it requires 5 hours to complete the FE simulation

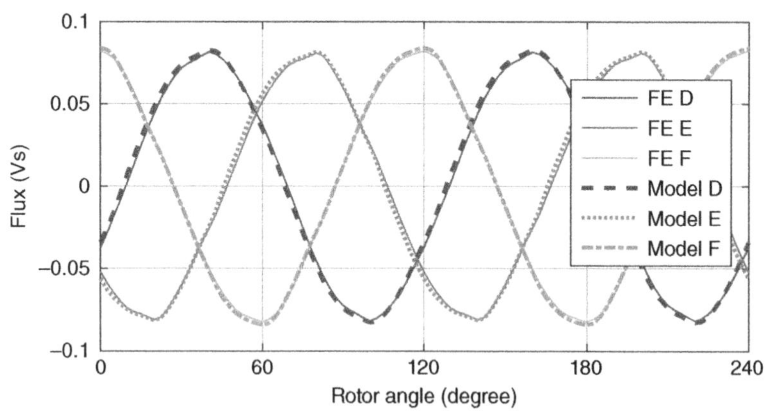

Figure 4.16 Comparison of predicted flux linkages in healthy DEF set by two models.

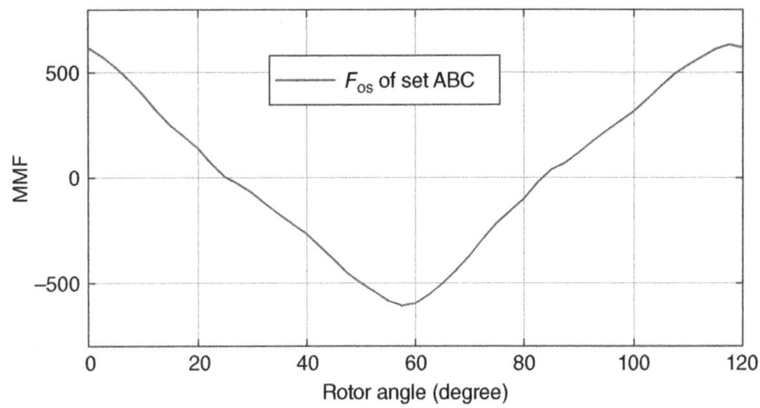

Figure 4.17 MMF offset component under short-circuit condition.

of the short-circuit fault. The prediction by the 4D table-based model only takes no more than 10 seconds. Further, the mismatches in the predictions by the 4D table-based and FE models are lower than those in the open-circuit case. This is indeed because the resultant MMF offset component of the fault set ABC as shown in Fig. 4.17 is lower compared to that in Fig. 4.12 of the open-circuit operation.

4.3.4 Unbalanced Current Operation

In practical operation, it is quite normal that the currents of different sets are slightly unbalanced due to limitations in current control bandwidth and inaccuracies in current sensors. This will result in an unbalanced current operation mode among the three 3-phase sets. Without loss of generality, the unbalanced current operation is assumed in that sets ABC, DEF, and GHI are fed by 1, 0.9, and 0.8 pu currents in magnitude, respectively. The predicted flux linkages and torque are then compared with the FE results as shown in Figs. 4.18 and 4.19. Since the flux linkages of the three sets exhibit similar behavior, only the flux linkages of set ABC are plotted and compared with the FE results. In general, the predicted torque and flux linkages by the two models agree very well. Furthermore, the 4D table-based model can capture the torque contributions from different 3-phase sets as depicted in Fig. 4.19. This provides more insights into the machine's behavior and allows for in-depth analysis of unbalanced operations of the machine drive because of position sensing and/or current control errors. The good matches between the predictions by the two models can be understood by the fact that the level of the current unbalance is relatively small and hence the resultant MMF offset components are quite low. For the FE predictions of the unbalanced operations, the full geometry model of the machine is also required.

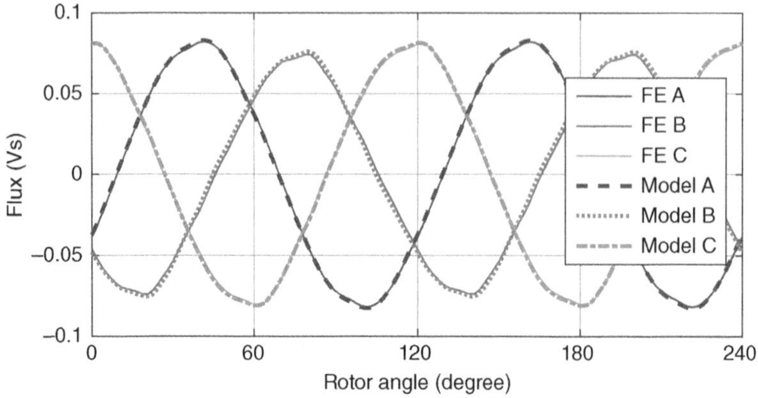

Figure 4.18 Comparison of predicted flux linkages of set ABC by two models.

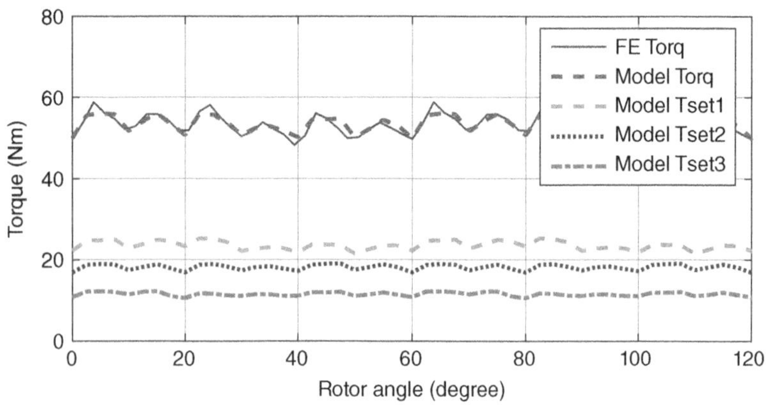

Figure 4.19 Comparison of predicted torque in unbalanced current operation by two models.

Hence, similar saving in computational time is gained using the 4D table-based model for predictions. It is worth noting that unbalanced operations caused by phase control errors in currents or asymmetries in both current magnitude and phase among the three 3-phase sets can be similarly simulated with the 4D table-based model in a computationally efficient manner.

4.3.5 Influence of Phase Resistances on Model Accuracy

As has been shown in (4.10) and (4.11), the predictions of the *dq*-axis currents in healthy and TSC conditions are obtained by the summation of all nonresistive voltages divided by the phase resistance. Since phase resistance is relatively small and dependent on temperature, it is important to assess with this modeling technique how changes in phase resistance influence the prediction of the physical quantities of the machine in healthy and fault conditions.

For this purpose, simulation studies have been performed first in the healthy condition when the machine operates at the rated speed and rated torque. By way of example, Fig. 4.20 compares the *dq*-axis flux linkages and their derivatives, and the *dq*-axis currents and voltages when there is 50% increase in phase resistance compared to its nominal value. This change represents a ~120 °C increase in temperature against the reference temperature at which the nominal phase resistance value is taken.

It is seen that there are very tiny changes which are invisible in the flux linkages, their derivatives and the currents when the phase resistance is increased by 50%. This is because although the predicted *dq*-axis currents in the model appear to be inversely proportional to the phase resistance and any initial change in the resistance will lead to a change in the currents, the change in the currents will in turn result in changes in the flux linkages and their derivatives. Consequently, the nonresistive voltage terms in the numerators of (4.10) will change accordingly, leading to a new equilibrium governed by the *dq*-axis voltage equations. The simulation study shows that the use of (4.10) for predicting *dq*-axis currents would not compromise the numerical accuracy and stability of the governing equations of machine operations. Since at the rated conditions, the resistive voltage drop is very small compared to the other voltage terms, the change in the phase resistance does not have much effect on the flux linkages and currents while a small change in the *dq*-axis voltages is seen in Fig. 4.20(d) as a result of the increase in the phase resistance.

Further simulations are performed with the same level of increase in the phase resistance when the ABC 3-phase set is terminally short circuited while the other two 3-phase sets are fed with the rated currents at the rated speed. The results for the ABC set are compared with those of the nominal phase resistance under the same conditions in Fig. 4.21. As can be seen, the resultant *dq*-axis flux linkages and their derivatives are close to zero under the short-circuit condition. The non-zero components are mainly caused by the phase leakage inductance. The changes in the flux linkages and their derivatives when the phase resistance is increased by 50% are noticeable because now the voltage across the phase resistance contributes a small portion to the total residual phase voltage. Similar changes in the nonresistive components of the *dq*-axis voltages ($u'_d = \omega\psi_q - \frac{d\psi_d}{dt}, u'_q = -\omega\psi_d - \frac{d\psi_q}{dt}$) are seen in Fig. 4.21(c). However, the predicted short-circuit currents are virtually not affected by the change in the phase resistance as is evident in Fig. 4.21(d). This is because, at the rated speed of operation, the inductive reactance (ωL_d) is dominant, and hence the short-circuit current is indeed governed by (ψ_d/L_d), or inductance limited.

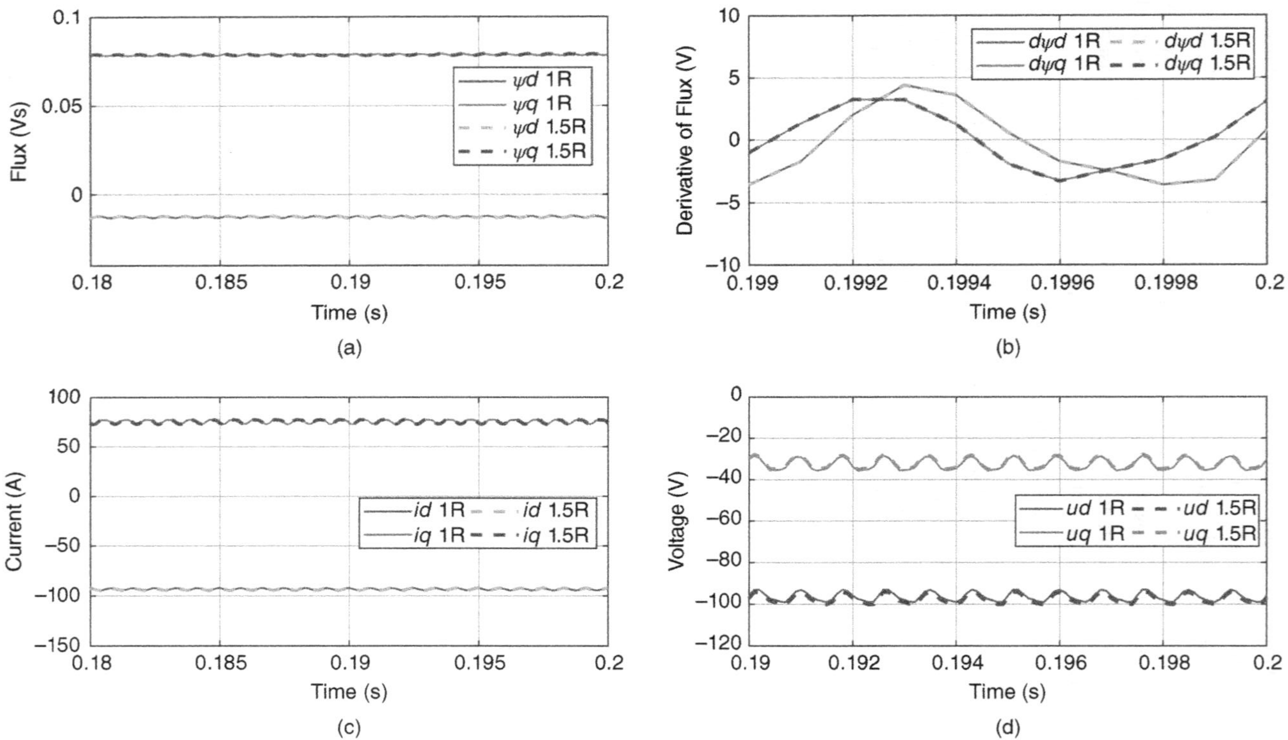

Figure 4.20 Comparisons of *dq*-axis quantities of (a) flux linkages and (b) their derivatives, and (c) currents and (d) voltages when the machine operates in healthy and the rated condition with 50% increase in phase resistance against the nominal value.

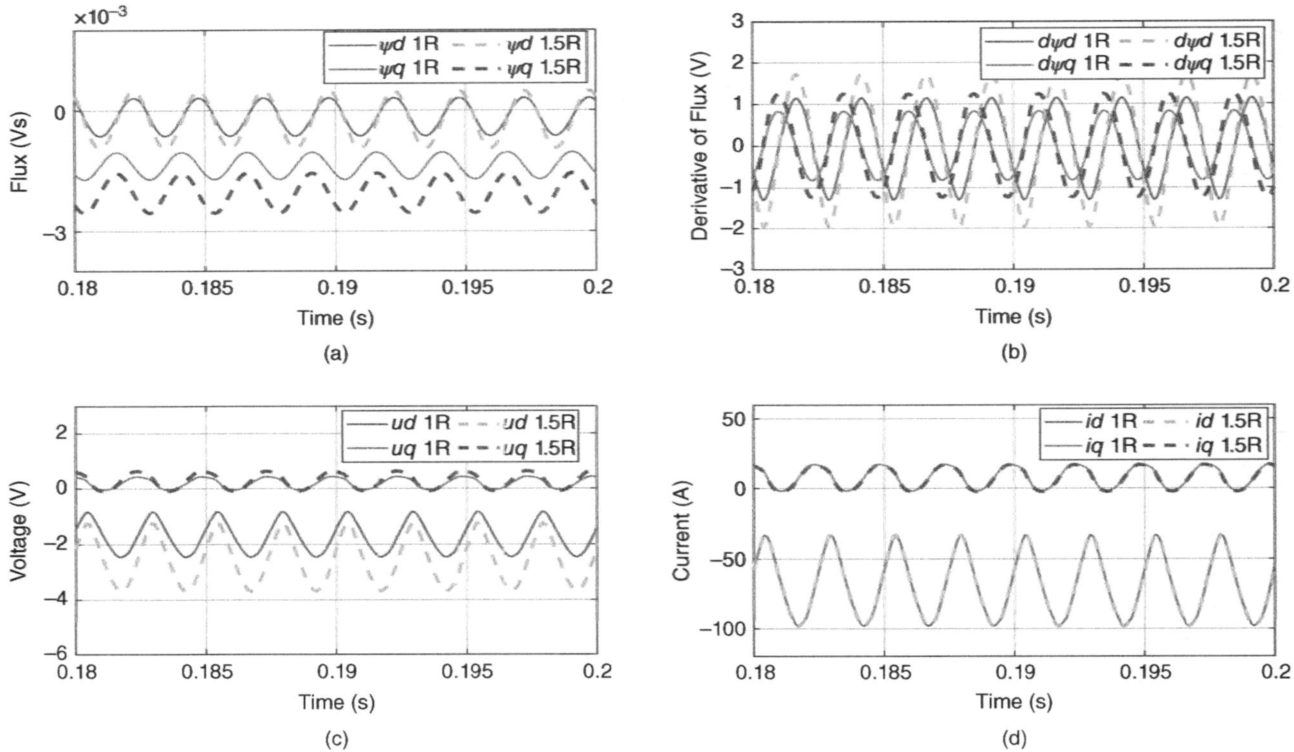

Figure 4.21 Comparisons of *dq-axis* quantities with 50% increase in phase resistance against nominal value when the ABC set is terminally short circuited while the other two 3-phase sets are fed with the rated current at the rated speed (a) *dq*-axis flux linkages, (b) their derivatives, (c) nonresistive components cf residual *dq*-axis voltages, and (d) *dq*-axis currents.

The foregoing simulations show that the current predictions based on (4.10) or (4.11) will not affect the numerical stability and accuracy of the model in both healthy and short-circuit conditions when there is a level of uncertainty in the phase resistance. The influence of the phase resistance on the machine operations is at a similar level to those seen in the conventional *dq*-axis-based machine model in which the currents are predicted from the integrations of the voltages across the *dq*-axis inductances.

4.4 General Modeling of Turn Fault

4.4.1 Turn Fault Description

As has been discussed and shown, the general modeling technique has proven to be flexible and accurate in predicting the flux linkage and torque of each 3-phase set for open- and short-circuit faults at the set level as well as for unbalanced current operations. However, it would be desirable to extend the model to predict the behavior of interturn short-circuit faults, denoted as turn faults. Therefore, a turn fault model is developed by adapting the general model, particularly for interturn fault conditions.

For discussion, it is assumed that a turn fault occurs in phase B of set ABC, specifically in coil B2. The schematic circuit for the turn fault condition is depicted in Fig. 4.22, where the phase B winding is divided into two parts, B_h to represent the healthy portion, and B_f to represent the faulty portion. The turn fault ratio μ is defined as the number of short-circuited turns N_f of the B_f portion divided by the total number of turns of the phase, $2N$. R_f and L_f are used to represent the external resistance and inductance associated with the fault current path, while i_f represents the current flowing through the faulty path.

4.4.2 Equivalent Input for the General Model Under a Turn Fault

As previously explained, the inputs for the 4D tables of one 3-phase set can be expressed as $(\theta, i_d, i_q, F_{os})$. To incorporate the turn fault into the model, the first step is to deduce the equivalent inputs for the 4D tables while accounting for the influence of the turn fault current i_f.

During the turn fault, additional fault current i_f flows through the short-circuited path, thereby impacting the MMF distributions of the machine. Furthermore, since i_f is present in the faulty part of coil B2, the MMF produced by the current in the B_f portion differs from that of phase B. This disparity can be accommodated by the phase shift between coil B2 and phase B as described in [5]. Under healthy conditions, the MMF of phase B is contributed by the two coils, as illustrated in

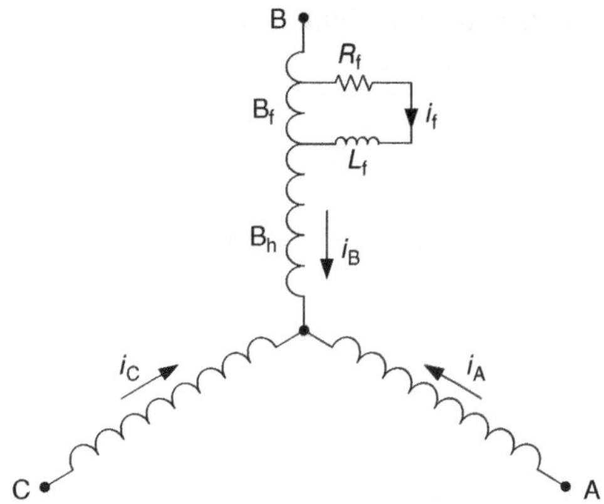

Figure 4.22 Schematic circuit for set ABC with turn fault in phase B.

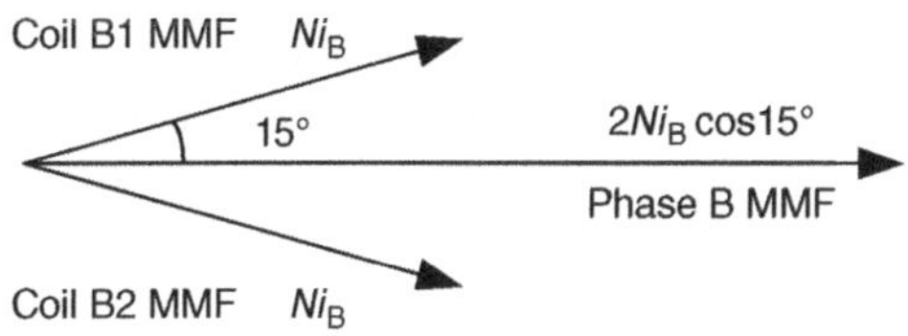

Figure 4.23 MMF phasors of phase B.

Fig. 4.23. It can be expressed in phasor form (4.12) when considering only the fundamental component, where 120° phase shift accounts for the phase shifts between phase A and phase B windings.

$$\begin{aligned} \mathrm{MMF_B} &= Ni_\mathrm{B}(e^{-j(\theta-120^\circ+15^\circ)} + e^{-j(\theta-120^\circ-15^\circ)}) \\ &= 2Ni_\mathrm{B}\cos 15^\circ e^{-j(\theta-120^\circ)} \end{aligned} \tag{4.12}$$

During a turn fault, the presence of fault current i_f will impact the MMF of phase B. Since the turn fault is assumed to take place in coil B2, the resultant MMF phasor denoted as $\mathrm{MMF_{Bf}}$ and given in (4.13) is directly proportional to the number of fault turns N_f. The additional phase shift of 15° represents the phase shift in space between the faulty coil B2 and phase B.

$$\mathrm{MMF_{Bf}} = -N_\mathrm{f} i_\mathrm{f} e^{-j(\theta-120^\circ-15^\circ)} \tag{4.13}$$

The fault current i_f and additional $\mathrm{MMF_{Bf}}$ caused by the fault current will influence both the MMF AC component of the ABC set and the MMF offset components of the other two 3-phase sets. This necessitates modifications to the *dq*-axis currents, i_{d1}, i_{q1}, of the ABC sets as a part of the inputs to the 4D table. Hence, the additional *dq*-axis components caused by i_f, denoted as $\Delta i_{d\mathrm{f}}, \Delta i_{q\mathrm{f}}$, can be expressed in (4.14) by the *abc*-to-*dq* axis transformation of the fault current in coil B2. By adding these two components to those flowing in the healthy parts of phases ABC denoted as i_{d1}, i_{q1}, the revised currents $i_{d\mathrm{f}}$ and $i_{q\mathrm{f}}$ are obtained for the 4D table inputs in the turn fault mode.

$$\Delta i_{df} = -\frac{2}{3} \times \frac{N_f}{2N} i_f \cos(\theta - 120^\circ - 15^\circ)$$
$$\Delta i_{qf} = \frac{2}{3} \times \frac{N_f}{2N} i_f \sin(\theta - 120^\circ - 15^\circ) \tag{4.14}$$

Meanwhile, the MMF offset components will also be affected by the fault current. The additional offset component, denoted as ΔF^1_{osf}, in set ABC region produced by the turn fault current i_f, can be expressed in equation (4.15) which is derived from (4.1) based on the turn function [6, 7] associated with the fault turns.

$$\Delta F^1_{osf} = -\frac{2}{3} N_f i_f \tag{4.15}$$

Consequently, the combined MMF offset, F^1_{osf}, over the set ABC region is obtained by summing ΔF^1_{osf} and F^1_{os} given in (4.17). Likewise, the offset components due to the fault current over the regions of the other two 3-phase sets can similarly be derived and given in equation (4.16).

$$\Delta F^2_{osf} = \Delta F^3_{osf} = \frac{1}{3} N_f i_f \tag{4.16}$$

$$F^1_{osf} = F^1_{os} + \Delta F^1_{osf},\ F^2_{osf} = F^2_{os} + \Delta F^2_{osf},\ F^3_{osf} = F^3_{os} + \Delta F^3_{osf} \tag{4.17}$$

By adding ΔF^2_{osf} and ΔF^3_{osf} to F^2_{os} and F^3_{os}, respectively, F^2_{osf} and F^3_{osf} can be calculated accordingly as a result of the turn fault. Subsequently, the adjusted variables $(\theta, i_{df}, i_{qf}, F^1_{osf})$ form the inputs for the 4D tables illustrated in Fig. 4.5 to determine the flux linkages and torque for the faulty set. Meanwhile, $(\theta, i_{d2}, i_{q2}, F^2_{osf})$ and $(\theta, i_{d3}, i_{q3}, F^3_{osf})$ are the inputs to the 4D lookup tables for the second and third 3-phase sets, respectively.

4.4.3 Flux Linkage of Fault Turn

By updating the inputs for the 4D tables, the flux linkages in d, q, 0 axes can be predicted for each 3-phase set under the turn fault condition. However, the flux linkage of the faulted turns has yet to be established. To accurately predict the fault current, it is essential to deduce the flux linkage of the fault turns from the $dq0$-axis flux linkages, $\varphi_{d1}, \varphi_{q1}, \varphi_{01}$, of set ABC, which are the flux linkage outputs of the 4D table depicted in Fig. 4.5.

The flow chart, illustrated in Fig. 4.24, outlines the procedure for deriving the flux linkage of the fault turns from the flux linkages of the faulty ABC set. The flux linkages of the faulty ABC set can be split into flux leakage and airgap flux linkage. However, only the airgap component is linked by the fault turns [8]. By utilizing equation (4.18), the $dq0$-axis components of the ABC set flux linkages can be separated into the flux leakage components and airgap flux linkage components,

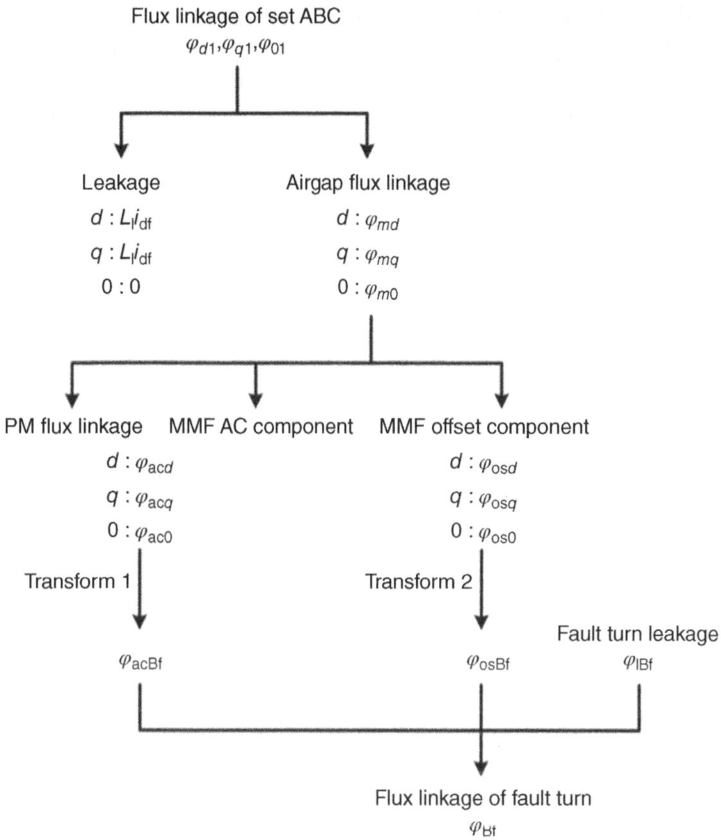

Figure 4.24 Derivation process of flux linkage of fault turn.

where φ_{mk} ($k = d, q$, and 0) represents the airgap flux linkage in the respective axes, and L_1 represents the phase leakage inductance. It is important to note that φ_{m0} is equal to φ_0 since the lookup table is generated without considering zero-sequence currents.

$$
\begin{aligned}
\varphi_{md} &= \varphi_{d1} - L_1 i_{df} \\
\varphi_{mq} &= \varphi_{q1} - L_1 i_{qf} \\
\varphi_{m0} &= \varphi_{01}
\end{aligned}
\tag{4.18}
$$

The airgap flux linkages originate from the rotor PM flux and the winding MMFs. When high-order harmonics are neglected, the PM flux linkage exhibits a sinusoidal distribution in the air gap. As described in Section 4.4.2, the MMF in the air gap encompasses both AC and offset components. The resultant flux linkages are directly proportional to the MMF. Consequently, the flux linkages generated by the AC component also exhibit a sinusoidal distribution in the air gap when the high-order harmonics are neglected. However, this is not true for the flux linkages generated by the offset components. Since the fault turns are affected by both the MMF AC and offset components, the resultant flux linkages will be treated separately due to their different distributions.

First, the flux linkages generated by the MMF offset component are derived based on the airgap permeance function, assuming infinitely permeable laminations. For instance, the flux linkage of phase A attributable to the offset component F^1_{osf} is determined using equation (4.19) [6], where φ_{osA1} and φ_{osA2} represent the flux linkages due to the offset component in coils A1 and A2, respectively. In addition, $n_{\mathrm{A1}}, n_{\mathrm{A2}}$ denote the turn functions for coils A1 and A2, while r represents the airgap radius and l denotes the axial length of the stator stack. $g^{-1}(\alpha)$ represents the inverse airgap function given in equation (4.20). The integration is performed over phase A winding span. Similarly, the induced flux linkages in phases B and C, φ_{osB} and φ_{osC}, can be calculated.

$$\begin{aligned}\varphi_{\mathrm{osA}} &= \varphi_{\mathrm{osA1}} + \varphi_{\mathrm{osA2}} \\ &= \mu_0 r l \int g^{-1}(\alpha)(n_{\mathrm{A1}} + n_{\mathrm{A2}}) F^1_{\mathrm{osf}} d\alpha \end{aligned} \tag{4.19}$$

$$g^{-1}(\alpha) = a + b\cos(2\alpha + \alpha_0) \tag{4.20}$$

Due to the full-pitched coils and space-invariance of F^1_{osf} across the ABC set region, it can be inferred that φ_{osA1} equals φ_{osA2} and both are independent of the rotor angle. Similarly, it can be inferred that φ_{osA} equals φ_{osB}. Because of the opposite polarity of the turn function for phase C with respect to phases A and B, the offset flux linkage of the phase C winding φ_{osC} is equal to that of phase A but in opposite polarity, as given in (4.21) and shown in Fig. 4.25.

$$\begin{aligned}&\varphi_{\mathrm{osA}} = \varphi_{\mathrm{osB}} = -\varphi_{\mathrm{osC}} \\ &\varphi_{\mathrm{osA1}} = \varphi_{\mathrm{osA2}} = \varphi_{\mathrm{osB1}} = \varphi_{\mathrm{osB2}} = -\varphi_{\mathrm{osC1}} = -\varphi_{\mathrm{osC2}}\end{aligned} \tag{4.21}$$

Since the turn fault is situated in coil B2, the flux linked by the fault turn is directly proportional to φ_{osB2}. Although the previous analysis provides a qualitative explanation of the relationships among the offset flux linkage components,

Figure 4.25 Flux linkage distribution induced by the MMF offset component.

there is a more convenient way to compute them using the 4D table established previously according to (4.22), where dq^{-1} denotes the inverse *abc*-to-*dq* axis transformation. Within the brackets on the right side of (4.22), the first column represents the flux linkages resulting from the combined MMF AC and offset components as well as the rotor PM field, while the second column denotes the flux linkages without the MMF offset component. By subtracting the two components, the flux linkages of the ABC set produced by the offset component only can be obtained, assuming the magnetic saturation is negligible. This condition is generally true, in the event of a turn fault because according to Lenz's law, the flux due to the fault current will oppose the original field, resulting in a lower flux density in the fault turn region. Consequently, the error introduced by equation (4.22) is relatively small.

$$\begin{pmatrix} \varphi_{osA} \\ \varphi_{osB} \\ \varphi_{osC} \end{pmatrix} = dq^{-1} \begin{pmatrix} \varphi_{osd} \\ \varphi_{osq} \\ \varphi_{os0} \end{pmatrix} = dq^{-1} \begin{pmatrix} \varphi_d\left(\theta, i_{df}, i_{qf}, F^1_{osf}\right) - \varphi_d(\theta, i_{df}, i_{qf}, 0) \\ \varphi_q\left(\theta, i_{df}, i_{qf}, F^1_{osf}\right) - \varphi_q(\theta, i_{df}, i_{qf}, 0) \\ \varphi_0\left(\theta, i_{df}, i_{qf}, F^1_{osf}\right) - \varphi_0(\theta, i_{df}, i_{qf}, 0) \end{pmatrix} \tag{4.22}$$

The flux linkage of the fault turns caused by the MMF offset component is proportional to the number of fault turns and can be evaluated by (4.23). $2N$ in the denominate corresponds to the total number of turns of a phase winding in the machine under study.

$$\varphi_{osBf} = \frac{N_f}{2N}\varphi_{osB} \tag{4.23}$$

On the other hand, the flux linkages caused by the rotor PM and MMF AC component exhibit a sinusoidal distribution in the air gap, represented by AC flux linkages (φ_{acA}, φ_{acB}, φ_{acC} for each phase). The AC flux linkage phasors for each coil and each phase are illustrated in Fig. 4.26. For each phase, the flux linkages are also the combined results of the two subcoils. The AC component of the flux linkage of the fault turns is, therefore, proportional to the AC flux linkage of coil

Figure 4.26 AC component phasors of flux linkages.

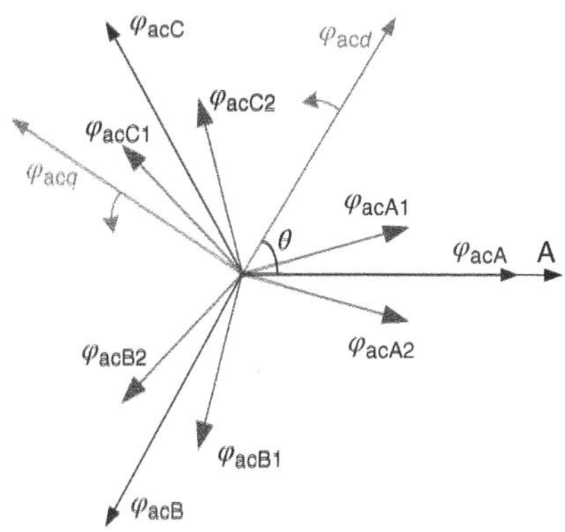

B2 φ_{acB2}, where the fault occurs. As shown in Fig. 4.26, the φ_{acB2} phasor differs from the φ_{acB} phasor. This difference should be accounted for when evaluating the flux linkage of the fault turns.

By subtracting the offset flux linkage components in equation (4.22) from the total airgap flux linkages in equation (4.18) in the *dq*-axis frame, the *dq0*-axis AC flux linkages of the ABC set, denoted as $\varphi_{\mathrm{ac}d}$, $\varphi_{\mathrm{ac}q}$, φ_{ac0}, can be evaluated in equation (4.24).

$$\begin{pmatrix} \varphi_{\mathrm{ac}d} \\ \varphi_{\mathrm{ac}q} \\ \varphi_{\mathrm{ac0}} \end{pmatrix} = \begin{pmatrix} \varphi_{md} \\ \varphi_{mq} \\ \varphi_{m0} \end{pmatrix} - \begin{pmatrix} \varphi_{\mathrm{os}d} \\ \varphi_{\mathrm{os}q} \\ \varphi_{\mathrm{os0}} \end{pmatrix} \tag{4.24}$$

These results are projected to the fault turn in coil B2 using the inverse *abc*-to-*dq* transform given in equation (4.25). As can be seen, the coil location of the fault turn in coil B2 is taken into account by the phase shift between the B2 coil and phase A winding axes.

$$\begin{aligned} \varphi_{\mathrm{acBf}} = \frac{N_{\mathrm{f}}}{2N\cos 15^\circ} \big(& \varphi_{\mathrm{ac}d}\cos(\theta - 120^\circ - 15^\circ) \\ & - \varphi_{\mathrm{ac}q}\sin(\theta - 120^\circ - 15^\circ) + \varphi_{\mathrm{ac0}} \big) \end{aligned} \tag{4.25}$$

Consequently, the airgap flux linkage of the fault turns can be estimated.

Furthermore, the slot leakage fluxes caused by the phase current and fault turn current need to be taken into account [2, 8]. As depicted in Fig. 4.27, the slot leakage flux and inductance associated with each turn is dependent on its location in the slot. The first turn encompasses the largest amount of the slot leakage flux and hence has the largest leakage inductance while the 4th turn has the smallest leakage inductance. This implies that the leakage inductance associated with short-circuited turns is dependent on their relative position in the slot. Nevertheless, once the position of the faulted turns is known, the associated leakage inductances can be quantified analytically or by FE analysis [8, 9]. Hence, the leakage flux linkage associated with the faulted turns can be estimated by using equation (4.26), where L_{lf} represents the self-leakage inductance of the fault turns and M_{lf} denotes the mutual leakage inductance between the healthy turns and the fault turns within the same slot. As a result, the influence of the slot position of the fault turns on the machine behavior can be accounted by slot-position dependent L_{lf} and M_{lf}. Finally, the total flux linkage of the fault turns can be obtained in equation (4.27).

$$\varphi_{\mathrm{lBf}} = M_{\mathrm{lf}} i_{\mathrm{B}} + L_{\mathrm{lf}}(i_{\mathrm{B}} - i_{\mathrm{f}}) \tag{4.26}$$

$$\varphi_{\mathrm{Bf}} = \varphi_{\mathrm{acBf}} + \varphi_{\mathrm{osBf}} + \varphi_{\mathrm{lBf}} \tag{4.27}$$

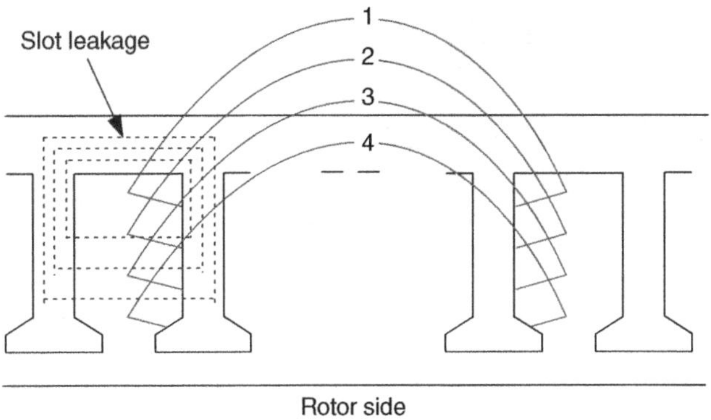

Figure 4.27 Illustration of the slot position and leakage flux of the fault turn.

4.4.4 Voltage Equations

By utilizing the obtained flux linkages of the faulted turns and the faulted ABC set, the voltage equations for both the fault turns and the entire 3-phase set in the dq-axis reference frame can be expressed by (4.28).

$$
\begin{aligned}
R_{\rm f}i_{\rm f} + L_{\rm f}\frac{di_{\rm f}}{dt} &= \frac{\mu}{2}R(i_{\rm B} - i_{\rm f}) + \frac{d\varphi_{\rm Bf}}{dt} \\
u_{d1} &= Ri_d + \frac{d\varphi_{d1}}{dt} - \omega\varphi_{q1} - \frac{\mu}{3}Ri_{\rm f}\cos(\theta - 120^\circ) \\
u_{q1} &= Ri_q + \frac{d\varphi_{q1}}{dt} + \omega\varphi_{d1} + \frac{\mu}{3}Ri_{\rm f}\sin(\theta - 120^\circ)
\end{aligned}
\tag{4.28}
$$

It is important to note that the voltage equations of the other two 3-phase sets remain the same as those in healthy mode since the effect of the fault current has already been taken into consideration by the changes in the MMF offset components in equation (4.16). As a result, the general model, shown in Fig. 4.28, has been adapted to accommodate the turn fault, accounting for variations in coil location, slot position, and the number of faulted turns. By evaluating the faulted turn flux linkage, $\varphi_{\rm Bf}$, using (4.22)–(4.27), the fault current $i_{\rm f}$ and its influence on the currents in set ABC can be predicted by (4.28). The influence of the turn fault current, $i_{\rm f}$, on the operation of the two healthy 3-phase sets is accounted for by the MMF offset components given in (4.17). Any change in the fault position in a slot can be represented by the resultant self- and mutual leakage inductances, $L_{\rm lf}$ and $M_{\rm lf}$, while the fault location in different coils can be accounted for by appropriate spatial phase shift with respect to phase A axis.

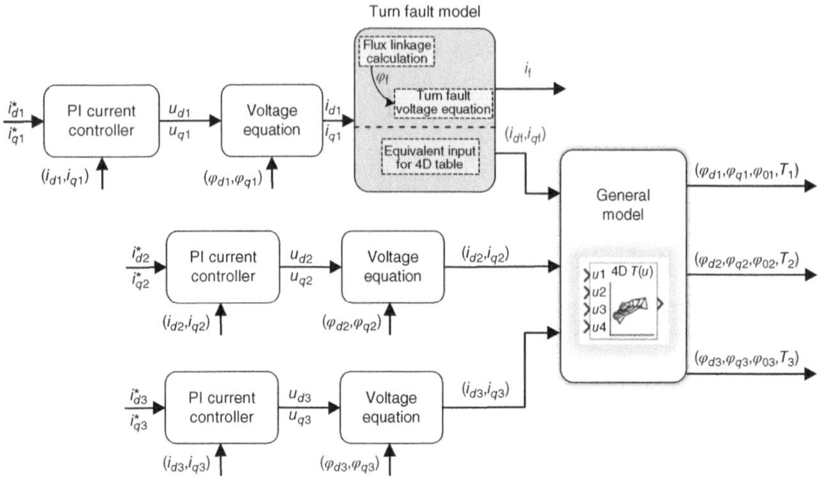

Figure 4.28 General model integrated with turn fault model.

4.5 Study on Model Accuracy and Computational Efficiency Under Turn Fault

The model capable of representing a turn fault, shown in Fig. 4.28, has been implemented in the Matlab/Simulink with the previously established 4D tables. To validate the developed turn fault model and access its merits over the FE-based model, comparisons were made between the two models in terms of accuracy and computation time under various fault conditions, including varying faulty coil locations, with and without TSC mitigation action. For each fault condition, the current control of the three 3-phase sets is assumed to be perfect, and hence reference currents and relevant parameters that describe the fault are fed into both models. This assumption will cause a small deviation in the prediction of fault currents, but significantly reduces the simulation time since the current control law and the voltage equations except that which governs the short-circuit current can be removed in both the FE and developed models.

To compare like-with-like, the same inputs and fault condition are specified in the FE model and in the developed turn fault model. In the FE model, the machine is fed by ideal current sources as shown in Fig. 4.29, where the turn fault without and with TSC application to the fault set terminals are illustrated in Fig. 4.29(a) and (b), respectively. The fault phase winding where the turn fault occurs is divided into two parts, namely, the healthy part BH and faulty part BF, and the external circuit condition of each part is separately defined in the FE model. Similarly, when TSC is not applied, the same reference currents for the

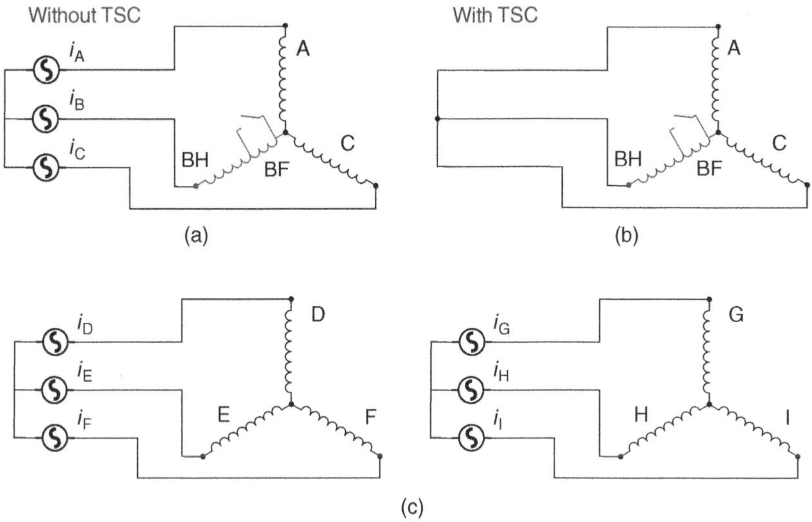

Figure 4.29 Illustration of turn fault simulation in FE: (a) set ABC without TSC, (b) set ABC with TSC, and (c) set DEF and set GHI with normal excitation.

three 3-phase sets used in the FE model are fed to the developed turn fault model shown in Fig. 4.28 with the PI current controllers and voltage equations being removed. The fault current is predicted in the turn fault submodel and its influence on the machine operation is reflected in the 4D table-based submodel via (i_{df}, i_{qf}). Under TSC application to the faulted 3-phase set, the short-circuit currents in the healthy part of the ABC phase windings are predicted by (4.28) with (u_{d1}, u_{q1}) being set to zero in the developed turn fault model. The other parts of the model are the same as those without TSC applications.

4.5.1 Turn Fault Behaviors Without Terminal Short Circuit

Initially, fault behaviors with a single-turn short circuit without TSC application are examined and the fault is assumed to take place in coil B2. All 3-phase sets are excited with balanced sinusoidal reference currents of 120 A in magnitude when the machine operates at 4000 rpm. Under this condition, the fault current predicted by the developed turn fault model matches well with the FE results, as depicted in Fig. 4.30. The relative rms error normalized to the FE prediction is 5.4%. It is seen that the fault current is nine times the rated value, resulting in significant distortion in the phase flux linkages, as shown in Fig. 4.31. The phase flux linkages of the ABC set predicted using the 4D lookup tables with the four equivalent inputs $\left(\theta, i_{df}, i_{qf}, F^1_{osf}\right)$ show very good agreements with the FE results. Additionally, it is observed that the flux linkage of phase B, where the faulty turn

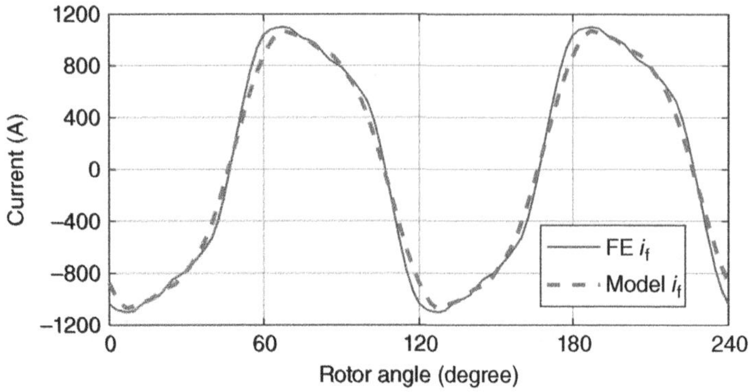

Figure 4.30 Comparison of predicted turn fault current when it occurs in coil B2 without TSC.

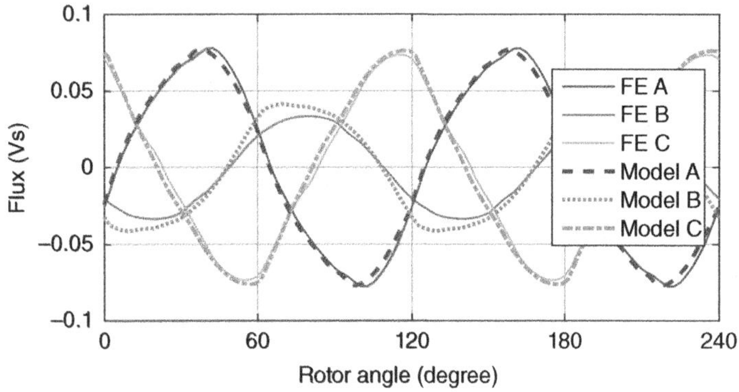

Figure 4.31 Comparison of predicted phase flux linkages when it occurs in coil B2 without TSC.

is located, is noticeably lower than those of phases A and C. This is understood from Lenz's law in that the induced short-circuit current tends to reduce the flux density in the fault region.

The model also predicts the torque accurately in comparison with FE analysis as demonstrated in Fig. 4.32. The torque components predicted by the model reveal that the torque contribution of the healthy sets remains relatively unchanged; however, a significant second harmonic is evident in the fault set. This observed second harmonic is primarily attributed to the distortion in the flux linkages resulting from the turn fault current. This understanding offers valuable insights for developing turn fault detection method, which will be explored and discussed in the subsequent chapter.

Furthermore, to account for a turn fault located in a different coil, the fault behaviors with a single-turn fault taking place in coil A1 are also evaluated with

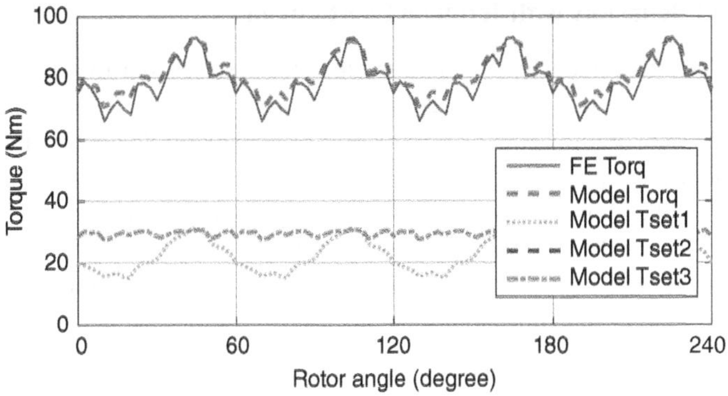

Figure 4.32 Comparison of predicted torques when it occurs in coil B2 without TSC.

the same reference currents at the speed stated previously. The resultant turn fault current predicted by the developed model, as depicted in Fig. 4.33, agrees closely with the FE result. The relative rms error is about 5.6%. Additionally, the model-predicted phase flux linkages and torque also agree well with the FE results, further demonstrating the flexibility of the model in specifying fault location. Their characteristic features are, though, like those shown in Figs. 4.31 and 4.32, and hence not repeated.

It is worth noting that the FE simulation of the machine behaviors under a turn fault also requires 10 fundamental electrical cycles to reach steady state in the short-circuit current. Thus, the required FE computation time is also about 5 hours for one specific fault scenario. In comparison, the developed model only takes 10 seconds to complete. The savings in time and computation efforts with the developed model are very significant.

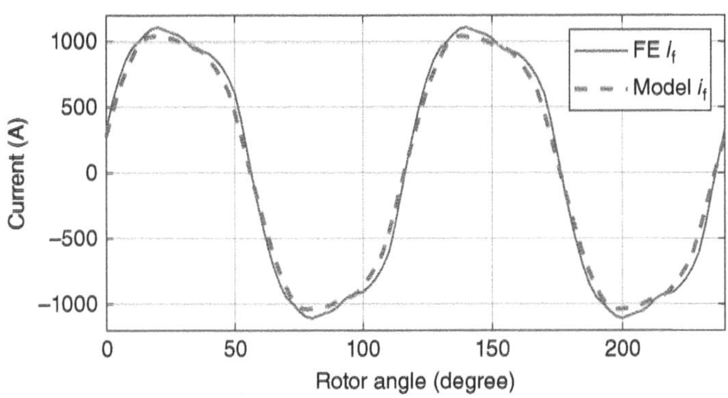

Figure 4.33 Comparison of predicted turn fault current when it occurs in coil A1 without TSC.

4.5.2 Turn Fault Behaviors with Terminal Short Circuit

As has been pointed out previously, the turn fault current is extraordinarily large if no mitigation measure is taken and may lead to catastrophic damages. Therefore, it is essential to employ appropriate fault mitigation actions upon detection of a turn fault. For the triple 3-phase PMA-SynRM under study, the turn fault can be mitigated by TSC application on the fault 3-phase set through control of inverter switches as has been described in Chapter 2. The circuit condition of the FE model representing the turn fault with TSC application has been illustrated in Fig. 4.29 (b), where the terminals of the fault 3-phase set are directly short circuited. In the developed turn fault model, the voltages given in (4.28) applied to the fault set are set to zero for predicting the short-circuit currents in both healthy and fault parts of the 3-phase windings.

First, the comparisons of the predictions by the turn fault model and the FE model are made when a single-turn fault is assumed to occur in coil B2 and TSC is applied to the fault ABC set. Meanwhile, the other two sets are still excited with the rated current at 4000 rpm. The predicted turn fault current and phase currents by the developed model are compared in Figs. 4.34 and 4.35 with the FE results. Good agreements between the two predictions are seen. However, a notable discrepancy is observed in the peaks and troughs of the turn fault current. This is attributed to the omission of high-order harmonics in PM flux linkage and in the MMF AC component and their influence on the fault turn flux linkage as described in Section 4.4.3. Nonetheless, the relative rms error of the turn fault current between the two predictions is only 4.5%. After the TSC application, the net fundamental flux linkage in the faulted turn becomes very small and consequently, neglecting high-order harmonics will lead to an increase in the prediction error. It is also evident that with the TSC application, the fault current decreases significantly to

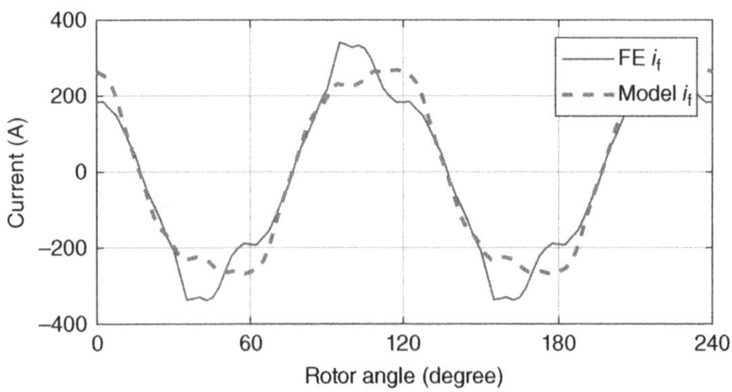

Figure 4.34 Comparison of predicted turn fault current when it occurs in coil B2 with TSC.

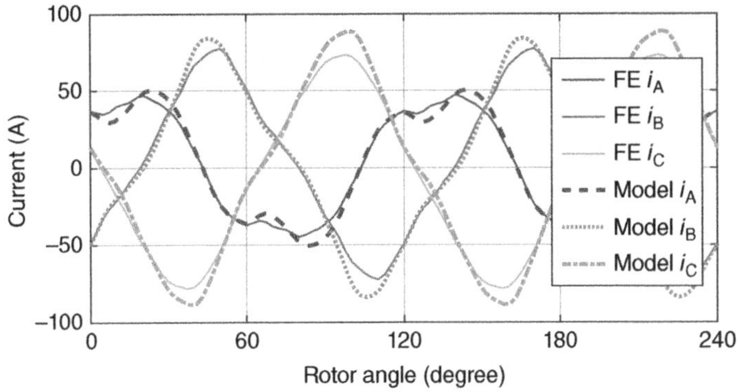

Figure 4.35 Comparison of predicted phase currents with turn fault in coil B2 with TSC.

approximately 2.4 pu, which becomes thermally sustainable for the machine since it is confined in the single fault turn while the currents in the healthy parts of the ABC phases are below the rated. In addition, the machine remains capable of delivering approximately 2/3 of its rated torque as illustrated in Fig. 4.36.

Likewise, the fault behaviors with TSC application are also assessed with the developed turn fault model and FE model when a single-turn short circuit is assumed in coil A1. Figures 4.37 and 4.38 compare the predicted turn fault current and phase currents from the two models. Again, a good agreement between the two predictions is demonstrated. The relative rms error of the turn fault current is 3.6%. It is observed that the turn fault current in coil A1 is notably lower than that in coil B2 under the same operating conditions. This observation is indeed consistent with those obtained in the previous fault analysis. In addition, similar computational efficiency is gained with the developed model over the FE predictions under the turn fault with TSC application to those without TSC application. Therefore, the detail is not repeated.

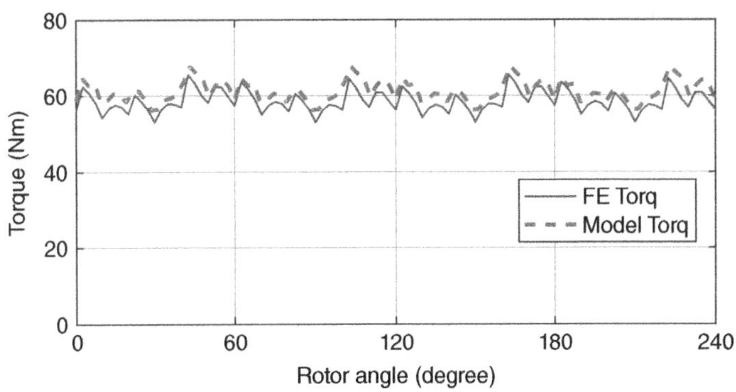

Figure 4.36 Comparison of predicted torques with turn fault in coil B2 with TSC.

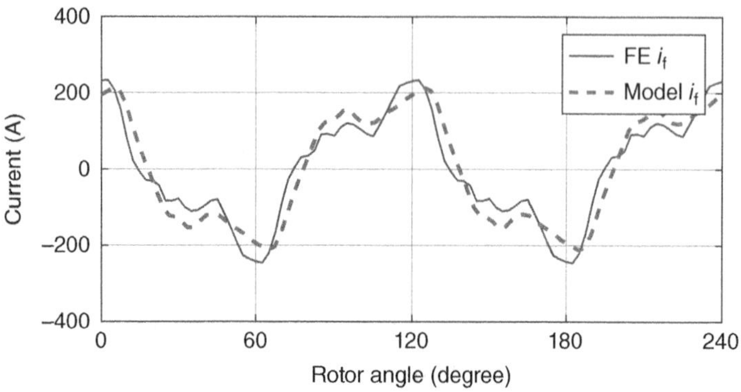

Figure 4.37 Comparison of predicted turn fault current when it occurs in coil A1 with TSC.

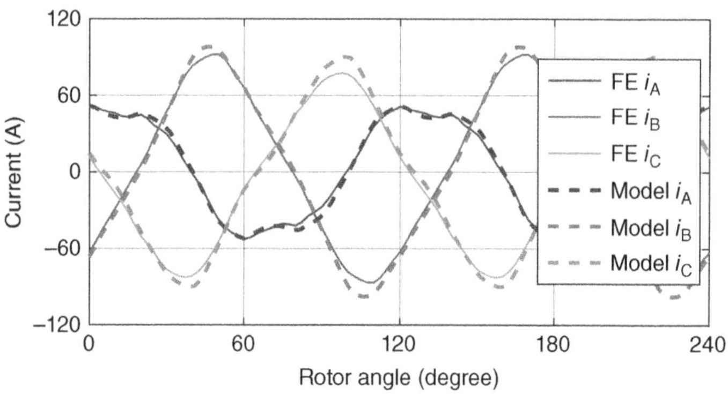

Figure 4.38 Comparison of predicted phase currents of faulty set with fault in coil A1 with TSC.

The predictions of turn fault current and the currents in the healthy parts of the phase windings with or without TSC are based on (4.28). The fault current and dq-axis currents in the healthy part of the 3-phase windings are determined from the sum of nonresistive voltage components divided by the resistance associated with the respective faulted and healthy turns. The influence of these resistances on prediction accuracy has also been studied by simulations under turn fault conditions with and without TSC applications and a similar conclusion to that given in Section 4.3.5 has been reached. That is, the influence of resistances on the currents and other physical quantities is governed by the voltage equations and the division by the resistance for predicting currents based on (4.28) does not affect the numerical stability and accuracy of the model. The details are omitted for the sake of conciseness.

4.6 Model Validation by Experimental Tests

This section describes experimental validations of the developed general model under healthy and different fault conditions, namely, open-circuit, short-circuit, unbalanced current operation, and interturn short-circuit faults with and without TSC application. The accuracy and effectiveness of the model are assessed through comparisons of the experimental measurements with the model predictions.

During tests, the triple redundant 3-phase machine operates at a given speed controlled by the dynamometer described previously and its currents in healthy operations or in healthy phases are controlled by independent PI current controllers with 200 Hz bandwidth. The sensor-sampled currents, and control output voltages in the digital signal processor (DSP) which are also the reference voltages for the inverter operation, have been exported for comparison purposes. In the predictions, the same current controllers and the voltage equations have been incorporated in the developed general models as shown in Figs. 4.7 and 4.28. Further, the same reference current commands used in the experiments have been fed to the general models for like-with-like simulation studies.

4.6.1 Healthy Condition

First, the machine undergoes tests in healthy operating conditions at 4000 rpm when each 3-phase set is supplied with *dq*-axis current commands of equal magnitude and phase angle for maximum torque per Ampere (MTPA) operation. Figure 4.39 compares the measured and predicted phase currents at 80 A in magnitude in phases ABC, where good agreements between the measurement and prediction are evident. Figure 4.40 shows the model-predicted and measured *dq*-axis voltages, where a close resemblance of the two is seen albeit

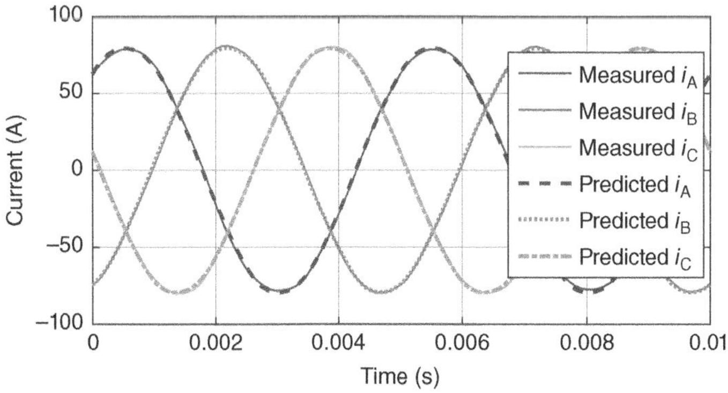

Figure 4.39 Comparison of measured and predicted currents in healthy mode.

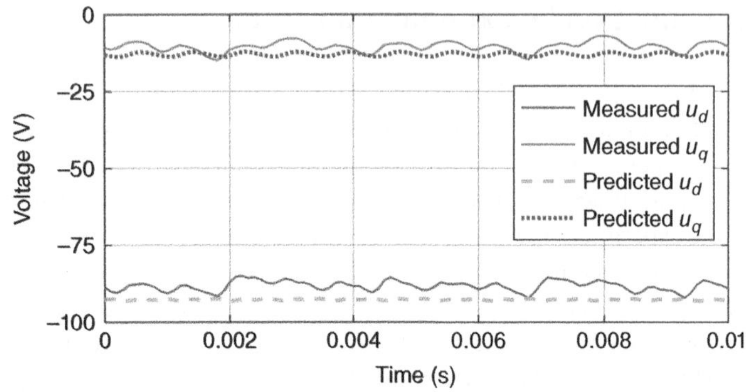

Figure 4.40 Comparison of measured and predicted voltages in healthy mode.

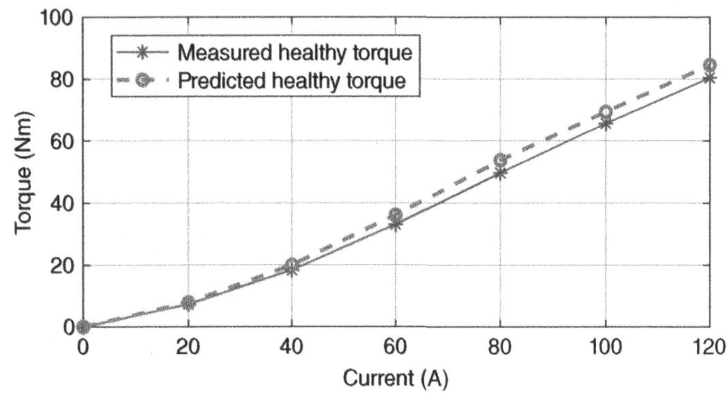

Figure 4.41 Comparison of measured and predicted torques in healthy condition.

with noticeable deviations in magnitude and harmonic contents. This is mainly because the measured *dq* voltages are obtained from the DSP as the reference voltages for the inverter modulation. Hence, a small discrepancy exists compared with the actual voltages applied to the machine windings due to the voltage drops across and nonlinearity of the inverters.

The predicted average torque variations are compared with the measurement in Fig. 4.41 when the phase current magnitude is varied from 0 to 120 A. It becomes evident that the measured torques are 4.8% lower than the predicted values. As previously shown, the flux linkages and torque predicted by the model match closely with FE calculations. Hence, the observed torque discrepancy can be primarily attributed to the inaccuracies within the 2D FE model, which is used to build the 4D tables as described in Section 4.2.3 and the omission of frictional and windage torque in the model.

4.6.2 One-Set Open Circuit

Next, one 3-phase set open-circuit operation is tested by intentionally deactivation of all the IGBTs in the ABC set inverter, while reference currents of 80 A in magnitude for MTPA operation are fed to the current controllers of the healthy sets DEF and GHI. In the developed model, the currents in the ABC phases are set to zero while the same current commands of 80 A in magnitude are given for the other two healthy sets. The measured currents and voltages from set DEF match well with the model predictions as shown in Figs. 4.42 and 4.43. Similar waveforms in set GHI are observed but not presented. As can be seen, under the open-circuit condition, the phase currents in the DEF set display visible imbalances and distortions, primarily due to the limited current-control bandwidth. These changes are reflected quite well in the predicted currents. Similarly, the measured control voltages also agree well with the model-predicted results, and the 2nd harmonics are present in both the measured and predicted *dq*-axis voltages although a small difference between the measurements and prediction exists due to the reasons explained earlier.

The measured and model-predicted torque variations with current magnitude are compared in Fig. 4.44, exhibiting a similar trend to that in the healthy mode. The difference can be attributed to the inaccuracies in the FE model and current prediction as well as neglection of friction and windage losses in the model.

4.6.3 One-Set Short Circuit

Subsequently, the machine drive undergoes tests with TSC in the ABC 3-phase set while the two remaining healthy sets are fed with the reference current of 80 A in magnitude under the MTPA condition. The same conditions are imposed in the prediction in which the *dq*-axis voltages applied to the ABC windings are

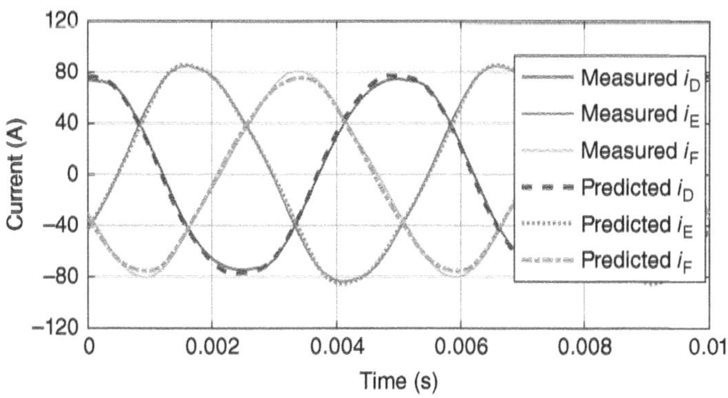

Figure 4.42 Comparison of measured and predicted currents in open-circuit mode.

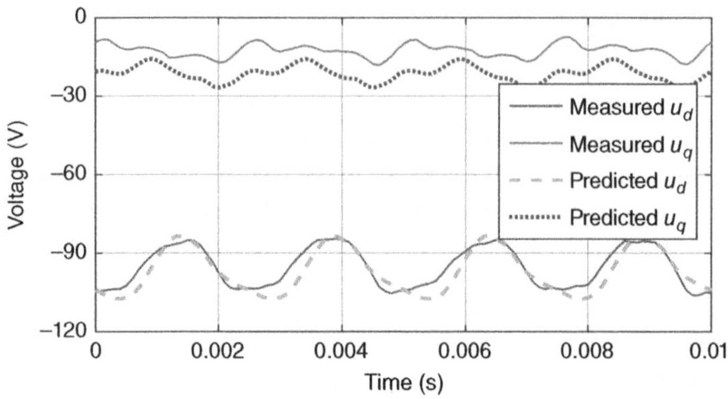

Figure 4.43 Comparison of measured and predicted *dq* voltages in open-circuit mode.

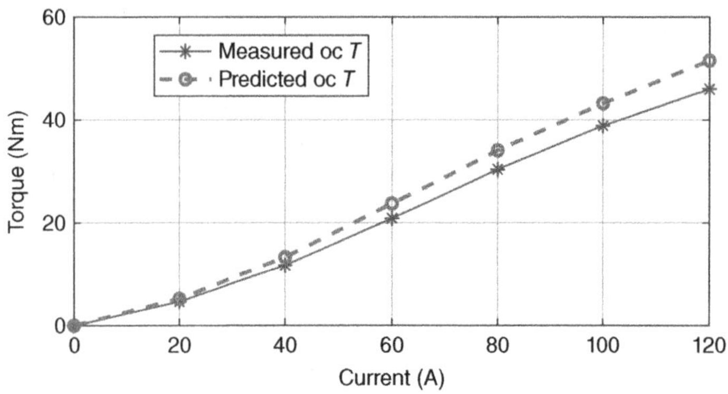

Figure 4.44 Comparison of measured and predicted torques in open-circuit condition.

set to zero while sets DEF and GHI are under current feedback control with the same reference currents. The measured and predicted short-circuit currents in set ABC are illustrated in Fig. 4.45. Again, the model can predict the primary characteristics of the unbalanced phase currents and their harmonic distortion, albeit with marginally higher peaks than the measurements. Such discrepancy is mainly caused by extra impedance of the connecting cables and the voltage drops across the switches which are not accounted for accurately in the model. Further, the measured currents and command voltages in the healthy sets are compared with the predictions in Figs. 4.46 and 4.47, demonstrating good fidelity of the developed model.

The measured and predicted torque variations with current magnitude in the healthy sets under this one set short-circuit mode are shown in Fig. 4.48. The lower measured torque is mainly attributed to the same factors as discussed previously.

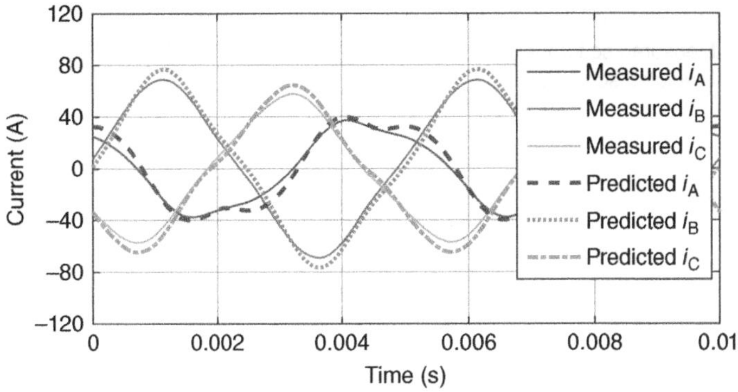

Figure 4.45 Comparison of measured and predicted short-circuit currents.

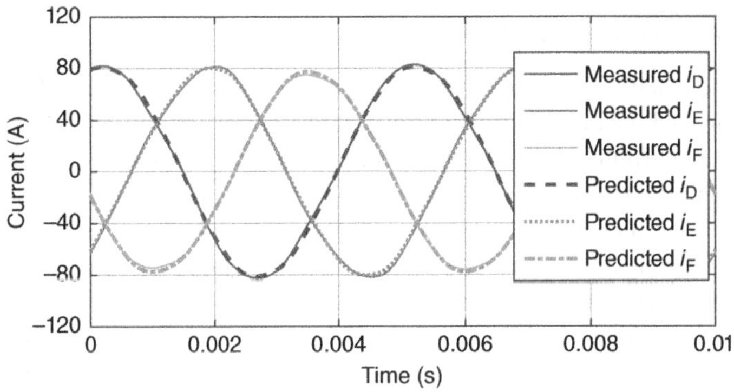

Figure 4.46 Comparison of measured and predicted currents in healthy DEF set when set ABC in short circuit mode.

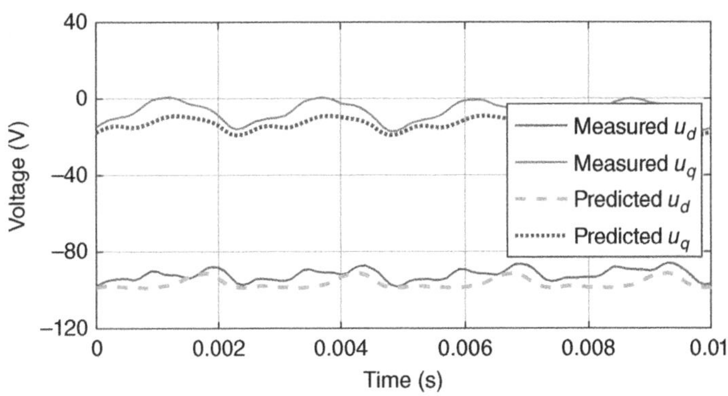

Figure 4.47 Comparison of measured and predicted voltages in healthy DEF set when set ABC in short-circuit mode.

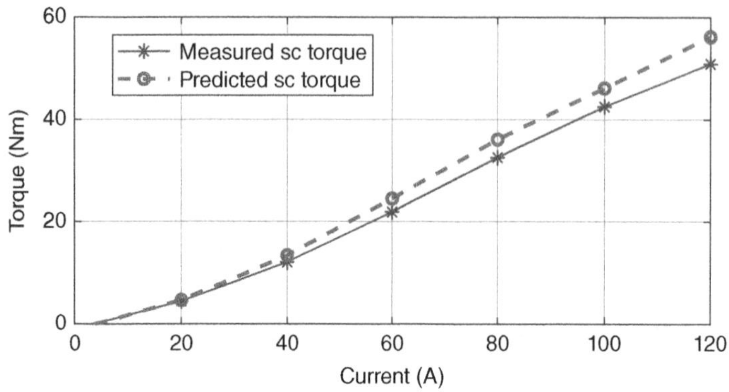

Figure 4.48 Comparison of measured and predicted torques in short-circuit condition.

4.6.4 Unbalanced Current Operation

Furthermore, the machine drive is tested under unbalanced current operations. Although unbalanced operations should be avoided in healthy conditions, slight variations in phase currents often exist due to inconsistencies across current sensors, inverters, and the three sets of phase windings. Current commands of 100, 80, and 60 A in magnitude under MTPA conditions are set in both the DSP controller and developed model for sets ABC, DEF, and GHI, respectively. These current references are tracked by the three independent current controllers for each 3-phase set. The measured and predicted currents and voltages of set ABC are plotted in Figs. 4.49 and 4.50 as examples. It is seen that due to the unbalance in the three 3-phase sets, the phase currents are slightly distorted, and this distortion causes noticeable second harmonics on the controller output voltages. These

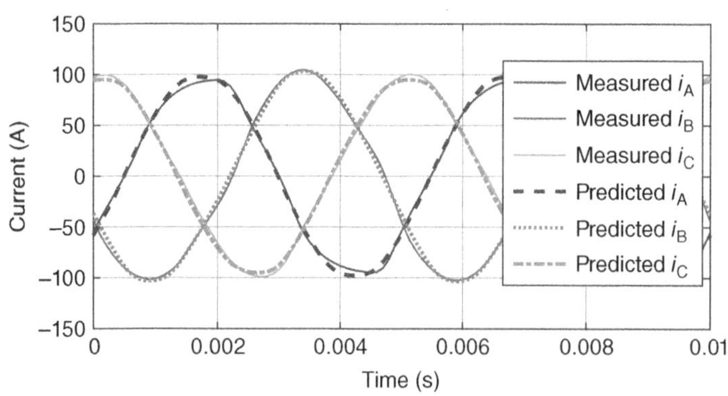

Figure 4.49 Comparison of measured and predicted phase currents in set ABC under unbalanced currents operation.

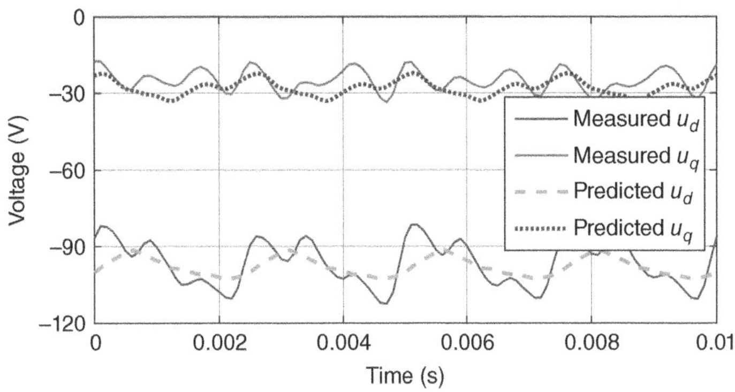

Figure 4.50 Comparison of measured and predicted voltages under unbalanced currents operation.

characteristics are captured by the predictions quite well, although the differences in the higher-order harmonics between the measured and predicted voltages are evident. The waveforms of the other two sets exhibit similar trends and, for conciseness, are not presented.

4.6.5 Turn Fault Without Terminal Short Circuit

As has been explained in Chapter 3, tests with one turn short circuit can be performed by the control relay connected to the fault emulation taps. The additional impedance caused by the relay and connecting leads is estimated to be 1.5 mΩ and 1 μH. This impedance is included in the developed turn fault model to replicate the experimental condition as closely as possible.

Prior analysis has demonstrated that turn fault current is considerably high without the TSC mitigation action. This underlines the necessity for experimental validation to establish whether the developed fault model is capable of accurately predicting such fault currents. For safe and repeatable tests, a single-turn fault within coil B2 is tested at 2000 rpm when each 3-phase set is loaded with 50 A current in magnitude under the MTPA condition. The single-turn fault is injected by the relay for a duration of 0.3 second. The measured turn fault current coincides closely with the model predictions as illustrated in Fig. 4.51. Throughout the fault injection period, the fault 3-phase set is still driven by the 3-phase inverter with the 50 A reference current. The acquired phase current and voltage readings are compared with the model predictions in Figs. 4.52 and 4.53. It is observed that the turn fault leads to minor distortions in phase currents, which are effectively captured by the model. Small discrepancies in the predicted voltages are evident, which are mainly attributed to the causes discussed previously. Meanwhile, the measured and predicted currents and voltages in the two healthy sets are

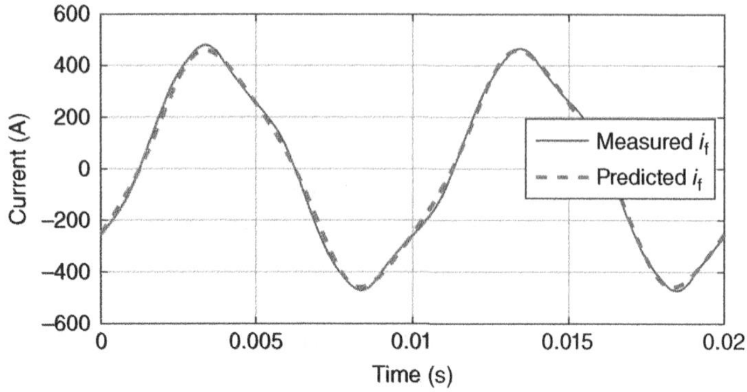

Figure 4.51 Comparison of measured and predicted turn fault currents when a turn fault occurs in B2 coil.

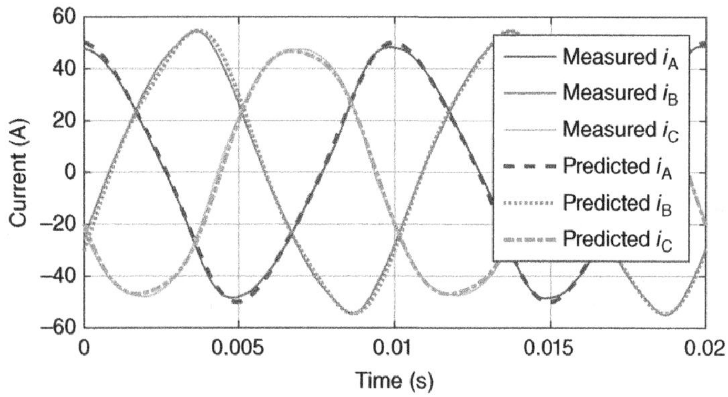

Figure 4.52 Comparison of measured and predicted phase currents in fault set when a turn fault occurs in B2 coil.

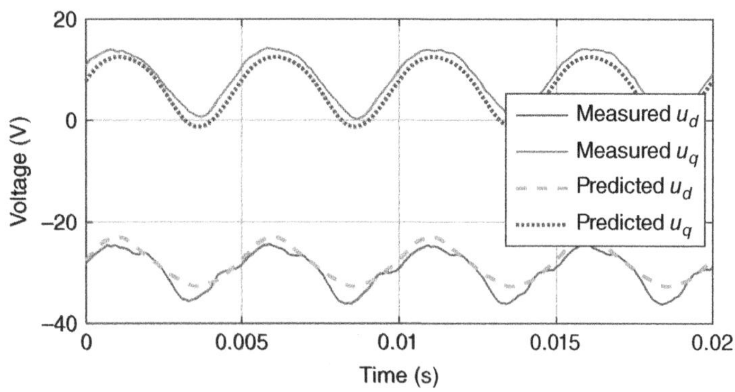

Figure 4.53 Comparison of measured and predicted dq-axis control voltages in fault 3-phase set when a turn fault occurs in B2 coil.

depicted in Figs. 4.54 and 4.55, demonstrating a high level of fidelity in the predictions.

Then, the single-turn fault tests are repeated by varying the phase current magnitude from 10 to 50 A at 2000 rpm with the phase angle kept at the MTPA condition for each magnitude. The measured variation of the rms fault current is compared with the prediction in Fig. 4.56. Again, a good agreement in the measurement and prediction is seen, indicating that the model can predict the fault currents with good accuracy in different operation conditions. Tests at higher speeds with larger current are not performed because high risk of damage to the machine.

In addition, the same tests of the turn fault operation are performed by rotating the rotor in the opposite direction. In this case, the fault occurs effectively in coil A1 with the same fault emulation setup. The measured fault currents and their rms

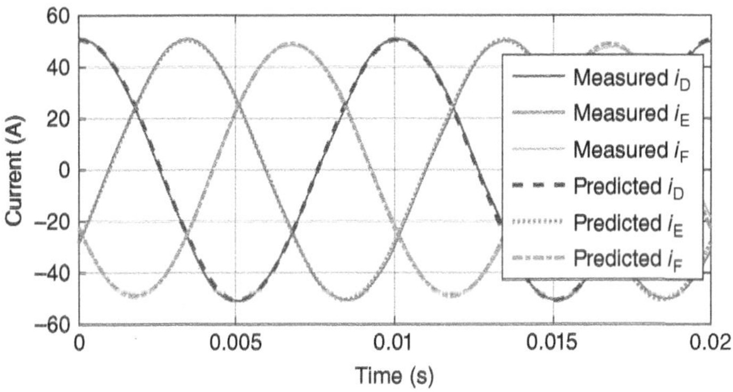

Figure 4.54 Comparison of measured and predicted phase currents in healthy DEF set when a turn fault occurs in B2 coil.

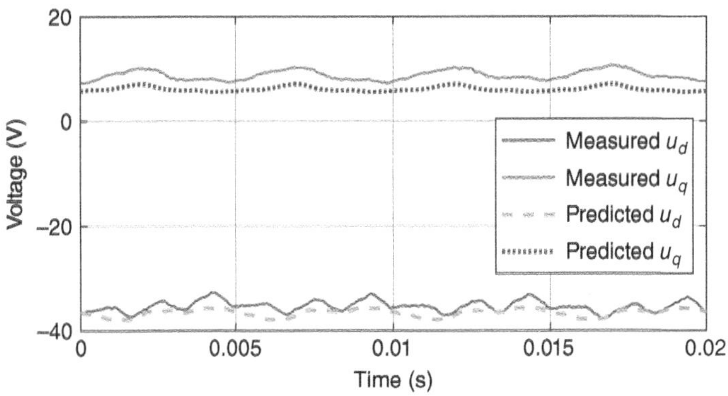

Figure 4.55 Comparison of measured and predicted *dq*-axis control voltages in healthy DEF set when a turn fault occurs in B2 coil.

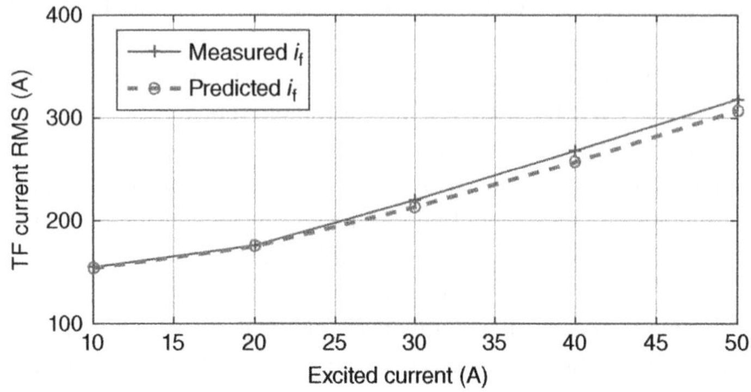

Figure 4.56 Measured and predicted turn fault current variations with healthy phase current magnitude when a turn fault occurs in B2 coil.

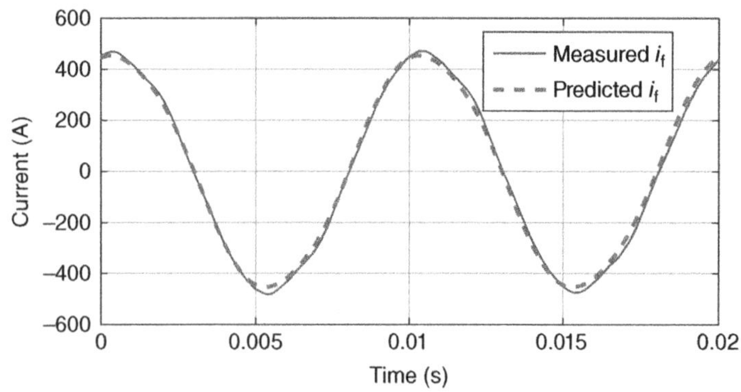

Figure 4.57 Comparison of measured and predicted turn fault currents at 50 A when a turn fault occurs in A1 coil.

variation are compared with the predicted values in Figs. 4.57 and 4.58, respectively. Other waveforms, like the phase currents and voltages, are similar to those shown in Figs. 4.52–4.55 and therefore not repeated. The close matches of the test results with the predictions demonstrate that the model is capable of predicting fault behaviors when a fault occurs in a different coil.

4.6.6 Turn Fault with Terminal Short Circuit

The machine drive under the single-turn short circuit with TSC application to the fault 3-phase set is also tested. With the TSC application, the resultant

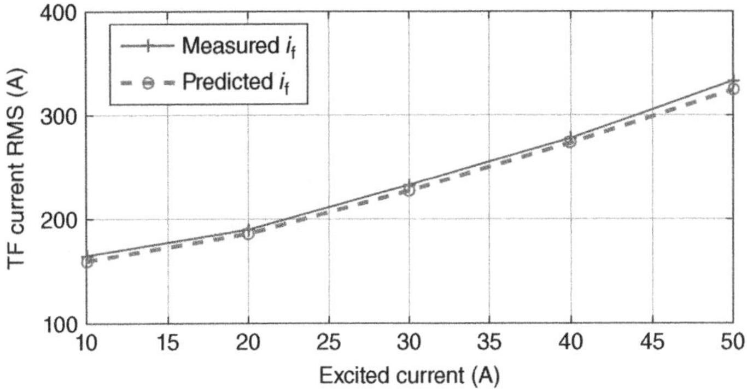

Figure 4.58 Comparison of measured and predicted turn fault current variations with healthy phase current magnitude when a turn fault occurs in A1 coil.

turn fault current is significantly reduced. This allows the turn fault test to be performed at the rated operation conditions (120 A, 4000 rpm). The measured and predicted fault currents are compared in Fig. 4.59. The relative rms difference between the measurement and prediction is 1.7% although the measured peak appears to be much larger. The difference is mainly caused by the neglection of high-order harmonics in the derivation of the fault turn flux linkage and the inaccuracy in the model for quantifying the extra impedance in the fault loop. Nonetheless, the rms values of both measured and model-predicted fault currents (Fig. 4.60) are in close agreement, which is a critical factor for thermal analysis. Further, the accuracy of predicting the currents and voltages of the healthy sets is

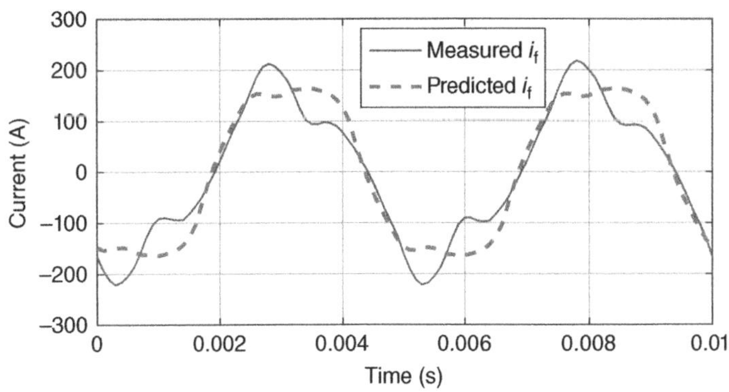

Figure 4.59 Comparison of measured and predicted turn fault currents with turn fault in B2 coil and TSC application.

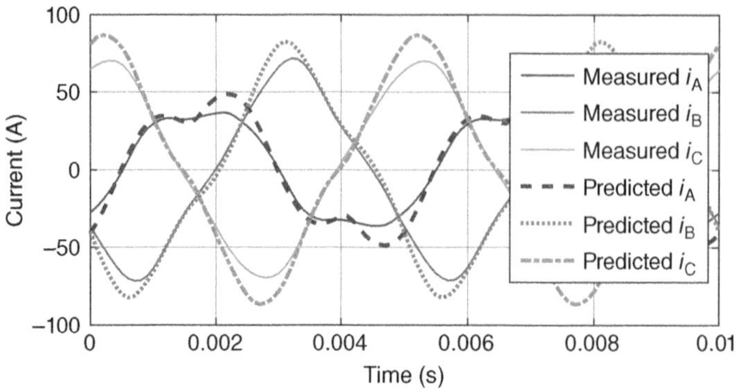

Figure 4.60 Comparison of measured and predicted short-circuit phase currents in ABC phases with turn fault in B2 coil and TSC application.

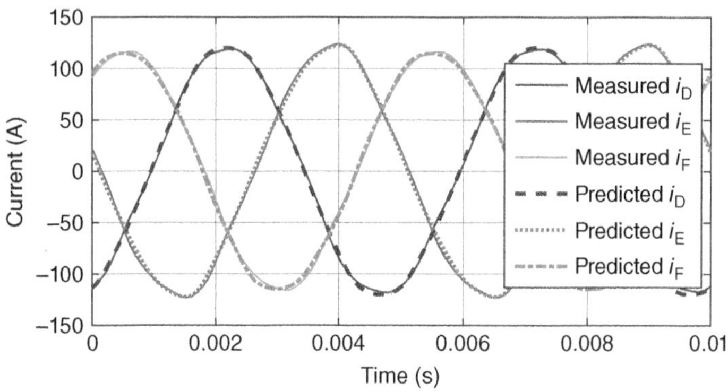

Figure 4.61 Comparison of measured and predicted phase current in DEF healthy set with turn fault in B2 coil with TSC application.

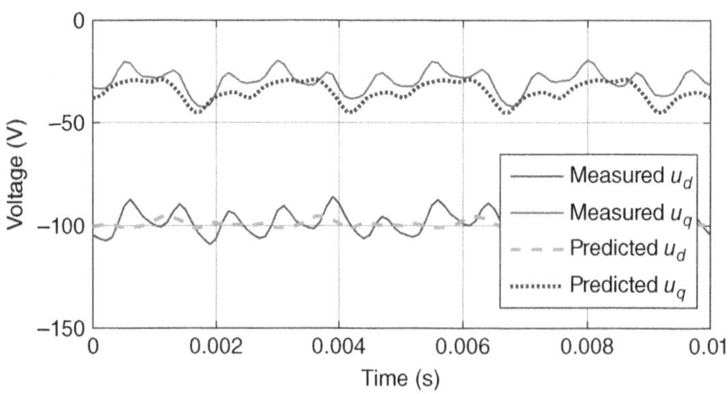

Figure 4.62 Comparison of measured and predicted control output voltages in DEF healthy set with turn fault in B2 coil with TSC application.

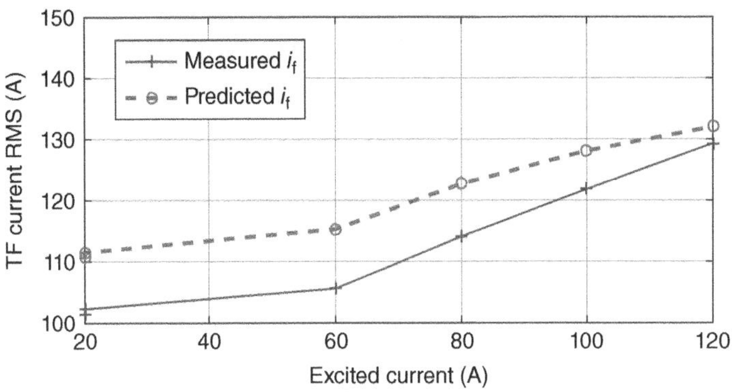

Figure 4.63 Comparison of measured and predicted rms turn fault current variations with healthy phase current magnitude when a turn fault occurs in B2 coil under TSC application.

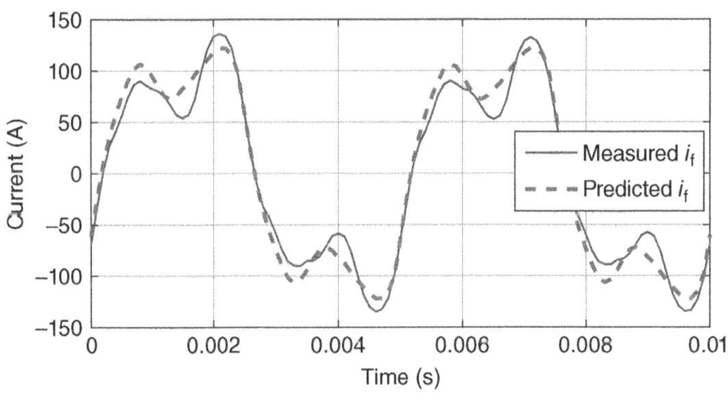

Figure 4.64 Comparison of measured and predicted turn fault currents when it occurs in A1 coil under TSC application.

demonstrated in Figs. 4.61 and 4.62. A comparison of the variations in rms turn fault current with the load currents in the healthy sets is presented in Fig. 4.63. It is shown that the maximum estimation error is under 10%.

Further turn fault tests with TSC application are conducted on coil A1 by rotating clockwise in motoring mode. Good agreements between the measured and predicted fault current waveforms and the rms fault current variation are seen in Fig. 4.64 and 4.65. This provides further confidence in model predictions under the different fault condition.

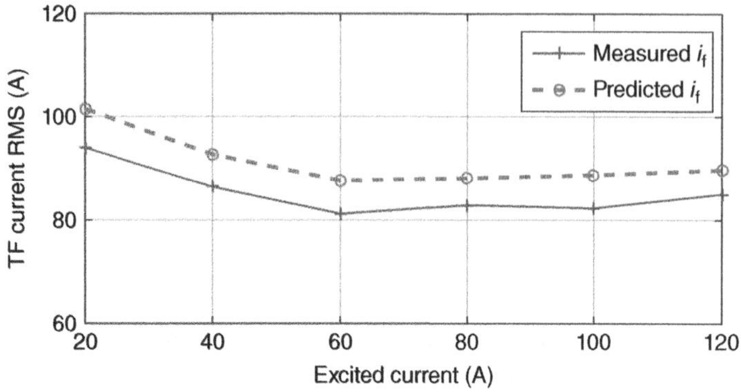

Figure 4.65 Comparison of measured and predicted rms turn fault current variations with phase current magnitude in healthy sets when it occurs in A1 coil under TSC application.

4.7 Summary

In this chapter, a general modeling technique for the triple redundant 3-phase PMA-SynRM drive has been developed. By decomposition of the MMFs associated with each 3-phase winding set into AC and offset components, the flux linkages and torque for each 3-phase set can be described by four variables, namely, the rotor angle θ, the dq-axis currents (i_d, i_q) of the 3-phase set, and an MMF offset component F_{os}. These relationships can then be determined by constructing 4D tables through FE simulations. The voltage equations and current control laws are subsequently integrated with the 4D table-based model for convenient time domain simulations. The key merits of the developed model lie in two important aspects for the development of a fault-tolerant drive. First, the model can represent the behaviors of the fault-tolerant machine drive across a wide range of operational modes, including healthy operations, open-circuit faults, short-circuit faults, unbalanced current operation, and postfault operations. Further, the model is computationally very efficient, being ~1500 times faster than the FE-based model for simulations of typical fault behaviors. The general model has been further extended to represent the behaviors of the machine drive with a turn fault in the phase windings and subsequent postfault operations after TSC application. The influential factors of a turn fault, such as the number of faulted turns, their locations in a slot, and their occurrence in different coils of the phase windings, can be accounted for in the model. Good accuracy and fidelity of the models have been demonstrated through a series of simulation studies and experimental tests. These models provide an effective and enabling tool for assessments of various fault behaviors and for the developments of fault detection techniques and postfault control strategies.

It is worth noting that the presented modeling approach is also applicable to other multiple 3-phase machines with similar segregated winding arrangements. The models developed with the concepts and techniques described in this chapter can provide as a useful tool for performance evaluations under healthy, unbalanced, and fault conditions.

It should be further understood that even for conventional 3-phase machine drives, the flux-linkage-based modeling techniques are far more advantageous over the classical *dq*-axis lumped parameter modeling approaches [4]. This is because a flux-linkage-based model is able to account for magnetic saturation, space, and time harmonics, which are often neglected in a classical *dq*-axis model, in a computationally effective manner and hence offers both accuracy and computational efficiency, and great potential for developing digital twins that provide real-time information on machine learning and condition monitoring.

References

1 D. C. Patel and M. C. Chandorkar, "Modeling and analysis of stator interturn fault location effects on induction machines," *IEEE Transactions on Industrial Electronics*, vol. 61, no. 9, pp. 4552–4564, 2014.

2 P. Arumugam, T. Hamiti, C. Brunson, and C. Gerada, "Analysis of vertical strip wound fault-tolerant permanent magnet synchronous machines," *IEEE Transactions on Industrial Electronics*, vol. 61, no. 3, pp. 1158–1168, 2014.

3 J. A. Haylock, B. C. Mecrow, A. G. Jack, and D. J. Atkinson, "Operation of fault tolerant machines with winding failures," *IEEE Transactions on Energy Conversion*, vol. 14, no. 4, pp. 1490–1495, 1999.

4 X. Chen, J. Wang, B. Sen, P. Lazari, and T. Sun, "A high-fidelity and computationally efficient model for interior permanent-magnet machines considering the magnetic saturation, spatial harmonics, and iron loss effect," *IEEE Transactions on Industrial Electronics*, vol. 62, no. 7, pp. 4044–4055, 2015.

5 B. Sen, J. Wang, and P. Lazari, "A detailed transient model of interior permanent magnet motor accounting for saturation under stator turn fault," in *Energy Conversion Congress and Exposition (ECCE) 2013 IEEE*, 2013, pp. 3548–3555.

6 J. Faiz and I. Tabatabaei, "Extension of winding function theory for nonuniform air gap in electric machinery," *IEEE Transactions on Magnetics*, vol. 38, no. 6, pp. 3654–3657, 2002.

7 J. Faiz, I. T. Ardekanei, and H. A. Toliyat, "An evaluation of inductances of a squirrel-cage induction motor under mixed eccentric conditions," *IEEE Transactions on Energy Conversion*, vol. 18, no. 2, pp. 252–258, 2003.

8 Z. Sun, J. Wang, D. Howe, and G. Jewell, "Analytical prediction of the short-circuit current in fault-tolerant permanent-magnet machines," *IEEE Transactions on Industrial Electronics*, vol. 55, no. 12, pp. 4210–4217, 2008.

9 B. Sen and J. Wang, "Analytical modelling of stator turn fault in surface mounted permanent magnet machines," in *Energy Conversion Congress and Exposition (ECCE) 2013 IEEE*, 2013, pp. 4445-4452.

5

Fault Detection Techniques for the Multiple 3-Phase Machine Drive

5.1 Introduction

Fast and reliable fault detection is indispensable for any fault-tolerant system as it enables timely application of suitable fault mitigation actions [1, 2] and postfault operations. Reliable fault detection is also of paramount importance for avoiding false alarms and associated unnecessary downtime. The key to fault detection in a machine drive system is to understand its fault behaviors and identify an optimal set of fault signatures that have the highest signal-to-noise ratio. These fault signatures are then monitored continuously for fault diagnosis. As previously discussed in Chapters 1 and 2, a switch open circuit and a winding interturn short circuit are two of the most frequent faults in a machine drive [3, 4]. Therefore, this chapter focuses on the detection methods of these two faults, respectively. The general model established in Chapter 4 will be utilized for the development of fault detection techniques. Appropriate simulations and experimental tests are performed to demonstrate the effectiveness of these fault detection methods.

First, detection techniques for the open-circuit fault in a machine drive are investigated. Open-circuit fault in inverter switches and machine windings is a common failure mode in an electrical machine drive [5, 6]. While it is relatively easy to detect a winding open-circuit fault with phase current sensors in a drive system, the detection of an inverter switch open circuit is more challenging because no direct measurement of currents in the switches is available. For purpose of discussion, a switch open circuit is defined as the power transistor in an inverter arm is open circuited such that no current flows through it, but the diode in the arm can still function. This is typical in an insulated gate bipolar transistor (IGBT)-based inverter drive since the diodes are separately integrated into each inverter arm and are usually more robust. The transistors are however more vulnerable to an open-circuit failure due to excessive current, thermal stress, thermal-mechanical cyclic aging, or faulty connection. It should be noted that while a transistor short circuit is also a common failure mode

Multiple 3-Phase Fault Tolerant Permanent Magnet Machine Drives: Design and Control, First Edition.
Bo Wang and Jiabin Wang.

in an inverter, such a fault would usually lead to an open-circuit failure in a voltage source inverter because of the excessive shooting-through current across the DC link. In a silicon carbide (SiC) metal-oxide-semiconductor field-effect transistor (MOSFET)-based inverter drive, the diode of an inverter arm is part of the MOSFET die. An open circuit in a SiC device typically means that both the MOSFET and diode in an inverter arm are open circuited. This failure mode bears similarity to the transistor-only open-circuit case although the resultant distortion in phase currents differs. To facilitate the detection of an open-circuit fault in inverter switches, the relative β-axis residual voltage signals associated with each phase during inverter operation will be described and utilized. The fault behavior of an inverter switch open circuit is analyzed by comparing the reference voltage and the estimated output voltage under both healthy and faulty conditions. The reference voltage is predicted by the real-time switching state of the inverter while the estimated output voltage is derived from the healthy machine model. It is found that the relative β-axis residual voltages associated with the fault phase do not change during an open-circuit fault in the inverter switch, but this is not the case for the healthy phases. This unique fault behavior is therefore exploited for detection of open-circuit fault in inverter switches in a machine drive. The voltage residuals are used to construct fault indicator and their polarities are employed for identification of the fault location within a 3-phase inverter.

Detection of the second common fault type, an interturn short-circuit fault, denoted as turn fault, in machine phase windings of a drive system, is far more challenging. This is because if the number of short-circuited turns is relatively small, the resultant fault signature is benign while the fault current can be excessively high, which requires an immediate remedial action. Reliable detection of the benign fault signature in an electronically noise environment is, however, technically very difficult. Several techniques have been developed for detection of such faults. The turn fault detection was investigated using the model-based approach by monitoring the current estimation errors [7–11]. This method required an accurate online model, which was usually achieved by adaptive state observer or online learning. As a turn fault gives rise to some forms of unbalance and harmonic distortion, specific harmonics arising from the fault are explored as a key fault indicator. The harmonic-based method, however, is only applicable in steady state, which takes a longer time to reach and hence may not be appropriated when immediate remedy action is required. The unbalance caused by a turn fault can also be monitored at the star-neutral point of 3-phase windings [12]. This, however, requires an additional connection to the neutral point, which is not available in most machines. In Ref. [13], the interturn short-circuit fault was alarmed by tracking estimated back electromotive force in a permanent magnet (PM) synchronous machine. Most of these detection techniques are based on monitoring fault signatures around the fundamental frequency or transients associated with the operational voltages and currents. However, they are dependent on operational conditions. Because the fault signatures due to a small number of short-circuit turns are benign compared with the fundamentals, it is very difficult to detect such faults with these techniques, particularly at low speeds of operation. Alternative methods have been developed for more reliable detection by utilizing high-frequency (HF) signals. Some of these techniques are discussed in detail in Section 5.8.

An alternative yet effective detection technique for winding interturn fault in the machine drive under study is proposed and described. Detailed analysis is performed to identify the behavior of a closed-loop controlled PM machine drive under interturn fault conditions. It is shown that significant 2nd harmonic components are generated in the *dq* voltages, currents, instantaneous active power (IAP), and instantaneous reactive power (IRP) during a turn fault condition. Further, the increase in the 2nd harmonic in IAP and IRP during fault conditions is comparatively higher than that of voltage and current, making them ideal candidates for turn fault detections. Thus, the turn fault detection technique based on the 2nd harmonic in IAP and IRP is implemented and demonstrated.

It should be noted that detection methods for the other faults discussed in Chapter 2 should also be established and implemented. Furthermore, upon detection of a fault, appropriate identification or classification of the fault type is also required to take correct mitigation measures against the fault [14, 15]. Comprehensive treatment of fault classification techniques is beyond the scope of this book and will not be elaborated. The detection methods for the open-circuit fault in inverter switches and winding turn fault serve as an example for the fault-tolerant triple redundant 3-phase permanent magnet-assisted synchronous reluctance machine (PMA-SynRM) drive.

5.2 Analysis of Fault Signal Under Open Circuit of an Inverter Switch

First, the fault behavior due to an open circuit in an inverter switch is studied to identify the suitable signal for fault diagnosis. The triple redundant 3-phase machine drive under study consists of three independent 3-phase modules, each operating as a typical 3-phase drive. Thus, the following analysis is focused on one 3-phase module, including the inverter and 3-phase windings for simplicity as shown in Fig. 5.1.

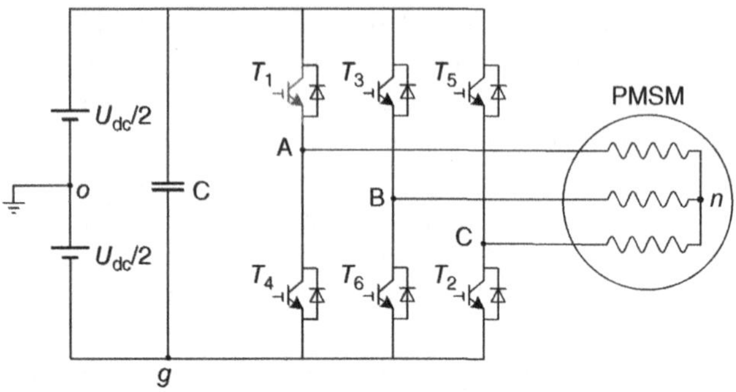

Figure 5.1 One 3-phase module with T_1 open-circuit fault.

5.2.1 Inverter Switching Function Definition

The operation of the 3-phase voltage source inverter is governed by the six gate drive commands for switches $T_1 - T_6$, denoted as $s_1 - s_6$, respectively, originating from the space vector pulse width modulation (SVPWM) unit within the digital signal processor (DSP) for drive control. These binary control signals determine the operating states for each switch, where a "0" value indicates switch-off state and a "1" indicates a switch-on state. It should be noted that the switching states for the two switches in the same leg must be complimentary to each other to avoid shooting through. For the convenience of discussions, the mid-point voltage (MPV) of the DC link is denoted as the zero-voltage reference. Consequently, the MPVs of the inverter legs with respect to the ground point as functions of the gating signals can be formulated as follows:

$$\begin{cases} u_{\mathrm{Ag}} = (s_1 + \overline{s_4})U_{\mathrm{dc}} \\ u_{\mathrm{Bg}} = (s_3 + \overline{s_6})U_{\mathrm{dc}} \\ u_{\mathrm{Cg}} = (s_5 + \overline{s_2})U_{\mathrm{dc}} \end{cases} \tag{5.1}$$

where U_{dc} is the DC-link voltage and u_{Ag}, u_{Bg}, u_{Cg} denote MPVs of the 3-phase legs A, B, and C, respectively. They are usually referred to as the inverter output voltages with respect to the negative end of the DC link.

5.2.2 Characteristics of Inverter Arm Mid-Point Voltages

In the event of an open-circuit fault within the inverter switches, a noticeable distortion in the phase currents is observed. This distortion is caused by the mismatch between the reference voltage and the actual voltage applied to the windings by the faulty inverter. Consequently, a postfault analysis of the actual voltage applied to the stator windings is of great importance for the fault detection.

Without loss of generality, the operation of the two switches in phase A is used for the illustration. As depicted in Fig. 5.2, the circulations of current within phase A in both the healthy and T_1 open-circuit fault scenarios differ. The current pathways are indicated by solid lines for the healthy state and dashed lines for the faulty state. In the healthy state, the MPVs of the 3-phase legs are determined by both the drive signal and the phase current polarity. When T_1 on, T_4 off and the phase current is positive, the MPV of phase A is positive DC-link voltage. By contrast, when T_1 off, T_4 on, and the phase current is positive, the MPV of phase A is zero, as shown in Fig. 5.2(a). Similarly, when the phase current is negative, there are also two different states of the MPV of phase A, as depicted in Fig. 5.2(b). It is important to note that the states when T_1 off and T_4 on with a positive phase current, and when T_1 on and T_4 off with a negative phase current, represent freewheeling conditions. Therefore, the duration of these states is normally very short.

In the event of an open circuit at T_1 with a positive phase current, the conduction path through T_1 is interrupted. Now the positive current has to flow through the diode D4 no matter whether the drive signal is high or low, as illustrated in Fig. 5.2(c). Consequently, the MPV of phase A is forced to be zero. It is noteworthy that since T_4 is operating normally during this condition, the conduction path for a negative phase current remains unaffected, as shown in Fig. 5.2(d). Thus, the resultant MPV is also not affected. In the extreme case, if both T_1 and D_1 are simultaneously open circuited, the freewheeling is impossible which will also cause more significant fault features.

The inference drawn from this analysis indicates that there is a deviation in the inverter's actual output voltage from the reference voltage due to a switch open-circuit fault. The complexity is further compounded when the switching operation is in a discontinuous conduction mode, where there is a short duration in which the phase current is zero and the phase voltage is indeed dependent on its induced voltage, consisting of the back electromotive force (back-emf) and the voltages from the mutual inductance effect. Nonetheless, the zero current interval duration is very brief, rendering its effect on the overall output voltage negligible.

5.2.3 Output Voltage Vector Analysis Under Fault Conditions

The inverter output voltages, u_{An}, u_{Bn}, and u_{Cn}, referred to as the star neutral n of the 3-phase windings, as shown in Fig. 5.1, are of more interests as they are essentially the phase voltages of machine operations. They can also be expressed

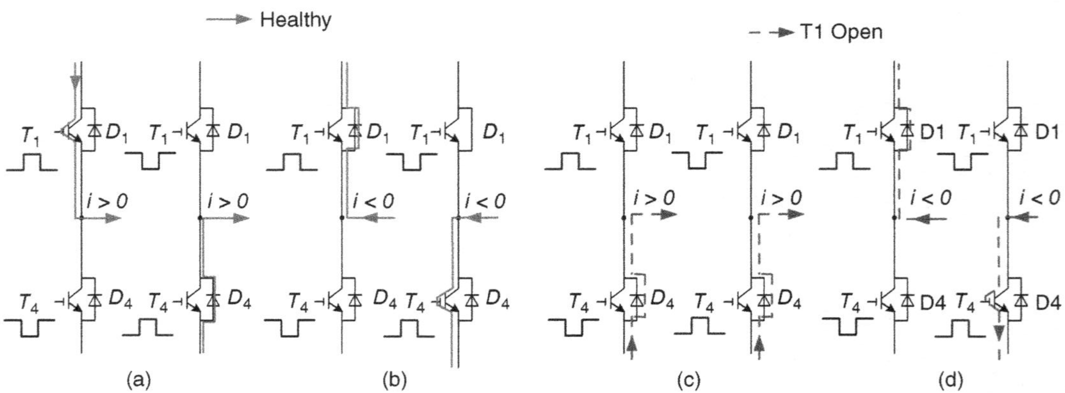

Figure 5.2 Current flow paths under healthy condition and T_1 open-circuit fault. (a) $i > 0$ under healthy condition. (b) $i < 0$ under healthy condition. (c) $i > 0$ under T_1 open-circuit fault. (d) $i < 0$ under T_1 open-circuit fault.

in the form of voltage space vectors u within the a-b-c stationary reference frame as follows:

$$u = u_{An}e^{j0} + u_{Bn}e^{j\frac{2}{3}\pi} + u_{Cn}e^{j\frac{4}{3}\pi} \tag{5.2}$$

According to the above analysis under the open-circuit condition, the phase voltages are also dependent on the polarities and state of currents. For notational convenience, the polarity of current is defined as 1 when it is positive and −1 when negative. Zero current is possible in the circuit; however, it only lasts for a very short period. Thus, it is disregarded in the following analysis. Figure 5.3 depicts these space vectors within the a-b-c frame under both healthy and T_1 open-circuit fault conditions, where the current polarities and state are indicated within the brackets. In healthy operational states, a set of eight space vectors, $u_0 - u_7$, form a hexagon as shown on the left of Fig. 5.3(a), where u_0 *and* u_7, known as zero vectors, are located at the center. The six active vectors, u_1 *and* u_6 divide the hexagon into six sectors, I to VI, as shown in Fig. 5.3(a). In the presence of the open-circuit fault, the space vectors are changed, which are also illustrated on the right of Fig. 5.3(b) and (c) for positive and negative currents, respectively.

When the current polarity is positive (denoted as 1), only half of the space vectors – $u_0'(1)$, $u_3'(1)$, $u_4'(1)$, and $u_5'(1)$ – coincide with u_0, u_3, u_4, and u_5 under the healthy condition. The four remaining space vectors $u_1'(1)$, $u_2'(1)$, $u_6'(1)$, and $u_7'(1)$ are distorted, with these differences depicted in Fig. 5.3(b). In contrast, with a negative current polarity (denoted as −1), the eight space vectors during a T_1 open-circuit fault align with those observed in normal operations as shown in Fig. 5.3(c).

The calculated output voltage vectors, derived from different switch states, are used to assess the combined voltage vectors for each SVPWM cycle as (5.2). To demonstrate the differences in voltage space vectors prefault and postfault conditions, Sector I of the hexagon is utilized as a representative case. During the healthy operation, the average reference voltage space vector over a PWM period can be evaluated as follows:

$$\begin{aligned} u &= u_\alpha^{\text{ref}} + ju_\beta^{\text{ref}} = k_1u_1 + k_2u_2 \\ &= t_1 \times \frac{2}{3}U_{\text{dc}}e^{j0^\circ} + t_2 \times \frac{2}{3}U_{\text{dc}}e^{j60^\circ} \\ &= \frac{2}{3}U_{\text{dc}}\left(\left(t_1 + \frac{1}{2}t_2\right) + j\frac{\sqrt{3}}{2}t_2\right) \end{aligned} \tag{5.3}$$

where u_α^{ref}, u_β^{ref} represent the α and β-axis reference voltage space vectors, respectively. t_1 and t_2 denote the active time duration of vectors u_1 and u_2, respectively. The reference space vector u_α^{ref} and u_β^{ref} can be further expressed

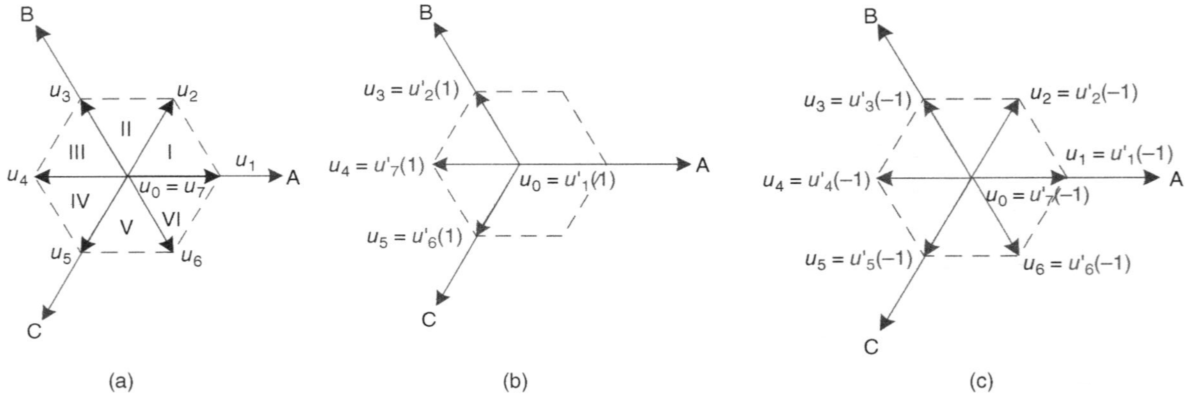

Figure 5.3 Space vectors in healthy condition and T_1 open circuit: (a) healthy conditions, (b) T_1 open circuit with positive current, and (c) T_1 open circuit with negative current.

in (5.4). Hence, equation (5.3) can be rewritten as (5.5). Combining equations (5.3) and (5.5), the values for t_1 and t_2 are deduced in (5.6).

$$\begin{aligned} u_\alpha^{\mathrm{ref}} &= U_{\mathrm{ref}} \cos(\theta + \gamma) \\ u_\beta^{\mathrm{ref}} &= U_{\mathrm{ref}} \sin(\theta + \gamma) \end{aligned} \tag{5.4}$$

$$u = u_\alpha^{\mathrm{ref}} + j u_\beta^{\mathrm{ref}} = U_{\mathrm{ref}}(\cos(\theta + \gamma) + j \sin(\theta + \gamma)) \tag{5.5}$$

$$t_1 = \frac{2u_\alpha^{\mathrm{ref}} - \sqrt{3} u_\beta^{\mathrm{ref}}}{2U_{\mathrm{dc}}}, \quad t_2 = \frac{\sqrt{3} u_\beta^{\mathrm{ref}}}{U_{\mathrm{dc}}} \tag{5.6}$$

where U_{ref} denotes the amplitude of the reference voltage, θ is the electric angle between the rotating d-axis of the rotor reference frame and phase A axis, and γ represents the angle of the reference voltage vector with respect to the d-axis.

When T_1 is open circuited, four non-zero space vectors emerge, $u_2'(1)$, $u_7'(1)$, $u_1'(-1)$, and $u_2'(-1)$. As a result, the combined fault voltage vector u' may be expressed by (5.7). Hence, the synthetic fault voltage vector u' can be derived by substituting (5.6) into (5.7) and is given by (5.8).

$$\begin{aligned} u' = {} & t_n \left(t_1 u_1'(-1) + t_2 u_2'(-1) \right) \\ & + (1 - t_n) \left(t_2 u_2'(1) + \frac{1 - t_1 - t_2}{2} u_7'(1) \right) \end{aligned} \tag{5.7}$$

$$\begin{aligned} u' = u_\alpha' + j u_\beta' = {} & \frac{1}{3} U_{\mathrm{ref}} \left(-1 + t_n + \sqrt{3}(1 + t_1) \cos\left(\theta + \gamma + \frac{\pi}{6} \right) \right. \\ & \left. + \sqrt{3} t_n \sin(\theta + \gamma) \right) + j U_{\mathrm{ref}} \sin(\theta + \gamma) \end{aligned} \tag{5.8}$$

where t_n denotes the ratio of negative current time against the switching period in sector I. The values of t_1 and t_2 are determined by the reference space vectors u_α^{ref} and u_β^{ref}.

Voltage vectors corresponding to the fault condition can be deduced in a similar way for the five remaining subsectors, and they are listed in Table 5.1. A consistent pattern is noticed in that the β-axis voltage shown in bold is not affected by the presence of the open-circuit fault. In contrast, the α-axis voltage experiences variations due to the fault occurrence. This pattern implies that alterations during the fault are confined to the α-axis voltage, which can be leveraged for the detection of the open-circuit fault.

The previous analysis assumes that the top switch (T_1) in phase A is in the open-circuit condition. It can be shown that if the bottom switch (T_4) is open circuited, the β-axis voltage is also not affected by the fault but the change in the α-axis voltage differs in polarity with respect to that of the open-circuit fault

Table 5.1 Faulty synthesized vector in six sub-sectors.

Sectors	$u' = u_\alpha + ju_\beta$; j is the imaginary operator
I	$\frac{1}{3}U_{\text{ref}}\left(-1+t_n+\sqrt{3}(1+t_n)\cos\left(\theta+\gamma+\frac{\pi}{6}\right)+\sqrt{3}t_n\sin(\theta+\gamma)\right)+j\mathbf{U}_{\mathbf{ref}}\mathbf{sin}(\theta+\gamma)$
II	$\frac{1}{3}U_{\text{ref}}(-1+t_n+3t_2\cos(\theta+\gamma))+j\mathbf{U}_{\mathbf{ref}}\mathbf{sin}(\theta+\gamma)$
III	$\frac{1}{3}U_{\text{ref}}\left(-1+t_n+\sqrt{3}(1+t_n)\cos\left(\theta+\gamma-\frac{\pi}{6}\right)-\sqrt{3}t_n\sin(\theta+\gamma)\right)+j\mathbf{U}_{\mathbf{ref}}\mathbf{sin}(\theta+\gamma)$
IV	$\frac{1}{3}U_{\text{ref}}\left(-1+t_n+\sqrt{3}(1+t_n)\cos\left(\theta+\gamma+\frac{\pi}{6}\right)+\sqrt{3}t_n\sin(\theta+\gamma)\right)+j\mathbf{U}_{\mathbf{ref}}\mathbf{sin}(\theta+\gamma)$
V	$\frac{1}{3}U_{\text{ref}}(-1+t_n+3t_n\cos(\theta+\gamma))+j\mathbf{U}_{\mathbf{ref}}\mathbf{sin}(\theta+\gamma)$
VI	$\frac{1}{3}U_{\text{ref}}\left(-1+t_n+\sqrt{3}(1+t_n)\cos\left(\theta+\gamma-\frac{\pi}{6}\right)-\sqrt{3}t_n\sin(\theta+\gamma)\right)+j\mathbf{U}_{\mathbf{ref}}\mathbf{sin}(\theta+\gamma)$

on (T_1). Furthermore, the derived formula for the voltage space vectors and the conclusion is also valid for switch open circuits in phases B and C if the α-axis is aligned with the axes of phases B and C, respectively, as illustrated in Fig. 5.4(a). Therefore, relative $\alpha\beta$-axis voltages for phases B and C can be computed by applying the Clark transformation with appropriate phase shift against phase A axis. In Fig. 5.4(a), α(A), α(B), and α(C) correspond to the relative α-axis associated with phases A, B, and C, respectively. Similarly, β(A), β(B), and β(C) pertain to

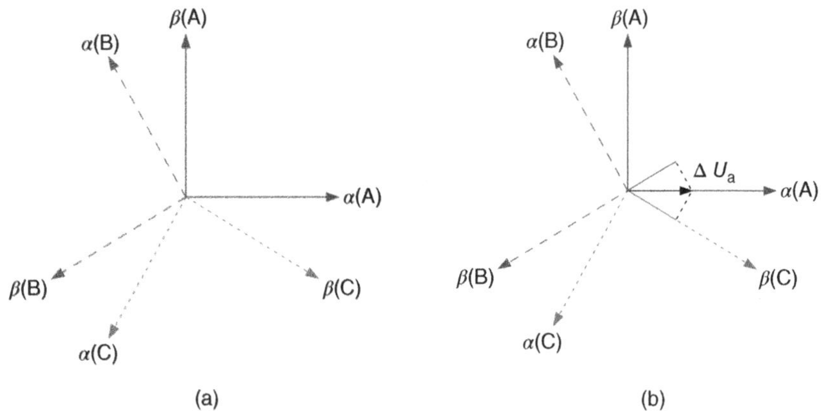

Figure 5.4 The relative $\alpha\beta$ coordinate systems with respect to the three phases: (a) relative $\alpha\beta$ coordinate systems associated with phases A, B and C and (b) projection of change in α(A)-axis voltage on the β(B) and β(C) axes.

the relative β-axis for each phase. If the change in the α(A)-axis voltage due to a switch open-circuit fault in phase A is projected on the β(B) and β(C) axes as shown in Fig. 5.4(b), the two projected changes will have equal magnitude but in opposite polarities. Hence, by continuously monitoring the changes in the voltages projected in β(A), β(B), and β(C) axes, a switch open-circuit fault can be detected and located.

5.3 Open-Circuit Fault Detection Design

Based on the prior analysis, notable differences in the relative $\alpha\beta$-axis voltages under healthy and switch open-circuit conditions have been identified. Thus, a fault detection scheme for switch open circuit can be devised by employing the relative β-axis voltages.

Initially, the voltage residual, which is defined as the difference between the reference voltage and the actual output voltage, is determined. To eliminate the need for voltage sampling circuit, the duty cycle of the switches is used to estimate the reference voltage, as described in (5.9).

$$\begin{cases} u_{\mathrm{Ag}} = D_{\mathrm{A}} U_{\mathrm{dc}} \\ u_{\mathrm{Bg}} = D_{\mathrm{B}} U_{\mathrm{dc}} \\ u_{\mathrm{Cg}} = D_{\mathrm{C}} U_{\mathrm{dc}} \end{cases} \tag{5.9}$$

where D_{A}, D_{B}, D_{C} are the duty cycles of A, B, and C phases. They are determined by the DSP control unit for a given PWM scheme assuming healthy inverter operations.

Referring to Fig. 5.1, it is straightforward to acquire (5.10) with the neutral-to-reference ground voltage, u_{ng}, as an intermediate variable. However, it can be eliminated by the fact that $u_{\mathrm{A}n} + u_{\mathrm{B}n} + u_{\mathrm{C}n} = 0$. Thus, the 3-phase reference voltages with respect to the star-neutral point n, $u^*_{\mathrm{A}n}$, $u^*_{\mathrm{B}n}$, and $u^*_{\mathrm{C}n}$ for phases A, B, and C, respectively, can be determined in (5.11).

$$\begin{cases} u_{\mathrm{A}n} = u_{\mathrm{Ag}} - u_{ng} \\ u_{\mathrm{B}n} = u_{\mathrm{Bg}} - u_{ng} \\ u_{\mathrm{C}n} = u_{\mathrm{Cg}} - u_{ng} \end{cases} \tag{5.10}$$

$$\begin{cases} u^*_{\mathrm{A}n} = \dfrac{2}{3} u_{\mathrm{Ag}} - \dfrac{1}{3} u_{\mathrm{Bg}} - \dfrac{1}{3} u_{\mathrm{Cg}} \\ u^*_{\mathrm{B}n} = -\dfrac{1}{3} u_{\mathrm{Ag}} + \dfrac{2}{3} u_{\mathrm{Bg}} - \dfrac{1}{3} u_{\mathrm{Cg}} \\ u^*_{\mathrm{C}n} = -\dfrac{1}{3} u_{\mathrm{Ag}} - \dfrac{1}{3} u_{\mathrm{Bg}} + \dfrac{2}{3} u_{\mathrm{Cg}} \end{cases} \tag{5.11}$$

Next, the actual output voltages are estimated by a simplified permanent magnet synchronous machine (PMSM) model using Euler approximation according to (5.12).

$$\begin{cases} u_d(t) = R_s i_d(t) + L_d \dfrac{(i_d(t) - i_d(t-\Delta t))}{\Delta t} - \omega_e L_q i_q(t) \\ u_q(t) = R_s i_q(t) + L_q \dfrac{(i_q(t) - i_q(t-\Delta t))}{\Delta t} + \omega_e L_d i_d(t) + \omega_e \psi_m \end{cases} \tag{5.12}$$

where R_s represents the stator phase resistance; u_d, u_q, i_d, i_q, L_d, and L_q denote the voltages, currents, and inductances in the *dq*-axis reference frame; ψ_m stands for the PM flux linkage; ω_e for the rotor electrical angular speed; and Δt for the sampling interval. With a predetermined set of the machine parameters, the *dq*-axis voltages can be estimated from measured *dq*-axis currents and rotor speed in (5.12).

To obtain the relative β-axis voltage across all three phases, the estimated voltages u_d and u_q by (5.12) are converted to the *abc* stationary reference frame using the inverse Park transformation. Further, the residual voltages for phases A, B, and C are normalized to the DC-link voltage in (5.13). Finally, the normalized relative β-axis voltage residuals for phases A, B, and C are derived and given in (5.14).

$$\Delta u_k = \frac{u_k^* - u_k}{U_{dc}}, \quad k = \text{A, B, C} \tag{5.13}$$

$$\begin{cases} \varepsilon_{\beta,A} = u_{\beta,A}^* - u_{\beta,A} = \dfrac{\sqrt{2}}{2}(\Delta u_B - \Delta u_C) \\ \varepsilon_{\beta,B} = u_{\beta,B}^* - u_{\beta,B} = \dfrac{\sqrt{2}}{2}(\Delta u_C - \Delta u_A) \\ \varepsilon_{\beta,C} = u_{\beta,C}^* - u_{\beta,C} = \dfrac{\sqrt{2}}{2}(\Delta u_A - \Delta u_B) \end{cases} \tag{5.14}$$

where $\varepsilon_{\beta,A}$, $\varepsilon_{\beta,B}$, and $\varepsilon_{\beta,C}$ are the normalized relative β-axis residual voltages for phases A, B, and C, respectively. $u_{\beta,A}^*$, $u_{\beta,B}^*$, and $u_{\beta,C}^*$ are the normalized reference voltages in the relative β-axes associated with phases A, B, and C, respectively, and $u_{\beta,A}$, $u_{\beta,B}$, and $u_{\beta,C}$ are the normalized estimated actual voltages in the relative β-axes associated with the three phases.

During healthy operation, the normalized relative β-axis residual voltages for the three phases should be equal or close to zero. Assuming a switch open-circuit fault in phase A, the β-axis residual voltage $\varepsilon_{\beta,A}$ remains close to zero. However, this does not apply to the residuals $\varepsilon_{\beta,B}$ and $\varepsilon_{\beta,C}$ associated with phases B and C, as indicated by (5.14).

As $\varepsilon_{\beta,A}$ maintains zero, it follows that the changes in Δu_B and Δu_C are equal, leading to $\varepsilon_{\beta,B}$ and $\varepsilon_{\beta,C}$ displaying one-sided variation and having opposite polarity

Table 5.2 Diagnostic rule.

Faulty component	F_A	F_B	F_C
No fault	0	0	0
T_1	0	−1	1
T_2	1	−1	0
T_3	1	0	−1
T_4	0	1	−1
T_5	−1	1	0
T_6	−1	0	1

to each other. Thus, the three relative β-axis residual voltages contain the key indicators of the location of the open-circuit fault and form an integral part of the fault diagnostic approach.

To refine the detection process, the residual outputs from equation (5.14) are conditioned by a low pass filter (LPF), which helps reduce the impact of differential operations in (5.12), and switching and sampling noises. Furthermore, the fault indicator for phase k (k = A, B, C), symbolized as F_k, is characterized by

$$F_k = \begin{cases} 1, & \mathrm{LPF}(\varepsilon_{\beta,k}) \geq \varepsilon_{\mathrm{th}} \\ 0, & |\mathrm{LPF}(\varepsilon_{\beta,k})| < \varepsilon_{\mathrm{th}} \\ -1, & \mathrm{LPF}(\varepsilon_{\beta,k}) \leq -\varepsilon_{\mathrm{th}} \end{cases} \tag{5.15}$$

where $\varepsilon_{\mathrm{th}}$ is a threshold value.

During online monitoring, when the processed residual voltage $\varepsilon_{\beta,k}$ exceed the predefined threshold $\varepsilon_{\mathrm{th}}$, the corresponding fault indicator F_k becomes active for that phase. A value of either 1 or −1 is assigned for that phase according to the residual voltage polarity. If an open-circuit fault takes place in the switch (T_1) of phase A, the residual voltage $\varepsilon_{\beta,\mathrm{A}}$ would not be affected; hence, F_A would continue to be zero. In contrast, $\varepsilon_{\beta,\mathrm{B}}$ and $\varepsilon_{\beta,\mathrm{C}}$ deviate from zero, as specified by equation (5.14), F_B turns to −1 and F_C turns to 1. Therefore, the sign of F_k serves as a diagnostic tool to pinpoint the fault location. Table 5.2 summarizes the correlation between fault locations and the fault indicators.

5.4 Experimental Validation on Open-Circuit Fault Detection

The proposed open-circuit detection technique is evaluated experimentally on the previous triple redundant 3-phase machine drive prototype. Since the three

3-phase sets are identical and independently controlled, one 3-phase module can be injected with a switch open-circuit fault for the purpose of demonstration. Consequently, the subsequent fault diagnosis experiments will be conducted on one 3-phase set as an example, and the current waveforms of the other two sets of 3-phase windings are not provided.

5.4.1 Single Switch Fault Diagnosis

The threshold value is a key parameter of the proposed method. Thus, it is determined before the fault test. The machine drive is commanded to operate in healthy condition, and the $|\mathrm{LPF}(\varepsilon_{\beta,k})|$ in (5.15) is recorded. After operating under the whole torque-speed range, the threshold value is defined as 0.1 pu. Then, the proposed fault detection method is tested at 2000 rpm with a phase current magnitude of 67 A. The open-circuit fault is injected in switch T_1 by deactivating T_1 drive signal as depicted in Fig. 5.5. After the fault occurrence, the positive phase A current reduces to zero since the voltage applied to the phase becomes zero. With the application of the proposed detection technique, the relative β-axis residual voltages for the three phases are determined. A noticeable surge

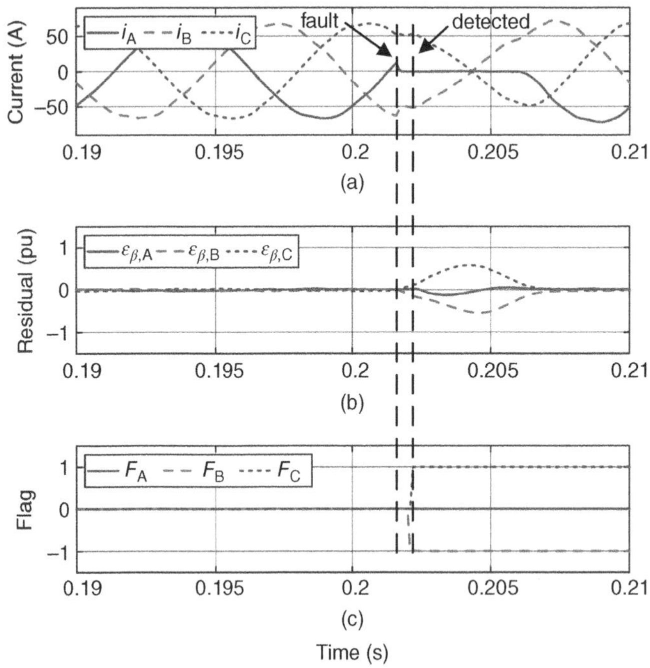

Figure 5.5 Fault detection waveforms when T_1 is open circuited: (a) phase currents, (b) voltage residuals, and (c) fault indicators.

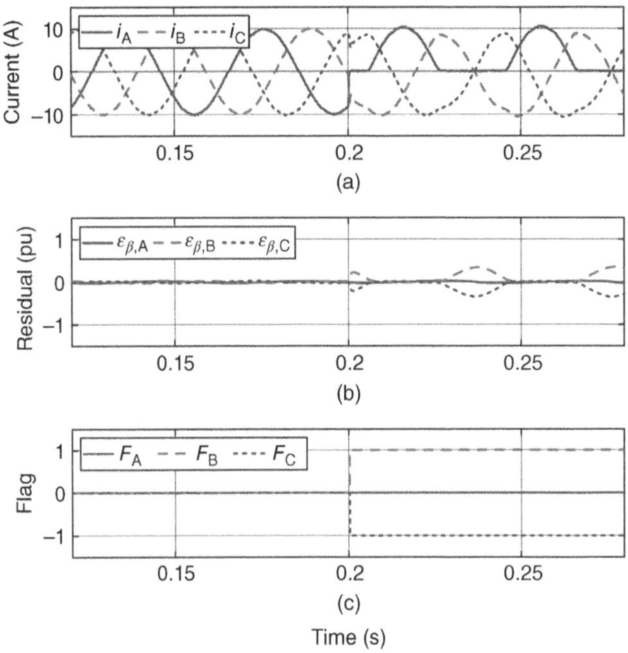

Figure 5.6 Fault detection waveforms when T_4 is open circuited: (a) phase currents, (b) voltage residuals, and (c) fault indicators.

is observed in the relative β-axis residual voltages, $\varepsilon_{\beta,B}$ and $\varepsilon_{\beta,C}$, of the healthy phases B and C, which are higher than the threshold value. In contrast, the value of phase A $\varepsilon_{\beta,A}$ is only slightly altered, which is consistent with the previous conclusion. As a result, the fault indicator signals F_A, F_B, and F_C are assigned values of 0, −1, and 1, respectively.

The proposed detection method is also examined in different power switch under the low speed and light load region. Figure 5.6 depicts test waveforms of switch T_4 open circuited when the machine operates at 500 rpm with 10 A current. The fault indicator, namely, the relative β axis voltage residuals, promptly responds to the fault. According to the polarity of F_A, F_B, and F_C, an open circuit can be identified in switch T_4. The test results confirm that the proposed method allows for the detection of open-circuit fault and accurate fault location identification according to Table 5.2.

5.4.2 Influence of Load Transients on the Detection Scheme

To eliminate any false-positive alarm in healthy operations, the presented fault detection algorithm is subjected to a robustness evaluation of step changes in load current as illustrated in Figs. 5.7 and 5.8. Initially, the motor operates in healthy

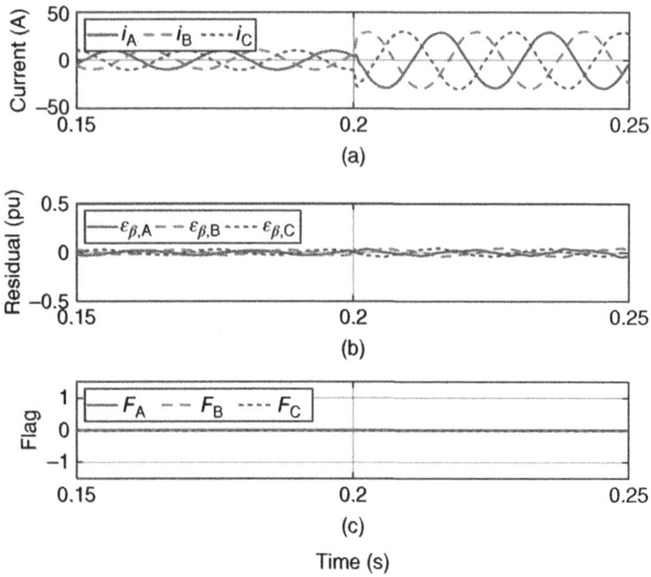

Figure 5.7 Fault detection waveforms during a step increase in load current: (a) phase currents, (b) voltage residuals, and (c) fault indicators.

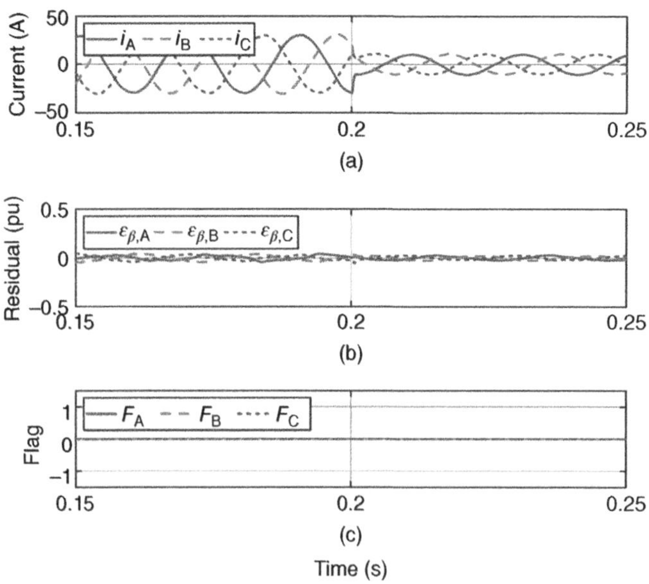

Figure 5.8 Fault detection waveforms during a step decrease in load current: (a) phase currents, (b) voltage residuals, and (c) fault indicators.

condition at 1000 rpm. At 0.2 second, the phase current experiences a step change in magnitude upwards from 10 to 30 A and downwards from 30 to 10 A. It is seen that the fault indicators are not affected by the change as the small deviations in the relative β-axis residual voltages are all within their thresholds. Such results affirm the introduced approach is robust to fluctuation in the load current.

5.4.3 Detection with Parameter Variation

The proposed detection technique employs voltage estimators instead of voltage sensors for evaluating the inverter output voltages. Therefore, it is necessary to assess the impact of parameter variations on fault detection. During operation, the machine parameters experience changes due to the load and speed variation, which typically vary within ±20%. Since voltage estimation involves four parameters, resistance, dq-axis inductance, and PM flux linkage, experimental tests are conducted with each of these parameters varying from 80% to 120%. Fig. 5.9 illustrates the experimental waveforms of fault detection when the machine operates under low-speed, light load and high-speed, heavy load conditions, respectively. The waveforms reveal a negligible impact on the relative β-axis residual voltages, no matter in healthy or fault mode for the four parameter variations. It can be observed that the impact of parameter variation on fault detection is minimal and can be essentially disregarded for both the scenarios. This is because the voltage residual caused by an open circuit fault is much greater than the error that caused by parameter variation. Therefore, the proposed detection method in this study demonstrates excellent robustness to parameter variations.

5.4.4 Fault Mitigation Test After Open-Circuit Fault

For the triple redundant 3-phase machine drive, an open-circuit fault in an inverter switch or phase winding can be effectively mitigated by deactivating the entire 3-phase windings when the fault has been detected. An integrated test is performed as shown in Fig. 5.10 including detection and mitigation action. The machine is initially operating at 1000 rpm with 30 A current of magnitude under healthy condition. Subsequently, an open-circuit fault is injected by deactivating switch T_1. As can be seen, the fault is rapidly detected after the injection, followed by the swift implementation of the mitigation strategy through the deactivation of the whole 3-phase set. This confirms the ability to quick detection of the fault and executing appropriate control measures to reduce the fault impact on the triple redundant 3-phase PMA-SynRM drive system. Meanwhile, it is evident that prefault and postfault operations of the other two healthy sets, DEF and GHI, are almost not affected.

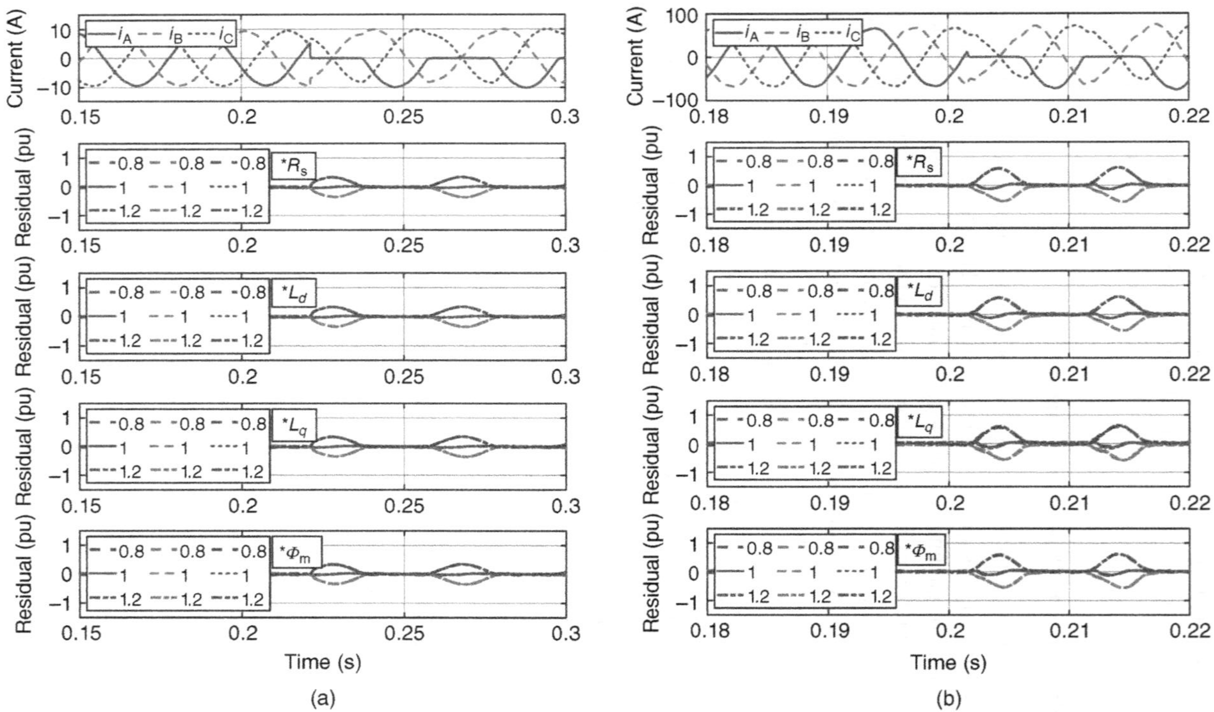

Figure 5.9 Influence of machine parameters on the voltage residuals during detection: (a) 500 rpm with 10 A load current and (b) 2000 rpm with 67 A load current.

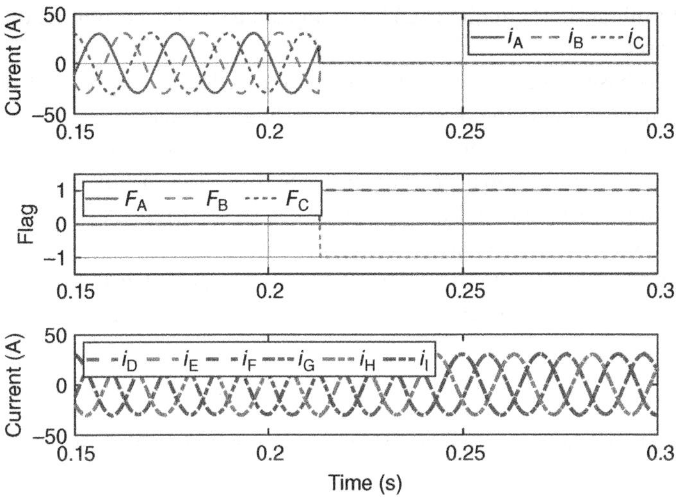

Figure 5.10 Integrated test with fault mitigation strategy: (a) phases ABC currents, (b) fault indicators, and (c) other phase currents.

5.5 Analysis of Turn Fault Signature

Before discussing techniques for detection of a turn fault in a phase winding, the machine drive behavior under a turn fault is analyzed for developing a suitable diagnostic method. For the sake of simplicity, the analysis is also based on a single 3-phase set of the triple-redundant 3-phase machine drive. As shown in Fig. 5.11, a single-turn fault in coil A1 of the ABC set is considered for its most benign fault signature. It has been shown that such a fault leads to a significant fault current i_f in the faulty part, which is roughly nine times the rated value. Consequently, the fault necessitates immediate detection to initiate a mitigation action, that is,

Figure 5.11 Schematic circuit for set ABC with turn fault in phase A.

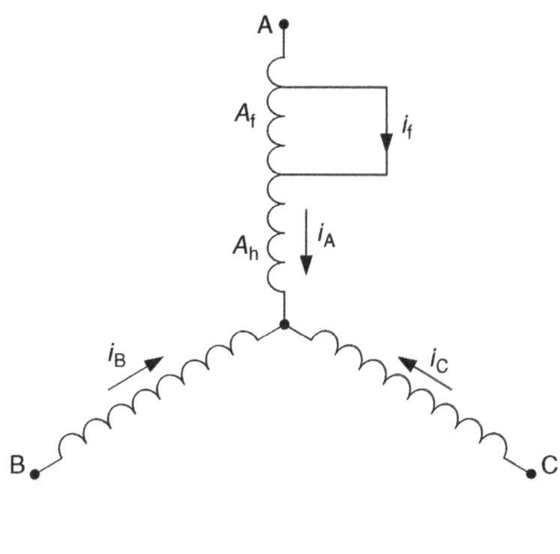

application of terminal short circuit (TSC) for, or injecting specific current to, the affected 3-phase set.

Given that a 3-phase drive operates typically with a feedback current control loop, the fault behavior under closed-loop control will be analyzed for selecting the most suitable fault indicator that provides optimal signal-to-noise ratio for detection purposes. As has been shown in Chapter 2, the distortion in flux linkages due to the fault introduces the 2nd harmonics in voltages, currents, and hence instantaneous power. These variables under closed-loop control will be further examined. During healthy operations, and neglecting higher-order harmonics, the flux linkage associated with a given number of turns in a phase winding can be described as follows:

$$\psi_{\text{th}} = \varphi \cos(\omega t + \theta_0) \tag{5.16}$$

where φ, ω, and θ_0 are the magnitude, electrical angular speed, and phase angle, respectively.

When a turn fault occurs, the resulting turn fault current is mainly reactive, generating a flux linkage that counteracts the flux linkage in the healthy operations. Therefore, the flux linkage generated by the fault current can be approximated as follows:

$$\psi_{\text{tf}} = -\varphi_{\text{f}} \cos(\omega t + \theta_0) \tag{5.17}$$

where φ_{f} denotes the magnitude of the flux linkage. This flux linkage can be expressed in the dq-axis reference frame as

$$\begin{aligned} \Delta\psi_d &= -\frac{1}{3}\varphi_{\text{f}}(\cos\theta_0 + \cos(2\omega t + \theta_0)) \\ \Delta\psi_q &= -\frac{1}{3}\varphi_{\text{f}}(\sin\theta_0 - \sin(2\omega t + \theta_0)) \end{aligned} \tag{5.18}$$

It is seen from (5.18) that the 2nd harmonics will be present in the dq-axis flux linkages, acting as perturbations to the current control loop. This leads to the generation of the 2nd harmonics in both dq-axis voltages and currents. To examine the emerging 2nd harmonic voltages and currents, the schematics of the dq-axis current control loop, as shown in Fig. 5.12, is considered. The proportional and integral (PI) current controllers are simplified as a proportional controller since the integral action has relatively low bandwidth, thus causing minimal immediate response to the 2nd harmonic components.

The dq-axis voltages of the machine drive can be expressed in (5.19) when the voltage drops on the winding resistive are neglected.

$$\begin{aligned} u_d &= \frac{d\psi_d}{dt} - \omega\psi_q \\ u_q &= \frac{d\psi_q}{dt} + \omega\psi_d \end{aligned} \tag{5.19}$$

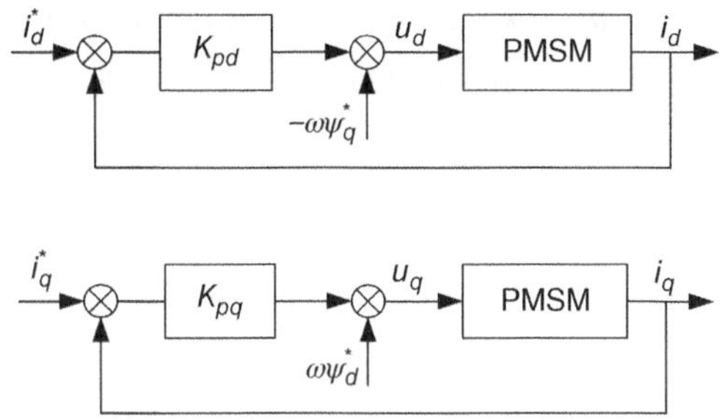

Figure 5.12 Simplified *dq*-axis current control loops in a 3-phase drive.

where ψ_d, ψ_q, L_d and L_q denote the *d*- and *q*-axis components of flux linkages and inductances, respectively. Further, ψ_d, ψ_q can be expressed as in (5.20) by accounting for the flux linkage induced by the turn fault current, where the DC-offset components in (5.18) associated with $\cos(\theta_0)$ are neglected since they do not cause 2nd harmonics in the *dq*-axis voltages and currents. ψ_{m} represents the PM flux linkage.

$$\begin{aligned}\psi_d &= \psi_{\mathrm{m}} + L_d i_d - \frac{1}{3}\varphi_{\mathrm{f}}\cos(2\omega t + \theta_0)\\ \psi_q &= L_q i_q + \frac{1}{3}\varphi_{\mathrm{f}}\sin(2\omega t + \theta_0)\end{aligned} \tag{5.20}$$

The decoupling feedforward compensations, $-\omega\psi_q^*$ and $\omega\psi_d^*$, in the current controllers of Fig. 5.12 are estimated typically using the *dq*-axis machine parameters as in (5.21).

$$\begin{aligned}\psi_d^* &= \psi_m + L_d i_d\\ \psi_q^* &= L_q i_q\end{aligned} \tag{5.21}$$

Hence, the voltage equations in (5.19) under the closed loop current control in Fig. 5.12 can be expressed as

$$\begin{aligned}-K_{pd} i_d - \frac{\omega}{3}\varphi_{\mathrm{f}}\sin(2\omega t + \theta_0) &= L_d\frac{di_d}{dt}\\ -K_{pq} i_q - \frac{\omega}{3}\varphi_{\mathrm{f}}\cos(2\omega t + \theta_0) &= L_q\frac{di_q}{dt}\end{aligned} \tag{5.22}$$

The gains K_{pd} and K_{pq} are typically designed as $\omega_{\mathrm{c}} L_d$ and $\omega_{\mathrm{c}} L_q$, respectively, based on pole-zero cancellation, where ω_{c} is the current controller bandwidth which is set to 200 Hz. By solving (5.22), the 2nd harmonic currents and voltages in steady state, denoted as i_{d2}, i_{q2} u_{d2}, u_{q2}, can be derived as

$$\begin{aligned}i_{d2} &= -\frac{k}{L_d}\sin(2\omega t + \theta_1)\\ i_{q2} &= -\frac{k}{L_q}\cos(2\omega t + \theta_1)\end{aligned}$$

$$
\begin{aligned}
u_{d2} &= k(\omega_c \sin(2\omega t + \theta_1) + \omega \cos(2\omega t + \theta_1)) \\
u_{q2} &= k(\omega_c \cos(2\omega t + \theta_1) - \omega \sin(2\omega t + \theta_1)) \\
k &= \frac{\omega \varphi_f}{\sqrt[3]{\omega_c^2 + 4\omega^2}}, \quad \theta_1 = \theta_0 - \tan^{-1}\frac{2\omega}{\omega_c}
\end{aligned} \tag{5.23}
$$

Defining i_{d0}, i_{q0} as the DC components of i_d and i_q, the dq-axis currents and voltages can be further expressed as

$$
\begin{aligned}
i_d &= i_{d0} - \frac{k}{L_d}\sin\delta \\
i_q &= i_{q0} - \frac{k}{L_q}\cos\delta \\
u_d &= -\omega L_q i_{q0} + k(\omega_c \sin\delta + \omega\cos\delta) \\
u_q &= \omega\psi_{d0} + k(\omega_c \cos\delta - \omega \sin\delta) \\
\psi_{d0} &= \psi_m + L_d i_{d0}, \quad \delta = 2\omega t + \theta_1
\end{aligned} \tag{5.24}
$$

It can be seen that the turn fault induces the 2nd harmonics in both the currents and voltages. The magnitudes of the 2nd harmonics are influenced by the current control bandwidth. Further, the (IAP and IRP, denoted as iap and iaq, can be evaluated as [16]

$$
\begin{aligned}
iap &= 1.5(u_d i_d + u_q i_q) \\
iaq &= 1.5(u_q i_d - u_d i_q)
\end{aligned} \tag{5.25}
$$

The 2nd harmonics of the IAP and IRP, denoted as p_2 and q_2, are extracted from (5.25) and given in (5.26).

$$
\begin{aligned}
p_2 &= 1.5k\left(\left(\frac{L_q}{L_d} - 1\right)\omega i_{q0} + \omega_c i_{d0}\right)\sin\delta \\
&\quad + 1.5k\left(\omega i_{d0} - \frac{\omega\psi_{d0}}{L_q} + \omega_c i_{q0}\right)\cos\delta \\
q_2 &= 1.5k\left(-\frac{\omega\psi_{d0}}{L_d} - \omega i_{d0} - \omega_c i_{q0}\right)\sin\delta \\
&\quad + 1.5k(\omega_c i_{d0} - 2\omega i_{q0})\cos\delta
\end{aligned} \tag{5.26}
$$

Given that the PM flux linkage ψ_m is notably small and a negative i_d is typically employed for maximum torque per Ampere (MTPA) operation in PMA-SynRM, ψ_{d0} remains small compared to other elements in (5.26). Therefore, the magnitude related to the *cosine* part of p_2 is roughly equivalent to that of the *sine* part of q_2. As a result, the magnitude of p_2 is predominantly determined by its *sine* part, whereas the magnitude of q_2 is primarily contributed from its *cosine* part. When assessing the *sine* part of p_2 against the *cosine* part of q_2, it becomes apparent that the magnitude of q_2 exceeds that of p_2 with a positive i_{q0}. Conversely, with

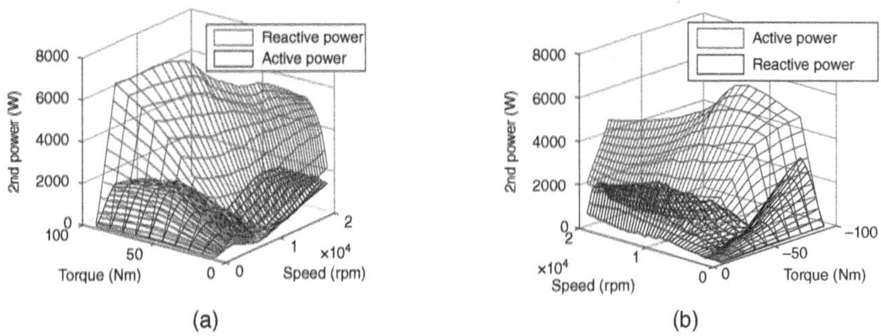

Figure 5.13 Comparison of 2nd harmonics in IAP and IRP: (a) motoring mode and (b) generating mode.

a negative i_{q0}, the magnitude of q_2 falls below that of p_2. This suggests that in motoring mode ($i_{q0} > 0$), the 2nd harmonic in the IRP surpasses the IAP, while in generating mode ($i_{q0} < 0$), it is lower than that of the IAP. Although this analysis is performed under the simplified condition by ignoring additional sources of the 2nd harmonics in IAP and IRP, such as natural phase imbalance and magnetic saturation, they are relatively small compared to those resulting from a turn fault. The foregoing analysis suggests that the 2nd harmonic of the IAP or IRP is the suitable fault indicator, depending on the modes of operation.

To capture the behavior of the 2nd harmonics in the IAP and IRP after the turn fault across the entire torque-speed operating range of the machine under study, the general model described in Chapter 4 has been utilized in simulations. Figure 5.13 illustrates the simulated 2nd harmonics in the fault 3-phase set under both motoring and generating modes of operation. It is noticed that during motoring, the 2nd harmonic generated in the IRP of the faulty set surpasses the IAP across the entire operation region, but the trend is reversed for the generating mode. This observation supports the findings discussed earlier. Additionally, the relationship between turn faults and the 2nd harmonic of IRP during motoring and IAP during generating displays a more uniform pattern throughout the operating range. Therefore, in terms of turn fault detection, the 2nd harmonic in the IRP is a more robust indicator for motoring mode, while the 2nd harmonic in the IAP becomes more suitable for generating mode. It is also observed that detection of turn fault at low speeds is more difficult, as the 2nd harmonics decrease fast with reduction in speed.

5.6 Turn Fault Detection Design

In this section, the proposed fault detection based on the 2nd harmonics in IAP and IRP is designed for the triple redundant 3-phase machine drive. It is important

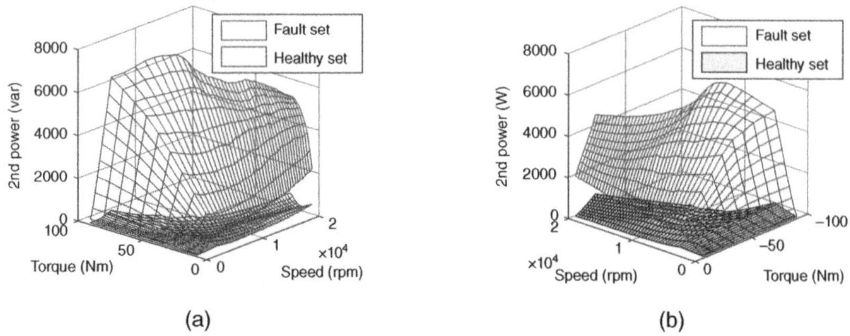

Figure 5.14 Comparison 2nd harmonics of faulty and healthy sets: (a) IRP in motoring mode and (b) IAP in generating mode.

to recognize that inherent asymmetries and imbalances within the machine are a natural consequence of production variances, the limited precision of current sensors [17], and disparities in inverter devices. These elements all contribute to the presence of small 2nd harmonics even under healthy operations. To account for these inherent 2nd harmonics, lookup tables for the 3-phase groups have been created to record the 2nd harmonics observed under healthy operations. This calibration can be executed automatically during healthy conditions of the drive, prior to an event of any fault.

The fault 3-phase set containing a turn fault obviously will show a pronounced increase in the 2nd harmonic within the fault indicator compared with the other two 3-phase sets. The 2nd harmonics between the healthy and fault-affected sets are compared in Fig. 5.14 across different modes of operation according to the foregoing analysis and simulations. It is evident that the 2nd harmonics generated by the faulty set are notably larger than those in the healthy sets. Consequently, the 2nd harmonic level of the three 3-phase sets serves as an internal benchmark for detecting the fault. Harmonics-based detection methods could be prone to variations in speed and load, potentially leading to false detection. Employing a comparison among the three 3-phase sets remarkably diminishes the impact of such transient changes on the accuracy of fault indicators, since these changes affect all sets equally. However, the increase in the 2nd harmonic in the fault set will be a clear indicator for detection, significantly lowering the likelihood of false alarms.

Figure 5.15 presents a schematic diagram of the detailed process for identifying turn faults in motoring mode. The 2nd harmonics within the IRP are labeled as q_2^k, $k = 1, 2, 3$. Initially, IRP values are computed for all 3-phase sets, utilizing the reference voltages in the DSP control unit and real-time current measurements as in (5.25). Subsequently, the amplitude of the 2nd harmonic is extracted by the process and low-pass filters shown at the bottom of Fig. 5.15. Reference data of

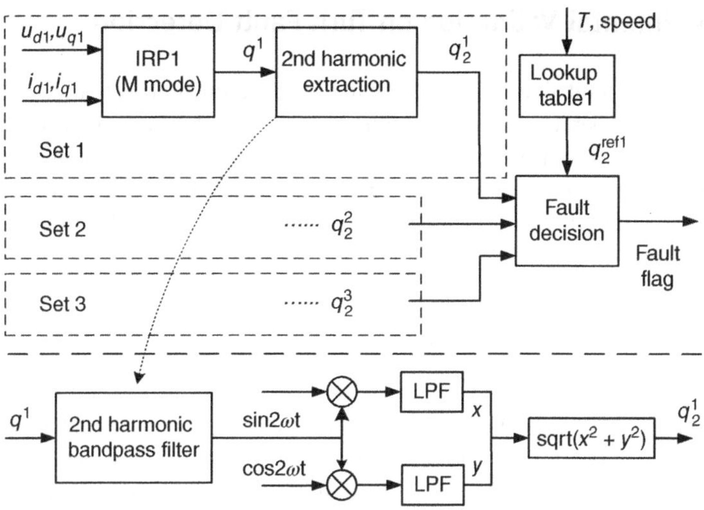

Figure 5.15 Fault detection diagram in motoring mode.

the 2nd harmonic under healthy operational conditions are stored in the lookup tables, which are generated through experimental testing. These tables account for the 2nd harmonics arising from intrinsic discrepancies in phase windings, such as phase misalignment, variance in resistance/inductance values, rotor eccentricity, and nonuniform magnetization of magnets, thereby minimizing the likelihood of false alarm. Finally, the fault indicators are processed within a decision block. The fault flag is triggered if

$$q_2^1 > 2q_2^{\text{ref1}} \quad \text{and} \quad q_2^1 > 1.5 \max\left(q_2^2, q_2^3\right) \tag{5.27}$$

where q_2^1, q_2^2, and q_2^3 represent the 2nd harmonic in IRP of the three 3-phase sets, and q_2^{ref1} is the reference data of the 2nd harmonic for set ABC in healthy condition at a given torque-speed combination. The assertion of the first inequality occurs when the 2nd harmonic surpasses twice the magnitude of its reference value under the healthy condition, while the second inequality is validated when this harmonic also exceeds the levels of the other two sets by a factor of 1.5. Adjustments to the two coefficients within these inequalities allow for enhanced sensitivity in detection and the prevention of false detection. Through the utilization of lookup tables and cross-referencing, the system effectively compensates for machine imbalances and transient effects, thus ensuring detection is both accurate and reliable. This detection method has been similarly applied to sets DEF and GHI. In terms of generating mode, the detection approach can be similarly designed by employing the 2nd harmonic in IAP.

5.7 Experimental Validation on Turn Fault Detection

The proposed turn fault detection is validated by experimental tests across a broad torque-speed operating range, including transient load. The fault current, *dq*-axis currents and voltages, and fault indicators, namely, 2nd harmonics in IAP and IRP are presented and discussed. In addition, an integrated test is conducted to demonstrate the developed fault detection and mitigation techniques, which enables the transition from healthy operations to sustainable postfault operations.

5.7.1 Typical Turn Fault Behavior

To mitigate the risk of damage due to excessive fault current, the turn fault behavior is evaluated at 1000 rpm. This is very representative as low-speed operation poses the greatest challenge for the fault detection. Initially, the motor drive operates at the rated current of 100 A in magnitude under MTPA condition and the turn fault is introduced at 0.06 second by closing the turn fault emulation relay for a short duration of 0.2 second. The resultant fault current is shown in Fig. 5.16, where the peak current reaches 480A (4 pu). The measured *dq*-axis currents and the resulting *dq*-axis voltages from the PI controllers are shown in Fig. 5.17 for the three 3-phase sets, where the subscripts 1, 2, and 3 correspond to the ABC, DEF, and GHI sets, respectively. The deduced IAP and IRP are depicted in Fig. 5.18(a). Prior to the fault where $t < 0.06$ second, the currents, voltages, IAP, and IRP signals appear consistent across all three 3-phase sets during healthy operation. Following the turn fault, the drive persists in operation through the current feedback control, but notable harmonics arise for all three sets, with the 2nd harmonic being particularly prominent. The amplitudes of the 2nd harmonic in both voltages and currents

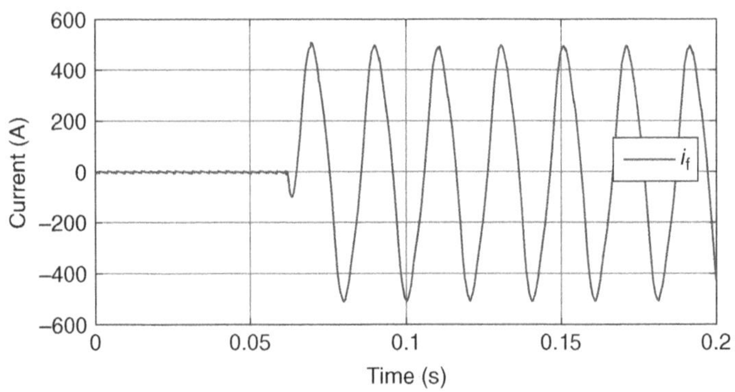

Figure 5.16 Turn fault current when the drive operates at 1000 rpm and 100 A.

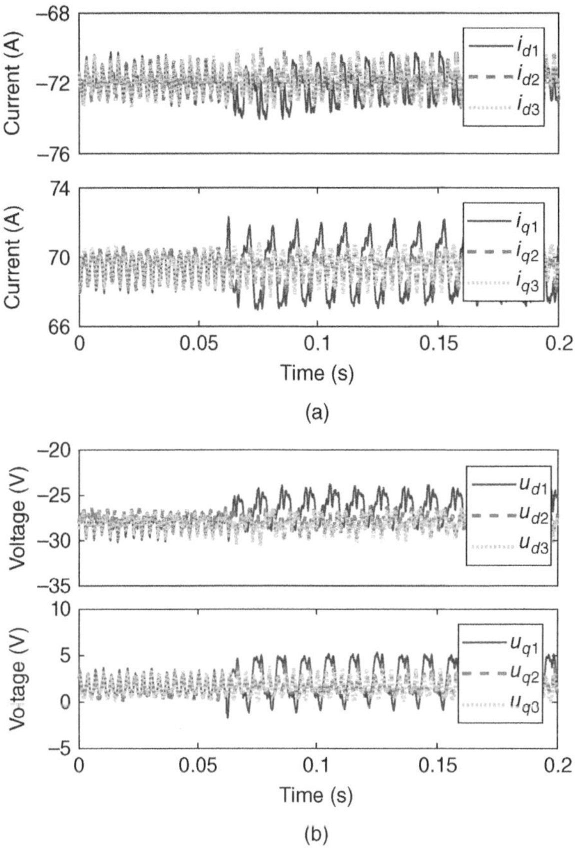

Figure 5.17 Measured currents and voltages before and after turn fault: (a) *dq*-axis currents and (b) *dq*-axis voltages.

are observed to be relatively small, being below 15.4% for the voltage and 2.9% for the current with respect to their amplitudes. Conversely, the 2nd harmonic in IAP and IRP are substantially higher, offering a much larger signal-to-noise ratio for the fault detection. Being consistent with the prior analysis and simulation, the 2nd harmonic in IRP extracted from the detector during the motoring operation surpasses that in IAP remarkably. For illustration, the extracted 2nd harmonics in IRP for the three 3-phase sets are compared in Fig. 5.18(b). It is evident that in the healthy operation, the magnitudes of q_2^1, q_2^2, and q_2^3 are small, so are the deviations between them. Following the turn fault, a significant rise in q_2^1 is observed, while q_2^2 and q_2^3 exhibit minor changes in opposite directions. This behavior is consistent with the theoretical study and simulations and has been exploited in the fault detection criteria outlined in (5.27) for improving reliability of the detection.

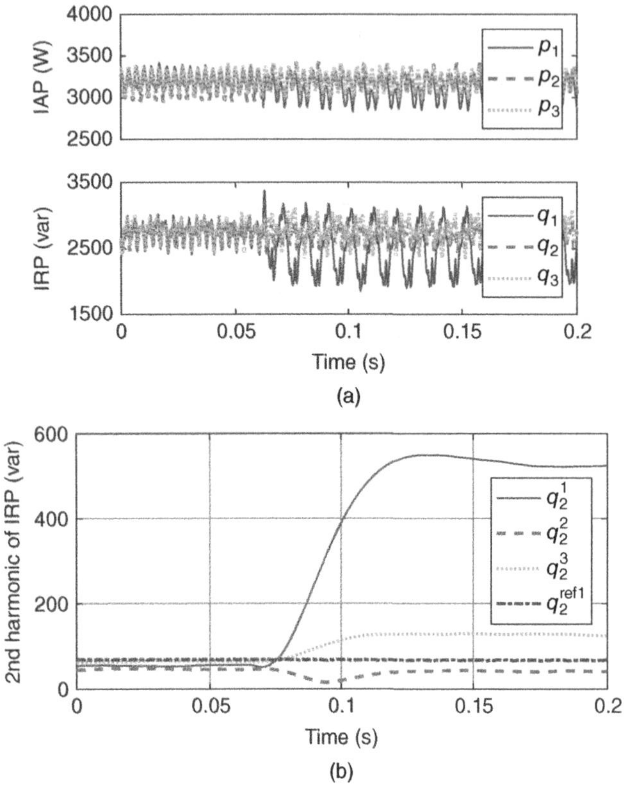

Figure 5.18 IAP and IRP before and after turn fault: (a) instantaneous values and (b) extracted 2nd harmonic magnitudes of IRP in three 3-phase sets.

Additional experiments are performed with 100 A load current in generating mode. The behaviors of the currents and voltages are similar to those shown in Figs. 5.16 and 5.17 except for the polarity of the q-axis current being reversed. Thus, only the graphs for the IAP and IRP before and after the fault are illustrated in Fig. 5.19(a). Unlike the motoring mode as expected, now the 2nd harmonic in IAP is evidently much greater than that in IRP. Consequently, for generating mode, the 2nd harmonic in IAP will be selected as the fault indicator. The filtered 2nd harmonics of IAP for the three 3-phase sets are depicted in Fig. 5.19(b). The relative patterns observed here are very close to those in Fig. 5.18(b), indicating that the turn fault can be reliably detected in this mode with the defined criterion.

5.7.2 Detectability over the Wide Operating Region

A critical aspect of assessing fault detection capability is the scope of detectable coverage within the operational torque-speed range for the machine drive. As

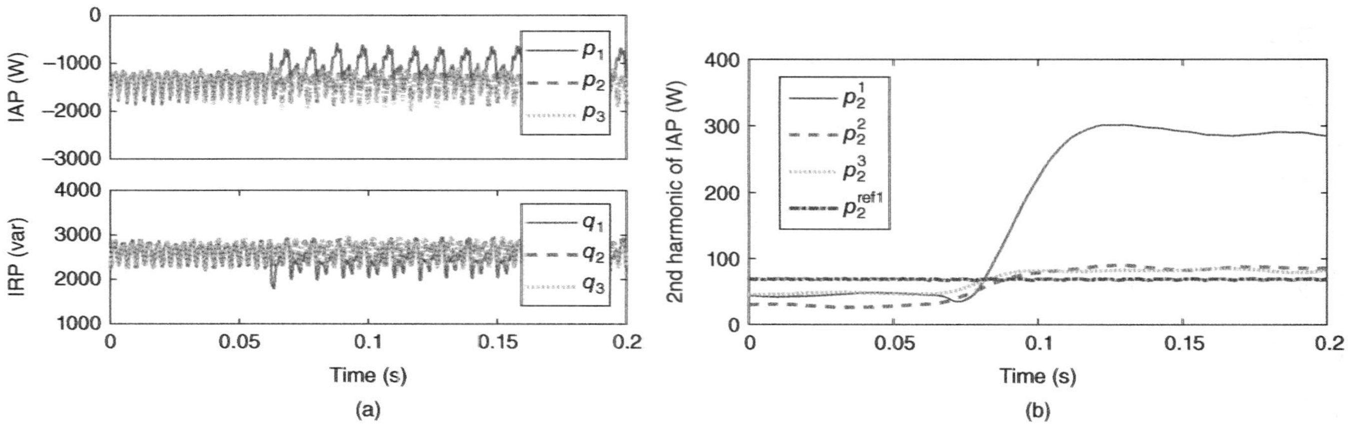

Figure 5.19 Measured IAP and IRP in generating mode: (a) instantaneous values and (b) extracted 2nd harmonic magnitudes of IAP in the three 3-phase sets.

depicted in Fig. 5.14, it is observed that the most challenging zone for the turn fault detection mainly locates at lower speeds due to the small 2nd harmonics in both IAP and IRP. Consequently, the turn fault detection algorithm is further examined at a lower speed of 500rpm. The proportions of the 2nd harmonics in IRP and IAP of the fault condition against their reference values under the healthy condition, defined as

$$\begin{aligned} \text{Ratio} &= q_2^1/q_2^{\text{ref}\,1} \quad \text{motoring mode} \\ \text{Ratio} &= p_2^1/p_2^{\text{ref}\,1} \quad \text{generating mode} \end{aligned} \tag{5.28}$$

have been measured in order to evaluate the detection criterion defined in (5.27). As depicted in Fig. 5.20, it is seen that the variations of the ratio with the load current consistently exceed 2 in the motoring mode, suggesting that the turn fault is detectable under all load current conditions at 500 rpm. As speed increases, the 2nd harmonic in IRP under a turn fault condition increases as is evident in Fig. 5.14, leading to a much greater detectability. Likewise, the ratio remains above 2 for speeds at 1000 rpm in the generating mode under all load conditions. At 500 rpm, however, the ratio falls below 2 when the load current surpasses 90 A, indicating that the turn fault may not be detected reliably with a threshold of 2 for the $p_2^1/p_2^{\text{ref}1}$ ratio when the speed is below 500 rpm. This phenomenon can be attributed to several factors, such as small turn fault current, and non-negligible voltage drop on the phase resistance and inverter switches. In the motoring mode, the reference voltage magnitude, $\sqrt{(u_d^*)^2 + (u_q^*)^2}$, is greater than that of the actual voltage due to the voltage drop on the inverter, but the opposite is true for the generating mode, resulting in lower 2nd IAP harmonics when evaluated with the voltage reference signals u_d^* and u_q^* in the DSP controller.

It should be noted that the above assessment is based on the experimental results with one-turn short-circuit fault. If the number of short-circuit turns is more than 1, the magnitude of the 2nd harmonics in IAP and IRP increases, leading to much larger fault signatures and even-better detectability.

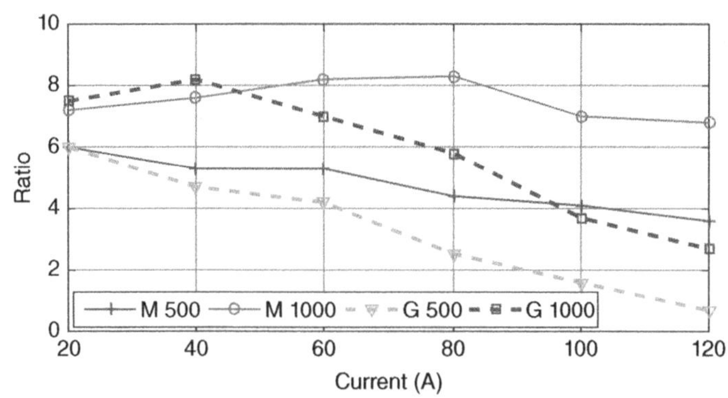

Figure 5.20 Detection ratio at 500 and 1000 rpm in motoring (M500, M1000) and generating (G500, G1000) modes.

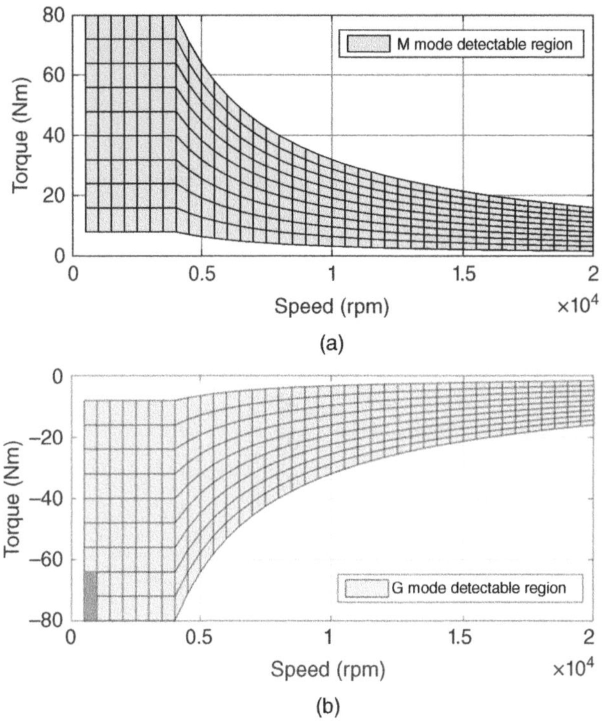

Figure 5.21 Detection zone under (a) motoring mode and (b) generating mode.

Based on the analysis above, the detection zones under the worst case of one-turn short circuit can be expressed as

$$\begin{aligned} &\text{abs}(T) > 0.1 \ \text{pu} \\ &\text{speed} > 500 \ \text{rpm} \end{aligned} \tag{5.29}$$

and they are illustrated in Fig. 5.21, where the undetectable regions are shown in gray. Thus, a turn fault can be detected in a very wide operational range under the motoring mode. While in the generating mode, the fault is also detectable when the speed is above 500 rpm and the load torque is lower than −60 Nm.

5.7.3 Effect of Transient Conditions on Detection

Harmonic-based fault detection may lead to unwanted alerts due to fluctuations in load torque (or current) and speed. For the machine drive under study, the online monitoring of the three 3-phase sets can provide a cross reference to avoid false alarms during transient. The robustness of the analyzed detection

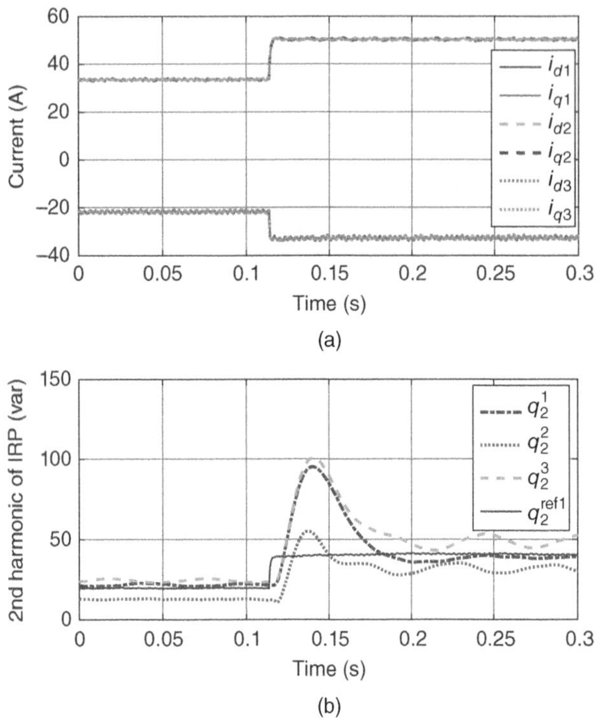

Figure 5.22 i_{dq} response and 2nd harmonic of IRP in transient (a) i_{dq} (b) 2nd harmonic of IRP.

method is assessed under such transient conditions and an example is illustrated in Fig. 5.22(a). Initially, the machine drive is operating with 40 A current at 1000 rpm, then a step increase in the load current to 60 A is applied at $t =$ 0.13 second. The resultant 2nd harmonics of the three 3-phase sets are extracted and displayed in Fig. 5.22(b). Temporary increases in the harmonics in all 3-phase sets are observed during the transient, and the peak of the 2nd harmonics is about three times higher than the reference value. However, the differences in these harmonics are below 1.5 times. Thus, no false alert will be activated as it does not satisfy the second inequality in (5.27).

Another test is performed when the same current step change takes place at 1000 rpm and a single-turn fault is injected simultaneously. The *dq* axis currents and the 2nd harmonic responses are shown in Fig. 5.23. In this case, although the 2nd harmonics of all three sets increase during the transient, the 2nd harmonic in the fault set is particularly higher. Thus, the fault can be detected during the transient.

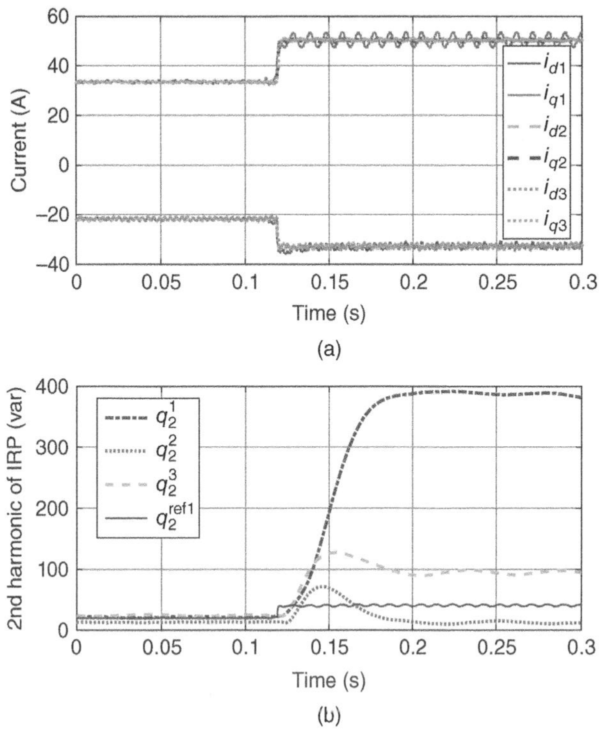

Figure 5.23 i_{dq} response and 2nd harmonic of IRP in transient with turn fault (a) i_{dq} and (b) 2nd harmonic of IRP.

To avoid the risk of damage in repeated tests with turn fault, simulations have been performed in various operating conditions with speed and torque changes. The results show that the fault criterion in (5.27) yields correct detection in all cases. It follows that by cross-referencing the 2nd harmonics of the three 3-phase sets the proposed fault detection is also effective during transient while false alarm can be avoided.

5.7.4 Fault Mitigation Test

Since a turn fault in a phase winding leads to excessive fault current and rapid increase in temperature in the vicinity of the fault conductors, an immediate action, that is, application of TSC through the inverter of the faulty 3-phase set, is necessary upon detection of the fault. Further tests are also performed to demonstrate the response of the triple redundant drive to a turn fault through fault detection and mitigation action to postfault operation. Initially, the machine drive is operating with 60 A load current at 1000 rpm in motoring mode. A single-

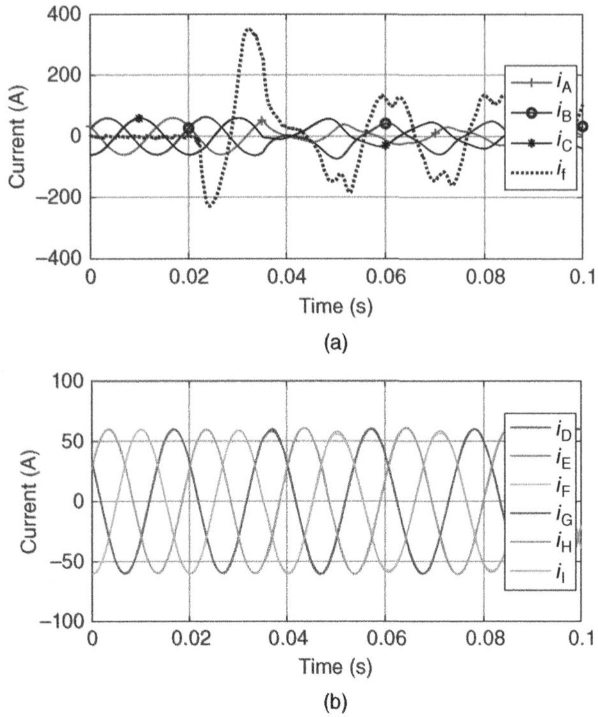

Figure 5.24 Current responses during fault detection and mitigation test: (a) currents in fault set and (b) currents in healthy sets.

turn fault is injected into phase B by triggering the relay across the turn fault emulation taps. It is seen in Fig. 5.24(a) that the fault current, denoted as i_f, soars to 340 A, which can cause serious damage to winding insulation if a mitigation action is not taken timely. With implementation of the fault detection technique described previously, the turn fault is detected after 15 ms within one fundamental electrical cycle and TSC is applied on the fault ABC set. With the mitigation, the rms fault current is reduced to a thermally sustainable level of 107 A while the rms phase currents in the healthy parts of the ABC windings are below the rated. Furthermore, the phase currents of the other two healthy 3-phase sets have not been affected noticeably following the mitigation action as seen in Fig. 5.24(b). The postfault operation continues with the two healthy 3-phase sets albeit with reduced torque production capacity. Thus, it confirms that the worst single-turn fault can be accommodated by the triple redundant 3-phase PMA-SynRM drive with proper detection and mitigation actions. It should be noticed that a short circuit across a single-turn represents the worst-case scenario in terms of fault current and detection. The integrated test provides further confidence in the ability of the triple redundant machine drive to operate against other possible turn faults.

5.8 High-Frequency Signal-Based Turn Fault Detection Techniques

As has been briefly discussed in Section 5.1, turn fault detection techniques based on signals around the fundamental frequency, denoted as low frequency (LF), are dependent on the machine operation condition. At low speed and light load conditions, the fault signature is very benign and reliable detection is of great challenge. The 2nd IRP and IAP-based detection technique described and tested in Sections 5.6 and 5.7 also falls into this category, whereby any detection at speed below 500 rpm is difficult. The direct physical changes due to a turn fault are (1) flux linkage which in turn affects induced voltage and current and (2) inductance. Since the first and second changes are proportional to the number of turns and the square of the number of turns, respectively, the change in inductance is much greater and even more pronounced at HF. It can be expected that HF signal-based detection methods will exhibit stronger capabilities and they have been widely investigated in the literature [2, 18–21].

Currently, there are two main types of detection methods based on HF signals. One method involves injecting HF voltages into machine windings, which is similar to the HF-injection based sensorless control of electric machine drives. The injected voltage is usually around a few kHz with a small magnitude. The resultant HF current response is then extracted and processed to detect interturn faults. As the injected signal is at a specific frequency, independent of machine operations, the detection is more robust and reliable even at low speed and in no-load operation. The techniques can be further enhanced by appropriate filtering. The HF signal-based detection was first analyzed in [22, 23] by measuring the HF negative sequence current and further investigated in [24] by monitoring the incremental inductances. In Ref. [25, 26], the HF voltage residual signals are extracted by comparing the injected HF voltage and estimated HF voltage. It is found that the fault signals are clearly reflected in the HF range, which are immune to the operation condition. The enhanced fault detectability of the HF injection-based technique can be further exploited to detect and classify turn fault and high-resistance connection fault as reported in [12]. However, HF injection introduces additional ripple in the current which, in turn, incurs additional HF losses in the machine windings and magnetic materials and may reduce voltage operating range at high speed.

The other method involves utilizing HF voltages generated by PWM switching action, which does not require additional HF injection. Therefore, the disadvantages of the HF injection-based techniques, such as increased noise and loss, are avoided. The PWM-induced HF voltages are usually concentrated around two times of the switching frequency, typically 20 kHz if the switching frequency is 10 kHz. In the presence of turn short-circuit faults, the asymmetry of the 3-phase windings leads to differences in the response of HF currents, enabling the detection of interturn short-circuit faults [27]. Further in Ref. [28], the equivalent HF impedance is employed to detect the turn fault with higher signal magnitude. However, the HF voltages and currents generated by PWM require a high sampling rate and processing power, which may increase the cost of the DSP unit of the drive. Alternatively, additional hardware processing circuits can be used to filter the HF voltages and currents.

The above two HF signal-based detection methods can be employed if accurate detection is required in the low speed and light load region. To illustrate the effectiveness of HF-signal based detection, the method in [25, 26] is exemplified as shown in Fig. 5.25. HF signals are injected to the phase voltages similar to the HF-injection based sensorless drive control [29]. Simulations are performed for a single-turn short circuit in phase B of the triple redundant machine under two operating conditions, light load at low speed (10 A current at 500 rpm), which is very challenging for detection with the technique described in Section 5.6, the other is heavy load at high speed (60 A current at 4000 rpm). The simulated turn fault currents and signals are presented in Fig. 5.26. It is seen that at low speed, the turn fault current is only 150 A while it is 550 A at high speed. As a result, the turn fault can be easily detected with LF-based techniques at high speed while the detection is very difficult in low-speed operation. In contrast, however, the captured HF components in the midpoint-to-ground-voltages u_{ng_ah}, u_{ng_bh}, u_{ng_ch} [25, 26] in the two operating conditions are almost the same, as shown in Fig. 5.26. Hence, the turn fault can be reliably detected for both operations from

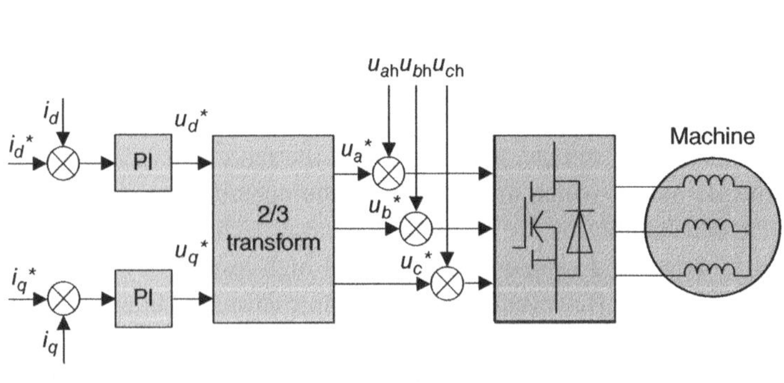

Figure 5.25 Scheme of HF voltage injection.

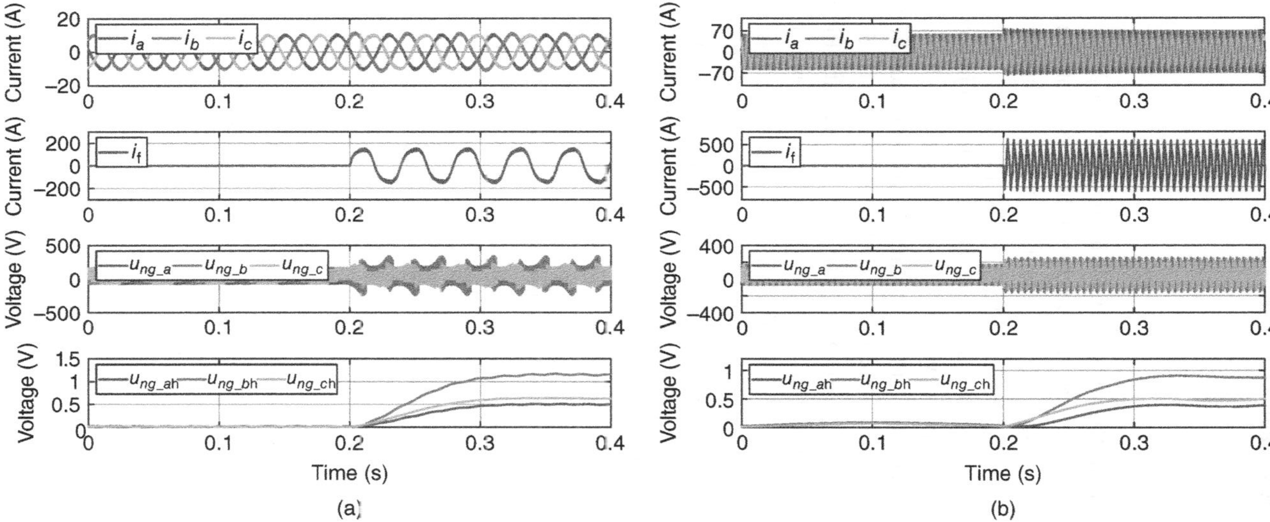

Figure 5.26 Diagnosis process of turn fault under different operating conditions: (a) 500 rpm with 10 A load and (b) 4000 rpm with 60 A load. (i_a, i_b, i_c, and i_f are the 3-phase currents and fault current and u_{ng_a}, u_{ng_b}, u_{ng_c} and u_{ng_ah}, u_{ng_bh}, u_{ng_ch} are the midpoint-to-ground-voltages and their high-frequency components.)

the HF signals. Thus, the HF-based method provides an alternative approach to detect the most challenging fault.

5.9 Summary

This chapter presents fault diagnosis methods for open-circuit faults in inverter switches and for interturn short-circuit faults in windings of a motor drive system. For open-circuit faults in power devices, a fault diagnosis method based on the relative β-axis residual voltages associated with each phase has been described, analyzed, and tested. It can not only detect an open-circuit fault in power devices but also identify the fault location within the 3-phase inverter. Through analysis of voltage and current behaviors under a winding turn fault condition, a fault detection method based on the 2nd harmonics in IAP and IRP has been developed and demonstrated. It has been shown that the 2nd harmonic in IRP exhibits a more significant fault signature during motoring mode, while the 2nd harmonic in IAP is more significant during generation mode. Thus, the two fault signatures are employed for turn fault detection in the two different operation modes. The robustness of the fault detection is further enhanced by cross-references of the fault signatures in three 3-phase sets. Extensive simulations and experiments have been conducted to validate the effectiveness and utilities of the two proposed detection methods for mitigation actions and sustained postfault operations.

In addition, the limitations of LF-based turn fault detection techniques are discussed and the merits of HF-based detection techniques are outlined. The latter is conducive not only for fault detection but also for fault classification.

References

1 S. Afrandideh, M. E. Milasi, F. Haghjoo, and S. M. A. Cruz, "Turn to turn fault detection, discrimination, and faulty region identification in the stator and rotor windings of synchronous machines based on the rotational magnetic field distortion," *IEEE Transactions on Energy Conversion*, vol. 35, no. 1, pp. 292–301, 2020.

2 B. Sen and J. Wang, "Stator interturn fault detection in permanent-magnet machines using PWM ripple current measurement," *IEEE Transactions on Industrial Electronics*, vol. 63, no. 5, pp. 3148–3157, 2016.

3 A. H. Bonnett and G. C. Soukup, "Cause and analysis of stator and rotor failures in three-phase squirrel-cage induction motors," *Industry Applications, IEEE Transactions on*, vol. 28, no. 4, pp. 921–937, 1992.

4 A. Gandhi, T. Corrigan, and L. Parsa, "Recent advances in modeling and online detection of stator interturn faults in electrical motors," *IEEE Transactions on Industrial Electronics*, vol. 58, no. 5, pp. 1564–1575, 2011.
5 P. Shi, X. Wang, X. Meng, M. He, Y. Mao, and Z. Wang, "Adaptive fault-tolerant control for open-circuit faults in dual three-phase PMSM drives," *IEEE Transactions on Power Electronics*, vol. 38, no. 3, pp. 3676–3688, 2023.
6 H. Mesai-Ahmed, I. Jlassi, A. J. M. Cardoso, and A. Bentaallah, "Multiple open-circuit faults diagnosis in six-phase induction motor drives using stator current analysis," *IEEE Transactions on Power Electronics*, vol. 37, no. 6, pp. 7275–7285, 2022.
7 C. H. D. Angelo, G. R. Bossio, S. J. Giaccone, M. I. Valla, J. A. Solsona, and G. O. Garcia, "Online model-based stator-fault detection and identification in induction motors," *IEEE Transactions on Industrial Electronics*, vol. 56, no. 11, pp. 4671–4680, 2009.
8 M. Yi Lu, M. A. Masrur, C. ZhiHang, and Z. Baifang, "Model-based fault diagnosis in electric drives using machine learning," *IEEE/ASME Transactions on Mechatronics*, vol. 11, no. 3, pp. 290–303, 2006.
9 B. Sen and J. Wang, "A fast detection technique for stator inter-turn fault in multi-phase permanent magnet machines using model based approach," in *7th IET International Conference on Power Electronics, Machines and Drives (PEMD 2014)*, 2014, pp. 1–6.
10 S. Nadarajan, S. K. Panda, B. Bhangu, and A. K. Gupta, "Online model-based condition monitoring for brushless wound-field synchronous generator to detect and diagnose stator windings turn-to-turn shorts using extended kalman filter," *IEEE Transactions on Industrial Electronics*, vol. 63, no. 5, pp. 3228–3241, 2016.
11 R. Hu, J. Wang, A. R. Mills, E. Chong, and Z. Sun, "Current-residual-based stator interturn fault detection in permanent magnet machines," *IEEE Transactions on Industrial Electronics*, vol. 68, no. 1, pp. 59–69, 2021.
12 R. Hu, J. Wang, A. R. Mills, E. Chong, and Z. Sun, "Detection and classification of turn fault and high resistance connection fault in permanent magnet machines based on zero sequence voltage," *IEEE Transactions on Power Electronics*, vol. 35, no. 2, pp. 1922–1933, 2020.
13 A. Sarikhani and O. A. Mohammed, "Inter-turn fault detection in PM synchronous machines by physics-based back electromotive force estimation," *IEEE Transactions on Industrial Electronics*, vol. 60, no. 8, pp. 3472–3484, 2013.
14 M. Drif and A. J. M. Cardoso, "Discriminating the simultaneous occurrence of three-phase induction motor rotor faults and mechanical load oscillations by the instantaneous active and reactive power media signature analyses," *IEEE Transactions on Industrial Electronics*, vol. 59, no. 3, pp. 1630–1639, 2012.

15 M. Drif and A. J. M. Cardoso, "On-line fault diagnostics in operating three-phase induction motors by the active and reactive power media," in *18th International Conference on Electrical Machines, 2008. ICEM 2008*, 2008, pp. 1–6.

16 H. Akagi, Y. Kanazawa, and A. Nabae, "Instantaneous reactive power compensators comprising switching devices without energy storage components," *IEEE Transactions on Industry Applications*, vol. IA-20, no. 3, pp. 625–630, 1984.

17 C. Dae-Woong and S. Seung-Ki, "Analysis and compensation of current measurement error in vector-controlled AC motor drives," *IEEE Transactions on Industry Applications*, vol. 34, no. 2, pp. 340–345, 1998.

18 J. Zhang, J. Hang, S. Ding, and M. Cheng, "Online diagnosis and localization of high-resistance connection in PMSM with improved fault indicator," *IEEE Transactions on Power Electronics*, vol. 32, no. 5, pp. 3585–3594, 2017.

19 S. C. Yang, "On-line turn fault detection of interior permanent magnet machines using the pulsating-type voltage injection," *IEEE Transactions on Industry Applications*, vol. PP, no. 99, pp. 1–1, 2016.

20 R. Hu, J. Wang, A. R. Mills, E. Chong, and Z. Sun, "PWM ripple currents based turn fault detection for 3-phase permanent magnet machines," *2017 IEEE International Electric Machines and Drives Conference (IEMDC)*, 2017, pp. 1–7.

21 R. Hu, J. Wang, A. R. Mills, E. Chong, and Z. Sun, "High-frequency voltage injection based stator interturn fault detection in permanent magnet machines," *IEEE Transactions on Power Electronics*, vol. 36, no. 1, pp. 785–794, 2021.

22 F. Briz, M. W. Degner, A. Zamarron, and J. M. Guerrero, "Online stator winding fault diagnosis in inverter-fed AC machines using high-frequency signal injection," *IEEE Transactions on Industry Applications*, vol. 39, no. 4, pp. 1109–1117, 2003.

23 F. Briz, M. W. Degner, P. Garcia, and A. B. Diez, "High-frequency carrier-signal voltage selection for stator winding fault diagnosis in inverter-fed AC machines," *IEEE Transactions on Industrial Electronics*, vol. 55, no. 12, pp. 4181–4190, 2008.

24 J. Arellano-Padilla, M. Sumner, and C. Gerada, "Winding condition monitoring scheme for a permanent magnet machine using high-frequency injection," *Electric Power Applications, IET*, vol. 5, no. 1, pp. 89–99, 2011.

25 B. Wang, X. Feng, J. Zeng, and Z. Wang, "A novel turn fault detection strategy based on high-frequency neutral-to-ground voltages," *IEEE Transactions on Power Electronics*, vol. 38, no. 12, pp. 15945–15955, 2023.

26 X. Feng, B. Wang, Z. Wang, W. Hua, M. Cheng, and B. Zhou, "Stator turn fault detection by relative voltage residual slopes with high frequency injection," *IEEE Transactions on Transportation Electrification*, vol. 10, no. 3, pp. 4809–4820, 2024.

27 B. Wang, L. Luo, W. Fu, W. Hua, G. Wang, and Z. Wang, "Study on the PWM ripple current based turn fault detection for interior PM machine," *IEEE Transactions on Transportation Electrification*, vol. 7, no. 3, pp. 1537–1547, 2021.

28 B. Wang, J. Hu, G. Wang, and W. Hua, "A novel stator turn fault detection technique by using equivalent high frequency impedance," *IEEE Access*, vol. 8, pp. 130540–130550, 2020.

29 G. Wang, M. Valla, and J. Solsona, "Position sensorless permanent magnet synchronous machine drives—a review," *IEEE Transactions on Industrial Electronics*, vol. 67, no. 7, pp. 5830–5842, 2020.

6

Postfault Control Strategies for Fault-Tolerant Machine Drives

6.1 Introduction

Postfault control is an important technique that enables continuous operation in a fault-tolerant machine drive. After the occurrence of a fault, the currents in the machine drive are usually distorted with increased voltage ripples and torque fluctuations. Excessive current will be induced in case of a short-circuit fault, particularly of an interturn short circuit. Thus, a suitable postfault control strategy should be in place to suppress the undesirable effects without further risk of fault propagation.

For multiphase fault tolerant machines with fractional slot concentrated windings (FSCWs) in which each phase is independently controlled by an H-bridge converter as shown in Fig. 6.1 [1, 2], significant torque ripple will result in case of open or short circuit in one phase. This will deteriorate the quality of postfault operation and may lead to many undesirable effects, such as excessive vibration and mechanical instability. This problem can be addressed by employing an optimal torque control strategy during the postfault operation [3–5].

For the multiple 3-phase permanent magnet-assisted synchronous reluctance machine (PMA-SynRM) under consideration, the fault in one module can be isolated by disabling the whole 3-phase set winding or applying terminal short circuit (TSC) on the fault set. However, as described in Chapter 2, in the case of interturn short circuit, the fault current in the worst-case scenario is about 3.2 pu even after TSC application to the faulted 3-phase set. This is because of the low impedance in the short-circuit loop [6, 7]. The localized high fault current may jeopardize the postfault operation of the machine drives although design measures have been taken to accommodate this fault as discussed in Chapter 3. Nevertheless, it would be desirable if the fault current could be further reduced.

TSC application is a common approach for fault-tolerant multiphase machines and multiple 3-phase machines to mitigate turn fault [1]. The reactive current produced by the TSC in the fault phase reduces the flux linkage in the fault turns

Multiple 3-Phase Fault Tolerant Permanent Magnet Machine Drives: Design and Control, First Edition.
Bo Wang and Jiabin Wang.

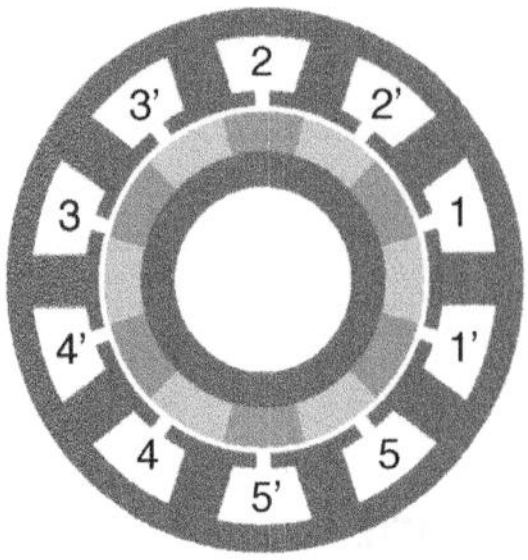

Figure 6.1 Schematic of 5-phase 10s12p fault-tolerant FSCW PM machine.

and hence the resultant fault current. However, the study in [8] showed that this method was only effective for a small wire wound machine and was not suitable for a bar wound machine. Therefore, an optimized current at 90° lagging the back electromotive was injected to the fault phase of a bar wound machine to limit the fault current. In [9], a magnet flux nullifying control was developed for a 3-phase interior PM (IPM) machine following a short-circuit fault either in the inverter or in an entire phase winding by employing open-end winding drives. It was demonstrated that the fault current was reduced to zero by injecting proper current to the healthy phases. In Ref. [10], the fault current was reduced for a 3-phase PM machine by activating field weakening before the base speed. The voltage of the fault turns is reduced by increasing the negative d-axis current. However, the effect of the fault current reduction is quite limited and this method also reduces the output torque. In addition, it is only applicable to operations below the base speed. In Ref. [11], 3-phase current injection was used to reduce the interturn short-circuit current in a PM aircraft starter/generator but leads to a complete loss of the torque capability of the machine. This mitigation action is still necessary to prevent excessive heat in the faulted turns and its propagation since the rotating PM field cannot be turned off during the engine operation. The conflict between the suppression of the fault current and the loss of the output torque is mainly due to the limited degrees of freedom in the conventional 3-phase machine.

This chapter describes the optimal torque control strategy for fault-tolerant multiphase FSCW PM machines and develops a turn fault mitigation strategy to minimize the fault current in the multiple 3-phase PMA-SynRM drive as two examples of postfault control. The optimal current control strategy is employed to deliver smooth output torque during postfault operation for both constant torque and constant power operating regions. It could be adapted for open-circuit, short-circuit, and interturn short-circuit faults. Current injection technique is employed without compromising output torque for the multiple 3-phase machine drive under study. The flux linkage of the faulted phase, where the fault turns are located, is estimated by accounting for the influence of the rotor PM field and the currents in the healthy 3-phase sets and the fault set while no precise information on the number of faulted turns and on the fault location in subcoils and slots is required. To minimize the flux linkage of the fault turns, specific currents are injected to the fault 3-phase set while the currents in the healthy sets are not affected by continuous torque output. The injected currents are estimated by two linear equations, which are independent of the rotor position. The proposed method is validated by extensive finite element (FE) simulations and experimental tests. It is demonstrated that the fault current is effectively reduced compared to the conventional TSC method while the torque is not compromised.

6.2 Postfault Optimal Torque Control Strategies for Multiphase FSCW PM Machines

Multiphase PM machines with single-layer FSCW are one of the classic forms of PM machine topology that offers fault-tolerant capability [1]. The number of phases employed is usually greater than 3, being 4 or 5 typically. Each phase is fed by an independently controlled H-bridge converter. Hence, the important machine design features to achieve fault tolerance, namely, electrical, thermal, and magnetic isolations can be easily realized with this machine topology. Further, by employing surface-mounted rotor topology, the machine can achieve high torque density at the expense of more magnetic material usage. Such a machine naturally has large phase inductance, which is conducive to achieving greater than 1.0 pu phase inductance for limiting short-circuit current below the rated value. In the event of an open-circuit or short-circuit fault in one phase, the remaining healthy phases can continue their operation to produce output torque or power. For a machine with m number of phases, the output torque capability under a fault condition is typically reduced to $(m-1)/m$ times the rated. This provides a degree of freedom in the selection of phase numbers to obtain the required torque capability in fault conditions.

It is generally assumed that such fault-tolerant machines have a sinusoidal back electromotive force (back-emf) waveform and the current controller for each phase is required to track a set of sinusoidal current commands proportional to a given torque demand. However, this simple control strategy inevitably results in an undesirable torque ripple during normal healthy operation, when the back-emf waveform is nonsinusoidal, and a large torque pulsation under a fault condition, such as an open-circuited or a short-circuited phase. In this chapter, a 5-phase, 10-slot, 12-pole fault-tolerant FSCW PM machine is considered to develop the optimal torque control whose schematic is shown in Fig. 6.1.

To reduce the torque ripple in healthy operations and alleviate large torque pulsation in fault conditions, an optimal current control strategy that aims to achieve a ripple-free electromagnetic torque under both healthy and faulted conditions with minimum copper loss can be employed. For a healthy m-phase machine of this type with a surface-mounted PM rotor, the reluctance torque is negligible. The electromagnetic torque, T_e, of the healthy machine can be expressed as

$$T_e = \sum_{j=1}^{m}(K_j(\theta)i_j(\theta)) \tag{6.1}$$

where $i_j(\theta)$ is the instantaneous current in phase j at a given rotor position of $\theta.K_j(\theta) = p(d\psi_{mj}/d\theta)$, p is the number of pole-pairs of the machine, and $\psi_{mj}(\theta)$ is the flux linkage of phase j due to the rotor PM field. If the machine develops an open-circuit or short-circuit fault in phase k, the resultant electromagnetic torque will be affected and given by

$$T_e = \sum_{j=1,\neq k}^{m}(K_j(\theta)i_j(\theta)) + T_f(\theta) \tag{6.2}$$

where $T_f(\theta)$ is the torque produced in the faulted phase and can be expressed by

$$T_f(\theta) = \begin{cases} 0 & \text{for an open circuit fault} \\ K_k(\theta)i_k(\theta) & \text{for a short circuit fault} \end{cases} \tag{6.3}$$

where $i_k(\theta)$ is the short-circuit current in the faulted phase and can be evaluated by

$$Ri_k(\theta) + \omega L\frac{di_k}{d\theta} = \omega\frac{d\psi_{mk}}{d\theta} \tag{6.4}$$

where R and L are phase resistance and phase inductance of the machine and ω is the electrical angular frequency. To produce the demand output torque T_e, the required current waveforms in the healthy phases can be obtained by minimizing the cost function F:

$$F = \sum_{j=1,\neq k}^{m}(i_j(\theta))^2 \tag{6.5}$$

subject to the output torque constraint given in (6.2). Since the cost function is proportional to the copper loss in the healthy phases, the resultant currents will minimize the copper loss while satisfying the required torque demand. The solution to the optimization problem can be derived by minimizing the augmented objective function $F(i_1,\ldots, i_j,\ldots, i_m, \lambda)$ with a Lagrange multiplier λ and $j \neq k$.

$$F = \sum_{j=1,\neq k}^{m}(i_j(\theta))^2 + \lambda\left\{\sum_{j=1,\neq k}^{m}(K_j(\theta)i_j(\theta)) - T_e + T_f(\theta)\right\} \tag{6.6}$$

Thus, the optimal currents in the healthy phases to produce the required output torque T_e with a minimum copper loss must satisfy

$$\frac{\partial F}{\partial i_j} = 0 \text{ and } \frac{\partial F}{\partial \lambda} = 0, j = 1, 2, \ldots, m, j \neq k \tag{6.7}$$

The instantaneous currents in the healthy phases can be derived from (6.6) and (6.7) and are given by

$$i_j(\theta) = \frac{K_j(\theta)[T_e - T_f(\theta)]}{\sum_{j=1}^{m} K_j^2(\theta)}, j = 1, 2, \ldots, m, j \neq k \tag{6.8}$$

Once the waveforms of the phase PM flux linkages and phase inductance are determined by either experimental measurements or FE analysis, the healthy phase currents at a given rotor position can be obtained from (6.8). These currents form the demand currents for the healthy phases, which are tracked by the independent current controllers, one for each phase, via their H-bridge converters.

In practical implementations, the maximum current in each phase is typically limited to $\pm I_{max}$. Depending on the demanded output torque T_e and the fault condition, the instantaneous current demands of the phases could exceed the current limit. In this case, the process of determining the optimal instantaneous currents may take several iterations. If the instantaneous current demand of phase l exceeds the current limit, it is clamped to

$$i_l = \pm I_{max} \tag{6.9}$$

The torque contribution from this phase becomes

$$T_l(\theta) = \pm I_{max} K_l(\theta) \tag{6.10}$$

where "+" and "−" signs correspond to positive and negative phase currents, respectively. The demand currents in the remaining healthy phases can be treated as a separate subset and determined by the same optimization technique to satisfy the new torque demand $(T_e + T_f(\theta) - T_l(\theta))$. The new current demands are given by

$$i_j(\theta) = \frac{K_j(\theta)[T_e - T_f(\theta) - T_l(\theta)]}{\sum_{j=1}^{m} K_j^2(\theta)}, j = 1, 2, \ldots, m, j \neq k \neq l \tag{6.11}$$

The new current demands should be checked to ensure that none of them exceeds the current limit. If the current demand for one or more phases from the new subset exceeds the limit, the above process is repeated until all the current demands satisfy the current limit. When there are more than one phase in which the current demands reach the limit, (6.10) should be the sum of the torque contributions from these phases and the index j in (6.11) should exclude these phases.

Cogging torque is usually present in a PM machine. In some applications, such as electric power steering and precision servo control, the presence of the cogging torque is undesirable. This cogging torque can be compensated by the electromagnetic torque via the same optimal current control strategy. In this case, the cogging torque $T_{cg}(\theta)$, predetermined by measurements or FE analysis, can be added in to the demand torque in (6.12):

$$T_e = \sum_{j=1,\neq k}^{m} (K_j(\theta)i_j(\theta)) + T_f(\theta) + T_{cg}(\theta) \tag{6.12}$$

The resultant optimal phase currents can be similarly derived and given by

$$i_j(\theta) = \frac{K_j(\theta)[T_e - T_f(\theta) - T_{cg}(\theta)]}{\sum_{j=1}^{m} K_j^2(\theta)}, j = 1, 2, \ldots, m, j \neq k \tag{6.13}$$

This concept can be further extended to compensate the other types of torque ripple if their variations with rotor position are known prior. However, it is important to point out that the effective compensation of the torque ripple is dependent on the current loop control bandwidth. In most applications, the torque ripple and torque pulsation due to a fault are more detrimental at low-speed operation. Consequently, the compensation can be quite effective even with a limited current control bandwidth.

By minimizing the copper loss, the current in each phase tends to be in phase with the phase back-emf. At high speed, therefore, the converter output voltage may reach the DC-link voltage of the H-bridge. This implies the optimal current control strategy is only applicable when the machine operates in healthy conditions below its base speed, ω_b, which is defined as the maximum speed at which the drive system can produce rated smooth torque without flux-weakening. In fault conditions, the current waveforms produced from the optimal control strategy are more distorted and yield higher operating voltages. Hence, the effective operation with the optimal current control in fault conditions will be below the base speed.

To extend the optimal current control operation above the base speed, the cost function of the optimization can be modified to that given in (6.14)

$$F = \sum_{j=1,\neq k}^{m} (Li_j(\theta) + w\psi_{mj})^2 \tag{6.14}$$

where w is a weighting factor used to adjust the flux-weakening effect at high speed. At speeds below the base speed, the weighting factor is set to zero. Therefore, the cost function becomes $L^2 \sum i_j^2$, which is equivalent to minimizing the copper loss $R^2 \sum i_j^2$. At speeds above the base speed, w is a non-zero positive number, which controls the degree of the flux-weakening effect and varies with the speed ω and torque demand T_e in the following form:

$$w = [(\omega - \omega_b)/\omega] \times \frac{T_e}{T_{rated}} \tag{6.15}$$

where T_{rated} is the rated torque of the machine. The increase in the weight factor on the PM flux linkage of each phase drives the optimal current profiles to advance the phase angles with respect to the phase back-emfs, resulting in lower phase voltages and, hence, flux-weakening effect.

In addition to open- or short-circuit fault in one phase, interturn short circuit that involves a few turns in a phase may also occur. In this case, a typical mitigation action is to apply TSC to the faulted phase via the H-bridge converter. This will reduce the interturn short-circuit current, and the resulting fault behavior is similar to that of a short-circuit fault in one phase. Hence, the optimal current control is also applicable after the mitigation action for the interturn fault.

6.3 Simulation Validation on the Optimal Torque Control Strategy

For the 5-phase fault-tolerant FSCW PM machine under consideration, the measured and predicted phase back-emf and cogging torque waveforms are shown in Fig. 6.2, from which the fundamental back-emf constant is obtained and given in Table 6.1 together with the phase resistance and inductance.

Each phase is fed by an independently controlled H-bridge converter with a DC-link voltage at 45 V. The current control bandwidth is set to 800 Hz, and the

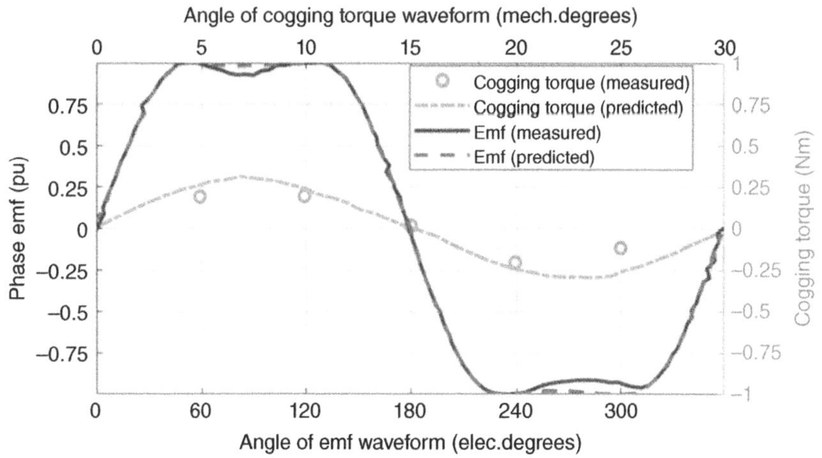

Figure 6.2 Measured and predicted waveforms of phase back-emf and cogging torque.

Table 6.1 Parameters of 5-phase fault-tolerant PM machine.

Parameters	Values
Emf constant (V/rad, fundamental)	0.104
Phase resistance (mΩ)	31.61
Phase inductance (mH)	0.155

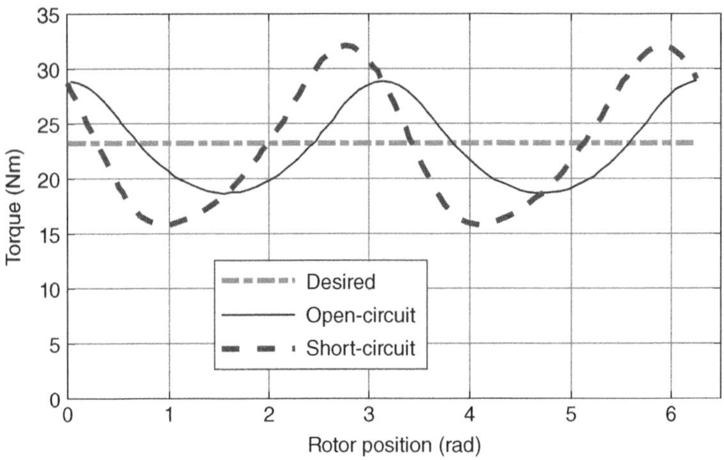

Figure 6.3 Output torque under open-circuit and short-circuit conditions with conventional control strategy.

converter is assumed to be ideal. Figure 6.3 shows the torque waveform when a conventional control is employed to track a set of sinusoidal current demands to produce the rated torque of 23.2 Nm at the rotor speed of 100 rad/s under one phase open-circuit or short-circuit condition.

6.3.1 Optimal Torque Control in the Constant Torque Region

It is seen that due to the emf harmonics and fault conditions, a torque ripple of 43.3% and 73.8% is present under the open-circuit or the short-circuit conditions, respectively. To eliminate the excessive torque ripple, the proposed optimal current control strategy is employed on the 5-phase prototype drive. By way of example, a test is performed under a short-circuit condition in the constant torque region at a rotor speed of 100 rad/s, the weighting factor w is set to 0. Figure 6.4 shows the current, voltage, emf, and torque waveforms, for the same torque demand and rotor

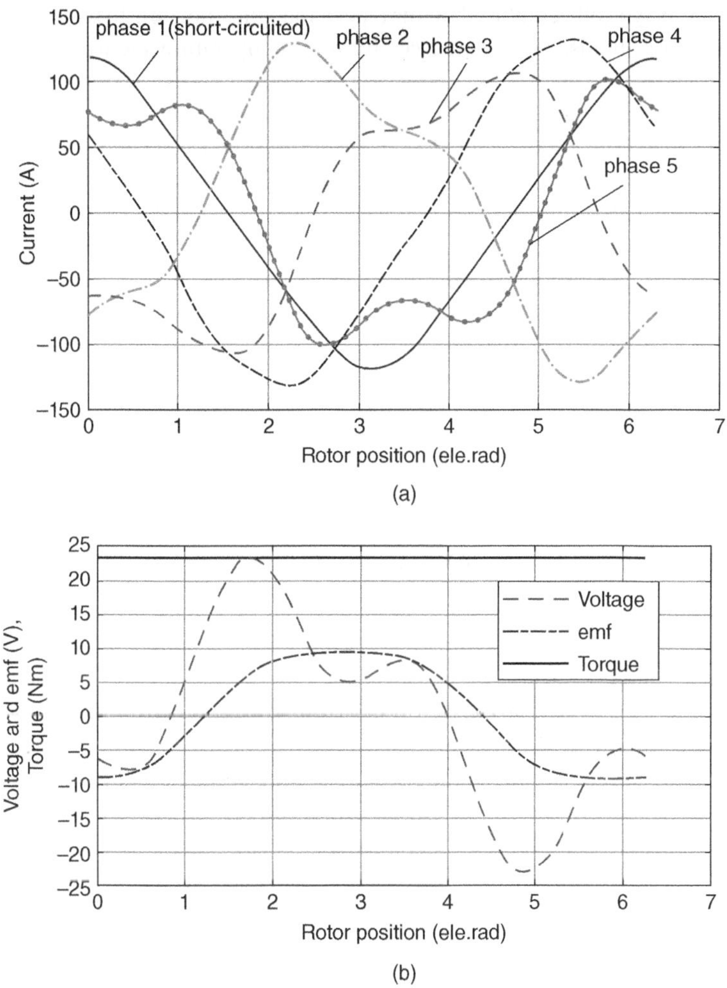

Figure 6.4 Current, voltage, back-emf, and torque waveforms under a short-circuit condition with the optimal control strategy at 100 rad/s: (a) phase currents and (b) voltage and back-emf in healthy phase 2 and resultant torque.

speed as in Fig. 6.3. It is evident that smooth torque is produced at the expense of increased current harmonics and *di/dt*, compared with sinewave phase currents. The required peak phase voltages to realize these current trajectories also appear to be higher than those for normal healthy operation.

The high-order harmonics in the phase currents that result from the optimal control strategy will increase eddy current loss in the rotor magnets [4], particularly at high-speed operations. Axial and circumferential segmentation of the rotor magnets is typically employed in this type of machine, not only to reduce the eddy current loss but also to prevent excessive rotor temperature and irreversible demagnetization.

Since the simulation is performed with ideal converters, pulse width modulation (PWM) switching effect is not seen in the current and voltage waveforms. Further under the current feedback control, the phase currents track very well their demands which results from the optimal current profiles at the relatively low operating speed. When the weighting factor w is set to zero, the optimal solution minimizes the copper loss, which leads to the phase currents being in-phase with their respective emfs. This will be evident from Fig. 6.4 by examining the current waveform in healthy phase 2 and its corresponding emf. Because of the high peak phase voltage seen in Fig. 6.4(b) required to track the optimal currents, for a fixed DC-link voltage, the maximum speed at which the rated ripple-free torque can be produced under a fault condition will be lower than that under healthy operating conditions.

6.3.2 Optimal Torque Control in the Constant Power Region

When the rotor speed increases, the available control voltage to realize the required current trajectories becomes progressively lower. Therefore, to achieve a smooth torque output, a degree of flux-weakening proportional to the speed and torque demand is necessary. Again, by way of example, a test is performed under a short-circuit condition in the constant power region at a rotor speed of 600 rad/s. The current, voltage, emf, and torque waveforms are shown in Fig. 6.5. As can be seen, even under the short-circuit condition, a ripple-free torque of 11.6 Nm is achieved in the flux-weakening mode. The flux-weakening effect is produced by increasing the weighting in the cost function given in (6.14). Consequently, the optimal current trajectories tend to suppress the flux linkage through phase advance, which is evident from the current and emf waveforms in Fig. 6.5 for healthy phase 2.

The voltage in phase 2 has reached the DC-link voltage over some periods in both the positive and negative polarities. Indeed, this illustrates that the flux-weakening effect is appropriate at the given speed and torque when the voltage reaches its limit while the currents are still under control.

It is also of interest to see that optimal currents under the flux-weakening operation are closer to sinewaves. This is because as the currents are more weighted to reduce the phase flux linkages, the effect of torque pulsation due to the fault on the currents in the healthy phases is also reduced.

It should be noted that the quality of the optimal current control relies on high-fidelity tracking of the current demands. In this regard, various advanced control techniques such as resonant control and repetitive control may be

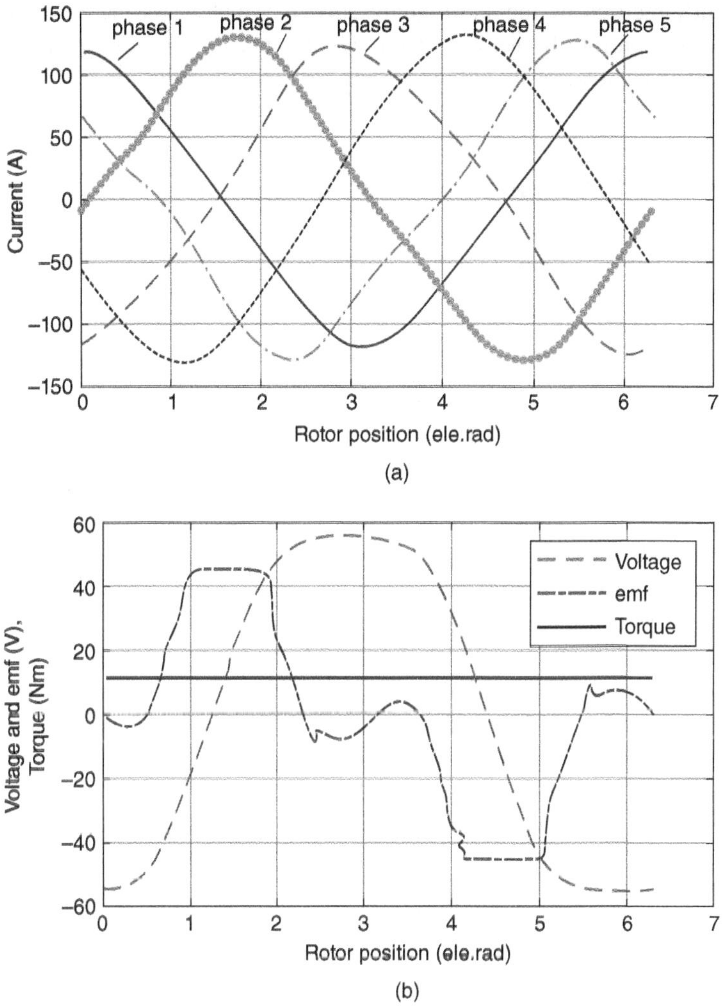

Figure 6.5 Current, voltage, emf, and torque waveforms under a short-circuit condition with the optimal control strategy at 600 rad/s: (a) phase current and (b) voltage and back-emf in phase 2 and resultant torque.

employed. Some of these techniques are described in [12]. Further, the flux-weakening effect will also be dependent on fault conditions. With an open-circuit fault, the tooth and back-iron of the fault phase are magnetically less saturated, and so are in the two neighboring phase regions. Hence, more phase advance effect is required to suppress the PM flux-linkage in these phases. In contrast, under a short-circuit fault, the PM flux linkage of the fault phase is significantly reduced by the short-circuit current, which eases the need for flux-weakening control. To achieve good performance in various operating conditions, the weight factor should ideally be adaptive, as described in [5].

6.4 Turn Fault Mitigation by Current Injection

For the fault-tolerant multiple 3-phase drive under study, it has been shown in Chapter 2 that the turn fault current after application of TSC is still much higher than the rated. The magnitude of the fault current is also dependent on fault location – being the highest in motoring mode if the fault occurs in coil B2, while in generating mode it is the highest in coil A1. To reduce the fault current further, the flux linkage of the fault turns should be minimized close to zero if possible. For the multiple 3-phase machine, it has more degrees of freedom than a conventional 3-phase machine in controlling the currents, which can be used to influence the flux linkage. Thus, instead of TSC application, specific currents can be injected to the fault set to minimize the flux linkage of the fault turns while the remaining healthy sets continue their postfault operation.

Due to the mutual magnetic coupling between the healthy sets and fault set, it is not possible to nullify all phase flux linkages of the fault set. However, only the flux linkage of the fault phase where a turn fault occurs needs to be minimized. With the fault detection techniques described in [13–15], it is possible to identify the fault phase but the subcoil in which a turn fault occurs may still be unknown if a phase winding contains more than one coil. For a machine with distributed windings, coils of the same phase and polarity are usually placed within a 60° phase belt. For example, the machine under consideration has two coils connected in series and they are displaced by 30° with respect to each other as shown in Fig. 6.6. The phase axis is displaced by 15° with respect to the two subcoils. Thus, if the total flux linkage of the fault phase is limited very close to zero, the flux linkage of the two subcoils will also be very low, leading to a small fault current.

To determine the injected currents, the flux linkage of the fault phase is first analyzed. According to the analysis in Chapter 4, the flux linkage of the fault phase is contributed by three sources, namely, the rotor PM field, the currents in the healthy 3-phase sets, and the currents in the fault set. The flux linkage of the fault

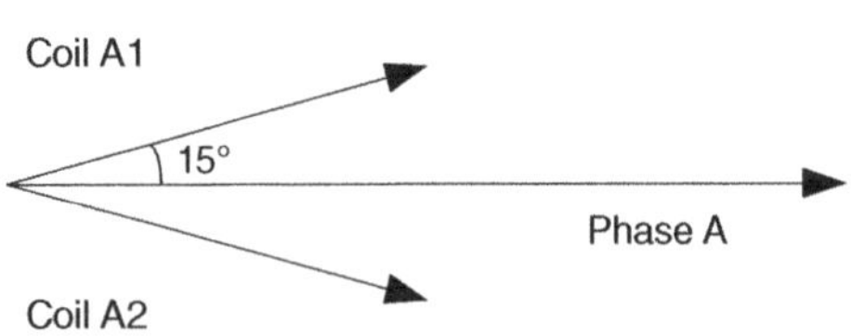

Figure 6.6 Phasors of the subcoils in phase A.

phase due to the three sources will be analyzed separately, assuming that they are independent from each other. While this condition is generally not true for an IPM machine in saturated condition, it will be shown that the injected currents render the fault 3-phase region unsaturated and the principle of superposition is valid.

6.4.1 Flux Linkage by Rotor PM Field

Assuming a turn fault occurs in a phase winding of the ABC set, the flux linkage in each phase produced by the rotor PM field will not be affected by the fault and can be evaluated by (6.16) when high-order harmonics are neglected.

$$\begin{aligned}\psi_{mA} &= \psi_m \cos(p\omega t)\\ \psi_{mB} &= \psi_m \cos(p\omega t - 120^\circ)\\ \psi_{mC} &= \psi_m \cos(p\omega t + 120^\circ)\end{aligned} \tag{6.16}$$

where ψ_m is the magnitude of the PM flux linkage of each phase, ω denotes the mechanical speed of the machine, and p is the number of pole pairs.

6.4.2 Flux Linkage by the Currents in Healthy Sets

Unlike the multiphase fault-tolerant PM machines in [7, 16] where mutual magnetic coupling between phases is negligible, the three 3-phase winding sets of the PMA-SynRM are mutually coupled via the magneto-motive force (MMF) offset component as explained in Chapters 2 and 4 [17]. Thus, the MMF distribution produced by the currents in the healthy DEF and GHI winding sets can be evaluated by multiplying the winding function of their coils with relevant phase currents, as detailed in Chapter 4. For example, Fig. 6.7 shows the MMF distribution at a given

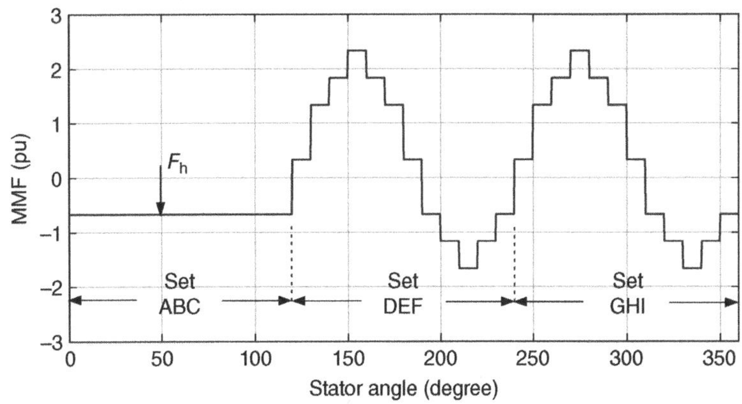

Figure 6.7 MMF produced by the healthy set currents.

instant of time over the three 3-phase sets due to the currents in the two healthy sets. It is seen that the currents in the two healthy 3-phase winding sets produce an MMF offset component in the faulted region with varying magnitude and phase angle. This MMF offset component produces a flux linkage component in the fault turns.

The MMF offset component F_h shown in Fig. 6.7 over (0°–120°) region can be estimated by (6.17). N is the number of turns per coil as stated in Chapter 4.

$$F_\mathrm{h} = -\frac{N}{3}(i_\mathrm{D} + i_\mathrm{E} - i_\mathrm{F}) - \frac{N}{3}(i_\mathrm{G} + i_\mathrm{H} - i_\mathrm{I}) \tag{6.17}$$

where i_D, i_E, ..., i_I are the phase currents which can be expressed in terms of the dq-axis currents ($i_{d\mathrm{h}}$, $i_{q\mathrm{h}}$) in (6.18).

$$\begin{aligned}
i_\mathrm{D} &= i_\mathrm{G} = i_{d\mathrm{h}}\cos(p\omega t) - i_{q\mathrm{h}}\sin(p\omega t)\\
i_\mathrm{E} &= i_\mathrm{H} = i_{d\mathrm{h}}\cos(p\omega t - 120^\circ) - i_{q\mathrm{h}}\sin(p\omega t - 120^\circ)\\
i_\mathrm{F} &= i_I = i_{d\mathrm{h}}\cos(p\omega t + 120^\circ) - i_{q\mathrm{h}}\sin(p\omega t + 120^\circ)
\end{aligned} \tag{6.18}$$

Substituting (6.18) into (6.17) results in (6.19):

$$F_\mathrm{h} = -\frac{4N}{3}(i_{d\mathrm{h}}\cos(p\omega t - 60^\circ) - i_{q\mathrm{h}}\sin(p\omega t - 60^\circ)) \tag{6.19}$$

It can be seen from (6.19) that when the dq-axis currents in the healthy 3-phase sets are kept constant under field-oriented control, the MMF offset component is time varying but not space varying over the ABC winding region. Hence, it will give rise to flux linkages of equal magnitude in phases A, B, and C of the fault set. They are given in (6.20) where the opposite polarity in phase C is due to the reversed turn function relative to those of phases A and B as has been explained in Chapter 4. k is a proportional constant associated with the airgap permeance. It is also of importance to see that these flux linkages are dependent on the dq-axis currents in the healthy phases.

$$\begin{aligned}
\psi_\mathrm{FhA} &= kF_\mathrm{h}\\
\psi_\mathrm{FhB} &= kF_\mathrm{h}\\
\psi_\mathrm{FhC} &= -kF_\mathrm{h}
\end{aligned} \tag{6.20}$$

Since the coils are full-pitched and F_h is space-invariant over the ABC winding region, the permeance k can be considered independent of rotor position when high-order permeance components are neglected [18]. Its value can be obtained via FE computation by injecting a specific MMF offset component over the fault set region. It is also possible to estimate the constant k through magnetic circuit analysis. However, for the rotor with three magnet layers, FE computation is more practical since the model is already available.

6.4.3 Flux Linkage by the Currents in Fault Set

Assuming specific currents (i_{df}, i_{qf}) are injected in the fault set to counteract the flux linkages caused by the PM field and healthy set currents in the fault phase, their flux linkages in the fault 3-phase set can also be evaluated by analyzing the MMF distribution.

The MMF caused by the injected currents (i_{df}, i_{qf}) in ABC set is plotted in Fig. 6.8 [17]. As analyzed in Chapter 4, the MMF contains a symmetrical AC component and an MMF offset component in the ABC winding region while it contributes to an MMF offset component over the region of the healthy sets. The MMF in the fault 3-phase set region can be decomposed to an AC component and an offset component as shown in Fig. 6.9.

The AC component exhibits the same form as that in healthy conditions when all three sets of windings are excited with (i_{df}, i_{qf}). Hence, the flux linkages induced by the MMF AC component can be calculated by the dq-axis inductance (L_d, L_q) in (6.21), which are further transformed to the phase quantities in (6.22).

$$\begin{aligned} \psi_{\mathrm{ACf}d} &= L_d i_{\mathrm{df}} \\ \psi_{\mathrm{ACf}q} &= L_q i_{\mathrm{qf}} \end{aligned} \tag{6.21}$$

$$\begin{aligned} \psi_{\mathrm{ACfA}} &= L_d i_{\mathrm{df}} \cos(p\omega t) - L_q i_{\mathrm{qf}} \sin(p\omega t) \\ \psi_{\mathrm{ACfB}} &= L_d i_{\mathrm{df}} \cos(p\omega t - 120^\circ) - L_q i_{\mathrm{qf}} \sin(p\omega t - 120^\circ) \\ \psi_{\mathrm{ACfC}} &= L_d i_{\mathrm{df}} \cos(p\omega t + 120^\circ) - L_q i_{\mathrm{qf}} \sin(p\omega t + 120^\circ) \end{aligned} \tag{6.22}$$

Further, the magnitude of the MMF offset component F_{f} can be estimated according to the turn functions and phase currents which are given in (6.23).

$$F_{\mathrm{f}} = \frac{2N}{3}(i_{\mathrm{A}} + i_{\mathrm{B}} - i_{\mathrm{C}}) = \frac{4N}{3}(i_{\mathrm{df}} \cos(p\omega t - 60^\circ) - i_{\mathrm{qf}} \sin(p\omega t - 60^\circ)) \tag{6.23}$$

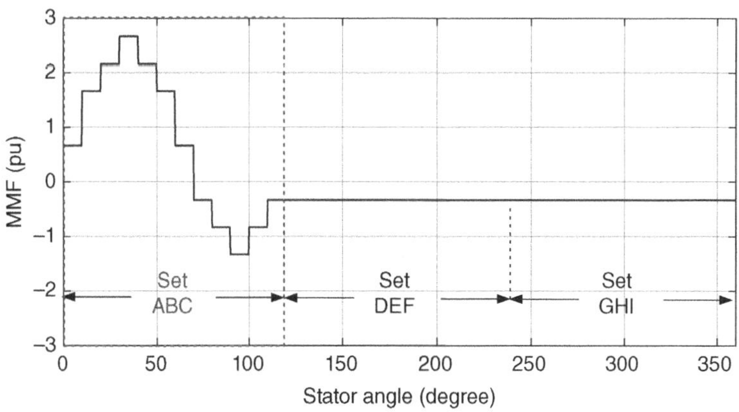

Figure 6.8 MMF produced by the fault set currents.

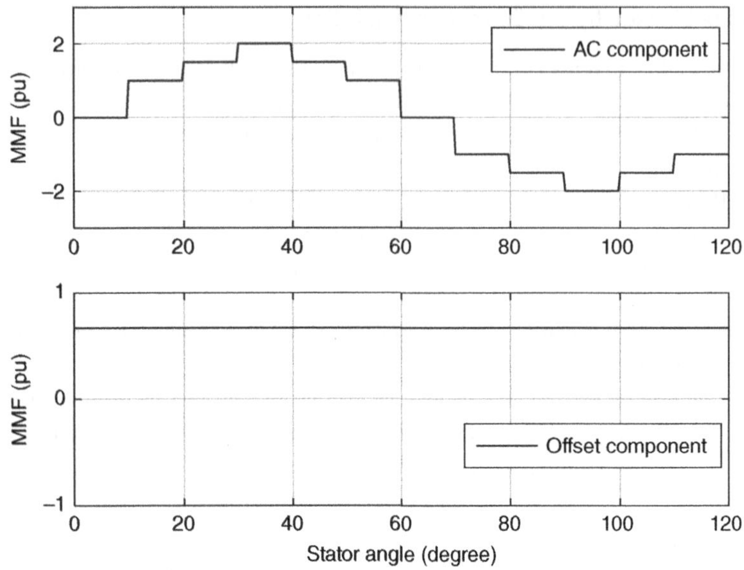

Figure 6.9 MMF AC and offset components in the fault set.

The flux linkages due to the MMF offset component can be derived similarly and they are expressed in (6.24).

$$\begin{aligned}\psi_{\mathrm{FfA}} &= kF_{\mathrm{f}}\\ \psi_{\mathrm{FfB}} &= kF_{\mathrm{f}}\\ \psi_{\mathrm{FfC}} &= -kF_{\mathrm{f}}\end{aligned} \tag{6.24}$$

Hence, the flux linkages in the fault ABC set produced by its own currents are obtained.

6.4.4 Total Flux Linkage in Fault Phase

From the foregoing analysis, the flux linkages of the fault phase can be obtained by summing equations (6.16), (6.20), (6.22), and (6.24). Without loss of generality, the fault turns are assumed in coil A1 of phase A. Thus, the flux linkage of the fault phase A is expressed in (6.25).

$$\begin{aligned}\psi_{\mathrm{A}} &= (\psi_m + L_d i_{\mathrm{df}})\cos p\omega t - L_q i_{\mathrm{qf}}\sin p\omega t\\ &+\frac{4k}{3}(i_{\mathrm{df}} - i_{\mathrm{dh}})\cos(p\omega t - 60^\circ) - \frac{4k}{3}(i_{\mathrm{qf}} - i_{\mathrm{qh}})\sin(p\omega t - 60^\circ)\Big)\end{aligned} \tag{6.25}$$

To minimize the turn fault current, the flux linkage of the fault phase should ideally be controlled close to zero. This requires that both the terms associated with cos $p\omega t$ and sin $p\omega t$ equal to zero. The above equation can be decomposed into two equations given in (6.26), associated with cos $p\omega t$ and sin $p\omega t$, respectively. Then, the cos $p\omega t$ and sin $p\omega t$ components can be omitted and the equations can be further written as in (6.27).

$$\begin{aligned}
&(\psi_m + L_d i_{df}) \cos p\omega t + \frac{4k}{3}(i_{df} - i_{dh}) \cos p\omega t \cos 60^\circ \\
&\quad + \frac{4k}{3}(i_{qf} - i_{qh}) \cos p\omega t \sin 60^\circ = 0 \\
&-L_q i_{qf} \sin p\omega t + \frac{4k}{3}(i_{df} - i_{dh}) \sin p\omega t \sin 60^\circ \\
&\quad - \frac{4k}{3}(i_{qf} - i_{qh}) \sin p\omega t \cos 60^\circ = 0
\end{aligned} \tag{6.26}$$

$$\begin{aligned}
&(\psi_m + L_d i_{df}) + \frac{2k}{3}(i_{df} - i_{dh}) + \frac{2k}{\sqrt{3}}(i_{qf} - i_{qh}) = 0 \\
&-L_q i_{qf} + \frac{2k}{\sqrt{3}}(i_{df} - i_{dh}) - \frac{2k}{3}(i_{qf} - i_{qh}) = 0
\end{aligned} \tag{6.27}$$

Hence, the currents (i_{df}, i_{qf}) to be injected to the fault set can be determined as functions of the currents (i_{dh}, i_{qh}) in the healthy sets based on the two equations. It is noticed that the injected currents (i_{df}, i_{qf}) are independent of the time and rotor angle, which is very beneficial for current control implementation. If the currents (i_{df}, i_{qf}) are injected to the fault 3-phase set, the flux linkages of the fault phase will be close to zero. Consequently, the flux linkage of the fault turns and the resultant fault current will be reduced to a very low level. Thus, the impact of the turn fault, namely, excessive fault current and resultant local hotspot, can be alleviated. It should be noted that the derivation of the injected currents from (6.26) does not require specific knowledge of the fault location in which subcoils of the fault phase and the number of fault turns. If the turn fault takes place in phases B and C, the flux linkages of phases B and C should be derived similarly as in (6.25) and they should be forced to be zero to minimize the fault current. Hence, the technique is applicable to various turn fault scenarios. The proposed current injection method will be examined by FE simulations and experimental tests.

6.5 Simulation Validation on the Current Injection Technique

The effectiveness of the proposed current injection method is evaluated by FE simulations. The influences of fault location in different phases and subcoils, slot position of faulted turns, and the number of fault turns on the turn fault current are assessed with the proposed current injection technique.

In the FE simulations, the machine is fed by ideal current sources. The fault turns in a coil are separated from the other healthy turns in the ABC phases as explained in Chapter 4 [19]. The turn fault is emulated by short circuiting the fault turns directly with zero external impedance. In contrast to the conventional TSC mitigation action in which the 3-phase terminals of the fault set are short circuited, the fault set windings are fed by ideal current sources which are determined from (6.26) under the proposed current injection mitigation method. The simulations, therefore, assume ideal 3-phase converters with infinite current control bandwidth.

6.5.1 Effectiveness of Current Injection

The simulation scheme is illustrated in Fig. 6.10. First, the turn fault under current injection control is examined at the base speed of 4000 rpm when the healthy sets DEF and GHI are excited with the rated currents (120 A, 51°) in motoring mode. A single turn fault occurring in all six different coil locations of set ABC has been examined. In each case, the injected currents are calculated by (6.26) for the fault 3-phase set. The injected currents in the fault set along with the resultant rms turn fault currents are listed in Table 6.2 and the waveforms of the turn fault currents are shown in Fig. 6.11 for the six scenarios. It is seen that the injected currents are different if the turn fault occurs in different phases and their magnitudes are lower than the rated value. This is because the injected currents aim to reduce the flux linkage of the fault phase only where the fault turns are located. The maximum rms value of the resultant turn fault current under the injection technique is 1.72 pu when the fault occurs in coil A2. In contrast, under TSC application, the maximum rms turn fault current is 3.2 pu when the fault occurs in coil B2 in motoring mode. Thus, the maximum fault current has been reduced by 45% with the proposed current injection technique. The 1.72 pu fault current only flows in one fault turn of coil A2, the incurred copper loss will not cause localized high temperature as the thermal loading in the rest A2 coil is much below the rated.

For the purpose of comparison, the total copper loss in the fault 3-phase set and the copper loss in the fault turn are listed in Table 6.3 under the conventional TSC mitigation and the proposed current injection method. It can be found that the turn fault will be safe under both mitigation measures if it takes place in phases A and C. However, if the turn fault occurs in coil B2 and is mitigated by TSC, the loss in the fault turn is about 10 times the thermal load under rated condition. This will cause irreversible damage to the whole machine drive. In comparison, by employing the proposed current injection, the loss in the fault turn has been reduced to about 2.3 times the rated thermal load. Though the total copper loss of the fault 3-phase set is higher than that of TSC, it is still within the rated thermal

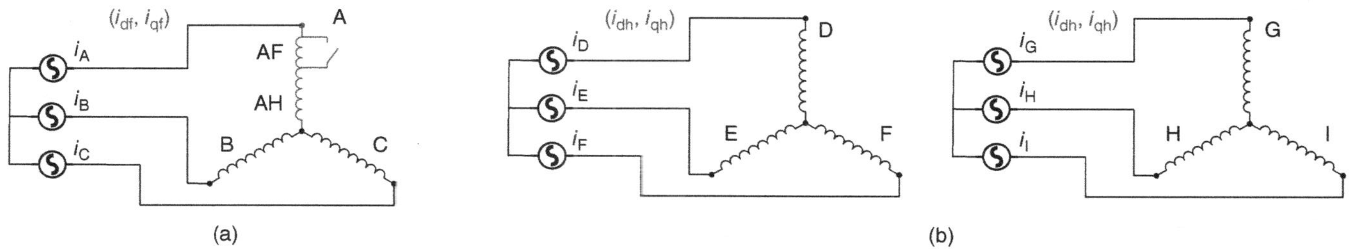

Figure 6.10 Illustration of current injection used in FE simulation of turn fault: (a) current injection in set ABC and (b) normal excitation in sets DEF and GHI.

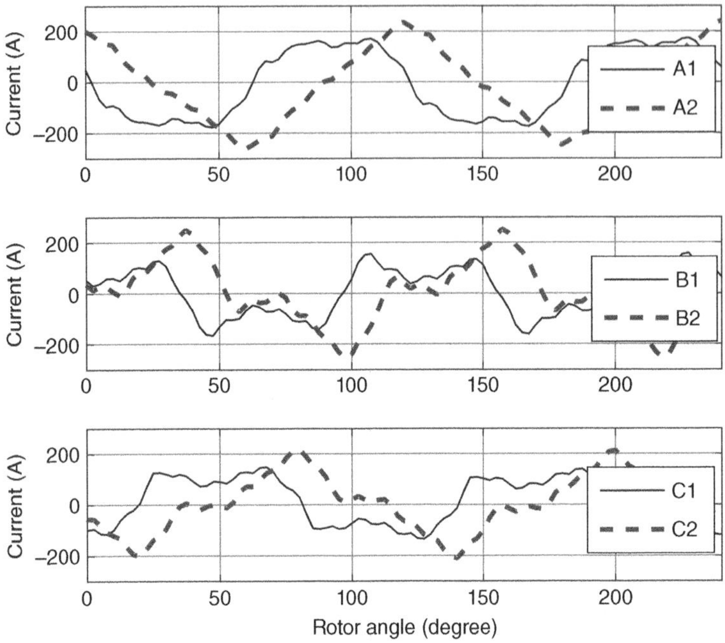

Figure 6.11 Fault current waveforms with current inject technique when a single turn fault occurs in six coil locations.

Table 6.2 Injected currents and resultant turn fault currents in six different coils.

Turn fault coil location	Injected *dq*-axis currents	rms value of turn fault current
A1	(−47.7 A, 24.8 A)	132 A, 1.58 pu
A2	(−47.7 A, 24.8 A)	144 A, 1.72 pu
B1	(−119 A, 19 A)	98 A, 1.18 pu
B2	(−119 A, 19 A)	127 A, 1.52 pu
C1	(−85 A, 19.7 A)	93 A, 1.11 pu
C2	(−85 A, 19.7 A)	108 A, 1.28 pu

load compared with healthy operation. Thus, it will not cause any further damage to the machine drive.

Previously, the saturation effect is not considered in the flux linkage estimation. This assumption is largely true because the injected current to the fault set has a large negative i_d and small i_q as shown in Table 6.2. Therefore, the fault 3-phase

Table 6.3 Copper loss comparison between current injection, denoted as CJ, and TSC.

Turn fault coil location	Total loss in the fault set (CJ)	Total loss in the fault set (TSC)	Loss in the fault turn (CJ)	Loss in the fault turn (TSC)
A1	115 W	192 W	28 W	27 W
A2	120 W	183 W	33 W	18.7 W
B1	451 W	175 W	15.4 W	10.3 W
B2	461 W	278 W	25.8 W	113 W
C1	242 W	199 W	14 W	34 W
C2	247 W	175 W	18.7 W	10.3 W

set operates in equivalent deep field weakening, which implies that the main flux path associated with the fault winding set is unsaturated.

The main objective of the proposed method is to minimize the flux linkage of the fault turns by injecting specific current to the fault set windings. However, the fault turn flux linkage cannot be completely nullified mainly due to two reasons. First, only the fundamental component is considered while the high-order harmonics of the faulted turn flux linkage will also induce short-circuit current in the fault turns. Second, the phase angle of the fault turn flux linkage is dependent on in which coil the fault occurs. However, no practical technique is capable of detecting the fault location to date. Therefore, in the proposed method, it is assumed that the axis of the fault turns coincides with that of the faulted phase. This will ensure, on average, effective reduction of the flux linkage in the faulted turns. The current is injected to minimize the residual flux along the phase axis. Hence, if a fault occurs in coil A1, the injected current is 15° leading while it will be 15° lagging in case the fault is in coil A2. Similar is true for a turn fault in phases B and C. Thus, fundamental component of the fault turn flux linkage will not be completely nullified and fault current still circulates in the fault turns. Nonetheless, it is much lower than that of the conventional TSC method. Owing to the above two factors, the injected currents affect the fundamental and high-order harmonic flux linkages of the faulted turn in a different manner when the fault occurs in different coils, and the resultant fault current waveforms shown in Fig. 6.11 are different.

6.5.2 Residual Flux Linkage of the Fault Turns

It has been shown that the proposed current injection technique is more effective than TSC application in the reduction of the turn fault current. This can be further explained by comparing the residual flux linkages of the fault turn and fault phase under the two mitigation strategies. A single turn fault, denoted as B2f, is assumed in coil B2 and the healthy sets of windings are excited with the rated current at the base speed. The resultant residual flux linkages following the application of the two mitigation measures have been compared in Figs. 6.12 and 6.13. After the TSC application, the flux linkages of phases ABC are reduced simultaneously with their magnitudes being very close. In contrast, when the current injection is employed, only the flux linkage of the fault phase B is reduced to a very low level while the flux linkages of phases A and C are not much affected.

The peak-to-peak residual flux linkages in phase B and fault turn B2f are compared in Table 6.4. It is evident that under the current injection, both the flux linkages of phase B and fault turn B2f are much lower than those with the TSC application, and consequently the fault current with the proposed method is much smaller.

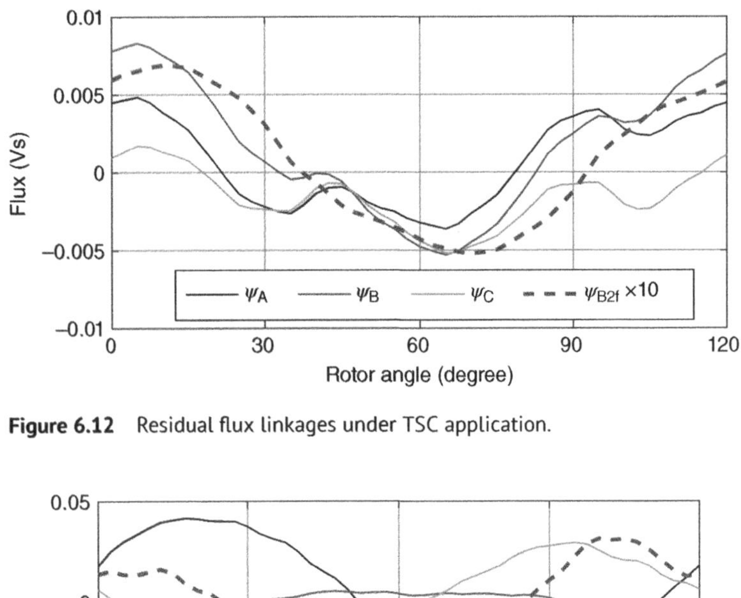

Figure 6.12 Residual flux linkages under TSC application.

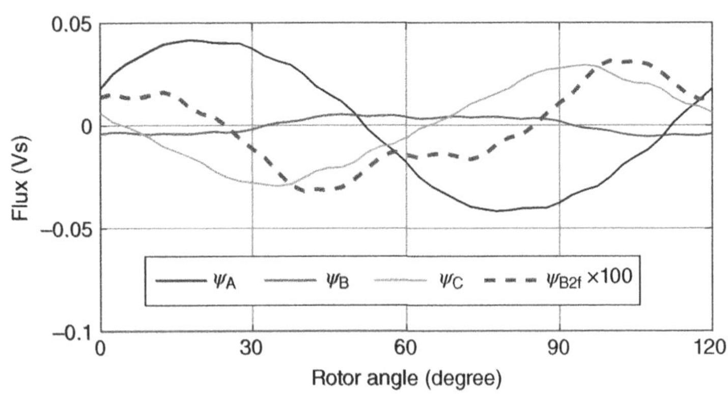

Figure 6.13 Residual flux linkages under current injection.

Table 6.4 Peak-to-peak value of residual flux linkage.

Residual flux linkage	TSC	Current injection
Fault phase B	1.36e-2 Vs	1.0e-3 Vs
Fault turn B2f	1.2e-3 Vs	0.63e-3 Vs

6.5.3 Influence of the Slot Position of Fault Turns

It is well known that the slot leakage inductance associated with faulted turns has a notable impact on the turn fault current. This leakage inductance is, however, dependent on the relative position of the faulted turns within a slot as shown in Fig. 6.14. Assuming a single-turn short circuit, the turn near the slot opening is denoted as "top turn" while that close to the stator yoke is "bottom turn." The analysis in [7, 20] shows that the turn fault current is highest if the fault occurs in the top turn while the fault current is much lower if the fault occurs in the bottom turn. An important cause of the difference is the slot leakage inductance. The top turn has a much lower slot leakage inductance than the bottom turn. Therefore, it is necessary to examine that if the proposed current injection technique is effective for arbitrary fault locations in a slot.

According to the proposed method, the injected currents derived from (6.26) are not dependent on the faulted turn location in a slot. Thus, the same currents are injected to the fault set windings irrespective of fault turn location. The effectiveness of the current injection when a single turn fault occurs in the top turn and in the bottom turn of coil B2 has been studied by FE simulation. For the two different fault locations, the same currents are injected to the fault 3-phase set while the healthy sets are fed with the rated currents at 4000 rpm. Fig. 6.15 compares the resultant turn fault currents when the fault occurs in the two different slot positions. It is shown that the fault turn at the top experiences a higher current (1.77 pu) than that at the bottom (1.21 pu) due to the screening effect of the healthy turns in the B2 coil with the injected current on the bottom

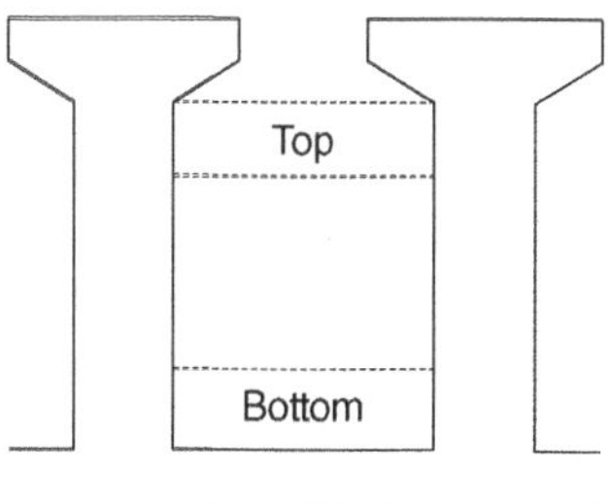

Figure 6.14 Illustration of the slot position of the fault turn.

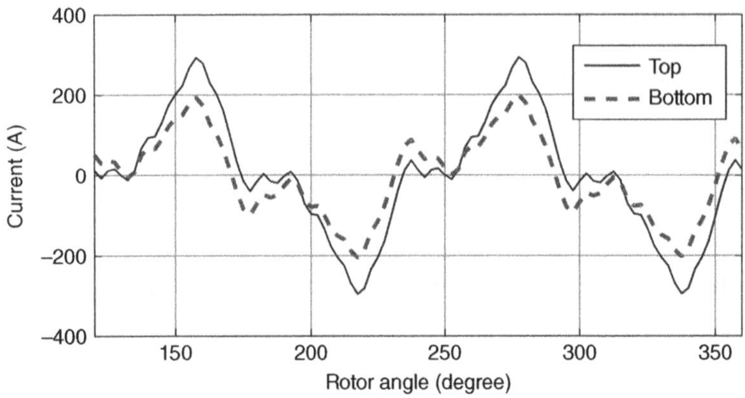

Figure 6.15 Turn fault currents when the fault occurs in two different slot positions.

turn [8] as well as relatively lower leakage inductance. Generally, the fault current when the injection technique is applied will be between the two extreme cases. However, the fault current in the top turn is still in the safe thermal range. Thus, the proposed mitigation strategy is effective in the worst-case scenario.

6.5.4 Influence of the Number of Faulted Turns

The fault current is also affected by the number of faulted turns. Generally, the turn fault current increases as the number of faulted turns decreases. This is because the inductance of the faulted turns is proportional to the square of the number of turns while the flux linkage of the faulted turns and the induced voltage are proportional to the number of turns. Therefore, the resultant short-circuit current which can be estimated as the ratio of the flux linkage to the inductance increases with the decrease in the number of faulted turns [1, 21].

For mitigation of a turn fault with the current injection, the flux linkage of the fault turns is reduced along with all turns of the fault phase regardless of the number of short-circuited turns. As a result, the real effect is expected to vary with the number of faulted turns. The influence of the number of faulted turns is examined by FE simulation when the number of faulted turns is set to 1, 2, 4, and 7 in coil B2 while the two healthy sets are fed by the rated currents at 4000 rpm. The same currents are injected to the fault set for the four cases of different numbers of faulted turns. The resultant turn fault currents are shown in Fig. 6.16. It is seen that the fault current decreases as the number of faulted turns increases. This result is consistent with other turn fault mitigation measures in that the fault current decreases with an increase in the number of faulted turns. Hence, the worst case is one-turn short circuit. Thus, the proposed method will be effective for different numbers

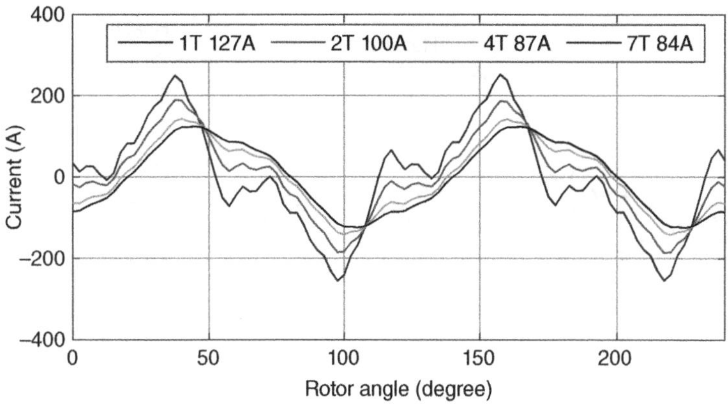

Figure 6.16 Turn fault currents with four different numbers of faulted turns.

of faulted turns if the fault current with single-turn short circuit is reduced to a thermally sustainable level.

6.5.5 Influence of Output Torque

As discussed in Chapter 2, when a turn fault is mitigated by TSC application, the torque of the triple redundant fault-tolerant machine will be reduced by ~1/3. For the purpose of comparison, the influence of the proposed current injection on the output torque has also been assessed when the two healthy sets are fed with the rated current and the machine operated at the rated speed. The torques under the healthy condition and turn fault conditions with different numbers of faulted turns when the two mitigation measures are applied are compared in Fig. 6.17. Note that these faulted turns are all very close to the slot opening. It is seen that the machine with different numbers of faulted turns exhibits almost the same torque waveforms as they coincide with each other and some are not visible.

Compared with TSC application, the output torque under the current injection method is ~10% higher. The torque is slightly boosted primarily because all the injected currents contain not only negative i_d with relatively large magnitude but also positive i_q as shown in Table 6.2. They will produce positive torque in the fault set region. Under the TSC method, only reactive currents flow in the fault set of windings without control, resulting in negative i_d and nearly zero i_q. The produced torque is negligible.

By employing the general model established in Chapter 4, the torque boosting mechanism can be explained by separating the torque contributions of each set region as shown in Fig. 6.18, where the operating conditions are the same as those in Fig. 6.17 except that one-turn fault is considered. It is seen that the healthy sets,

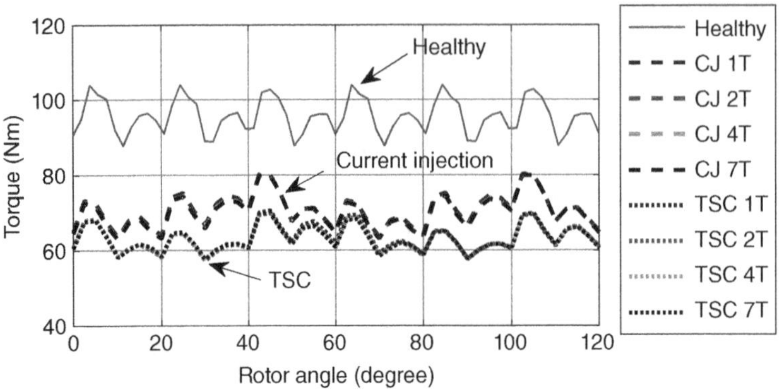

Figure 6.17 Output torques under different mitigation strategies considering different fault turns.

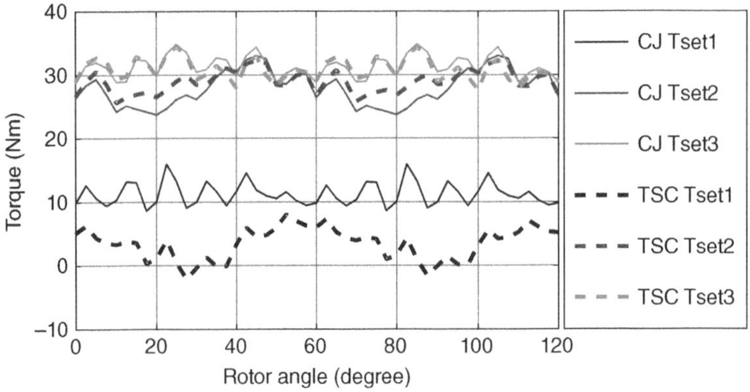

Figure 6.18 Torque contribution of different set regions under two mitigation strategies with turn fault.

namely, set 2(DEF) and set 3(GHI) produce similar torque under the two mitigation strategies. However, the fault set 1(ABC) under current injection contributes higher torque than that of TSC application. Hence, the current injection method for turn fault mitigation has an additional benefit of a few percent higher output torque.

If the machine operates in a generating mode, the healthy sets windings will be fed with negative i_d and negative i_q. The derived injected currents in the fault set will also contain negative i_d and negative i_q. Consequently, it will also produce higher torque (breaking torque) than that of the TSC method.

6.5.6 Effectiveness at High Speed

The effectiveness of the current injection method at high-speed operations is investigated. Generally, when the speed of a machine operation is sufficiently high, the main limiting factor of a turn fault current will be the inductance of the short-circuit loop. This condition is known as "inductance limited." Thus, when the inductance-limited condition is true, the fault current will not increase with speed. Hence, when the machine operates above the base speed, flux-weakening operation will be required and, consequently, the magnetic circuit of the machine will be less saturated with an increase in the magnitude of the negative *d*-axis current. This condition increases the inductance of the faulted turns and leads to a lower fault current. To further prove this analysis, one-turn fault in coil A1 has been simulated in FE at the maximum speed 19,200 rpm. The currents in the healthy sets of windings are (60 A, 70°) and the injected currents in the fault set windings are calculated as (i_{df} = −54.9 A, i_{qf} = 3.8 A). The simulated turn fault current is shown in Fig. 6.19. It is seen that the fault current is much lower than that under the rated operating condition given in Table 6.2. This is because all the three 3-phase sets operate in the deep field weakening region. As a result, the machine is less saturated and the inductance of the faulted turn increases. At the same time, the flux linkage estimation error for derivation of the injected current decreases. Both lead to a lower fault current. Nevertheless, the simulation confirms that the proposed method is very effective in reducing fault current at high speed.

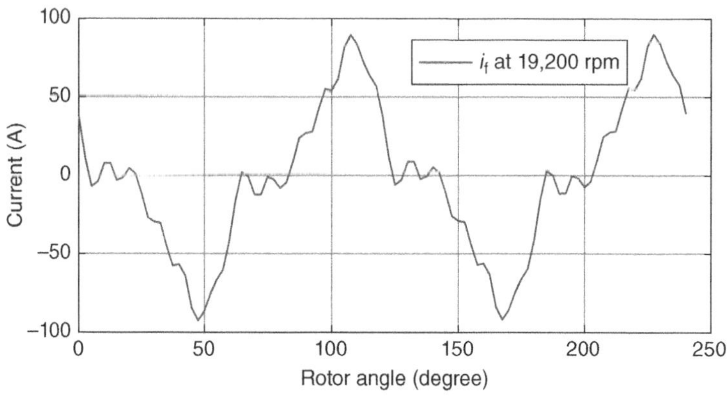

Figure 6.19 Simulated turn fault current in coil A1 at 19,200 rpm under current injection.

6.6 Experimental Validation on the Current Injection Technique

The proposed current injection method has been validated on the triple redundant 3-phase PMA-SynRM prototype drive. The schematic of the proposed current injection control is shown in Fig. 6.20. The calculation block provides the current commands based on equation (6.26) which will be injected into the fault set windings with information on the faulted phase and current feedback from the healthy sets of windings. The current commands for the two healthy sets are given according to a specific operating condition. These current commands are then tracked by the three *dq*-axis current controllers independently.

6.6.1 Turn Fault in Coil B2

First, the effectiveness of the two mitigation methods, namely, TSC and the proposed current injection method is evaluated. The turn fault is injected into coil B2 during the test when the healthy sets are fed with the rated currents of 120 A in magnitude at 4000 rpm under maximum torque per Ampere (MTPA) condition. The resultant turn fault current and phase currents under the two mitigation measures are shown in Figs. 6.21 and 6.22. Under the TSC application, the phase

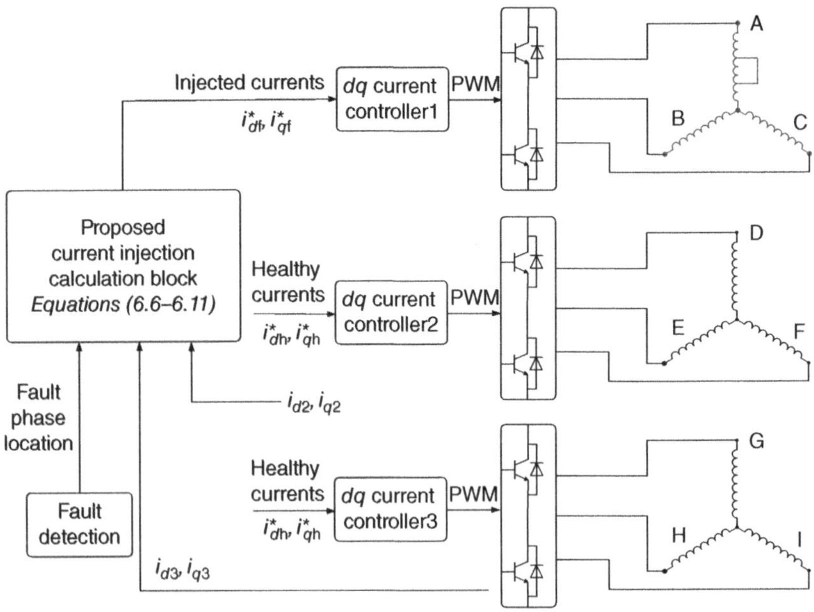

Figure 6.20 Schematic of the proposed current injection control diagram.

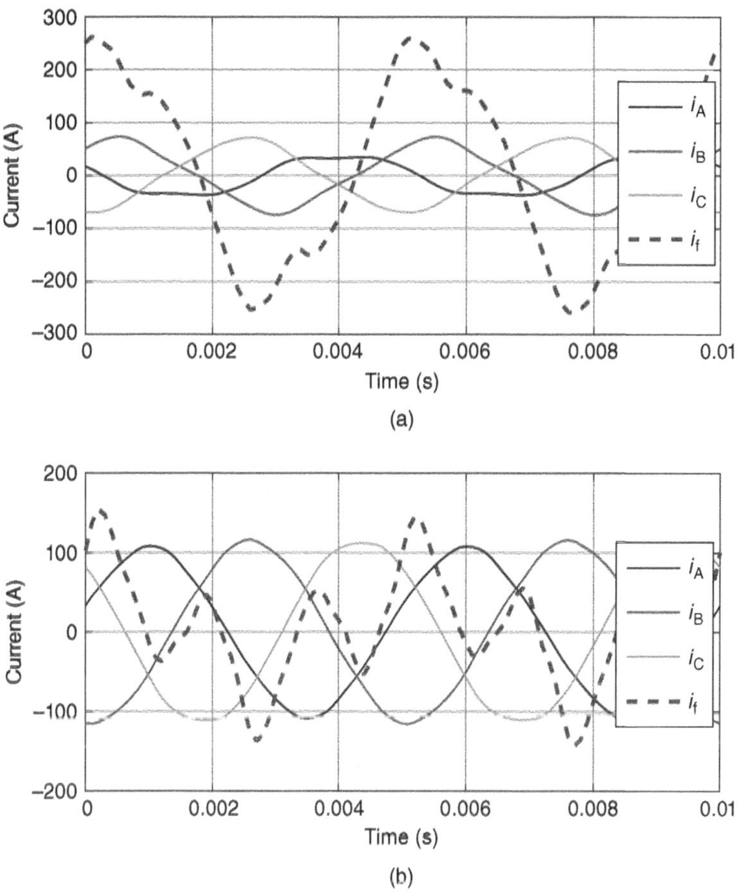

Figure 6.21 Fault current and phase currents in set ABC (a) with TSC and (b) with current injection.

currents in the fault set are reactive and unbalanced due to the mutual coupling with the healthy sets as seen in Fig. 6.21(a). Consequently, the rms fault current is relatively higher at 166.6 A albeit the currents in the healthy sets are well below rated as shown in Fig. 6.22(a). With the current injection technique, the phase currents of the fault 3-phase set are controlled and exhibit a reasonable level of symmetry. Since the flux linkages of the fault phase and the fault turn are minimized by the injected currents, the resultant rms fault current is 68 A, much lower than that of the TSC scheme. The fundamental component in the fault current is effectively suppressed, and the third harmonic becomes the dominant component. Further, the currents in the healthy sets are also well-controlled with both mitigation measures as is evident in Figs. 6.21(b) and 6.22(b).

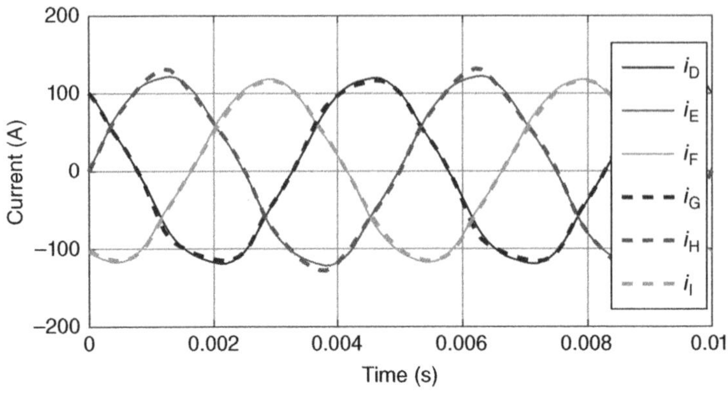

(a)

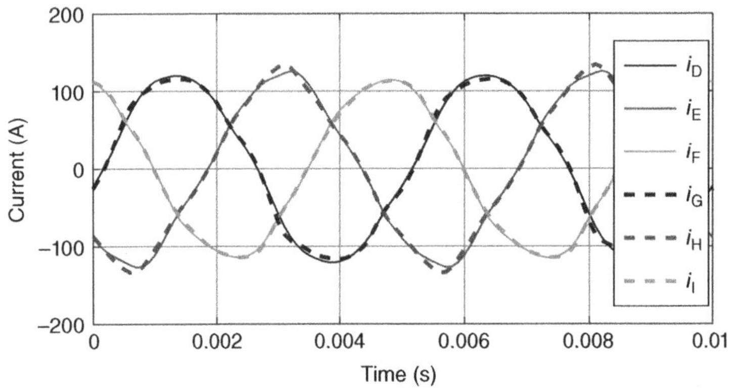

(b)

Figure 6.22 Phase currents in sets DEF and GHI (a) with TSC and (b) with current injection.

The effectiveness of the proposed method is also validated in a wide operation range by varying the magnitude of the currents in the healthy sets from 20 to 120 A in six steps. At each current magnitude, the phase angle is determined for MTPA condition. Since the fault current of one-turn short circuit without mitigation is excessive at 4000 rpm, it may cause permanent damage to the machine. The tests without mitigation action are performed at a low speed of 1000 rpm. It is seen from Fig. 6.23 that the fault current without mitigation action increases with the current in the healthy sets and is about four times the rated value when the current in the healthy sets reaches the rated. The fault current will be even larger at higher speed before the inductance-limited condition is met. Thus, the effective mitigation method is essential to limit the excessive fault current. The resultant turn fault current under the TSC and current injection at 4000 rpm are compared in

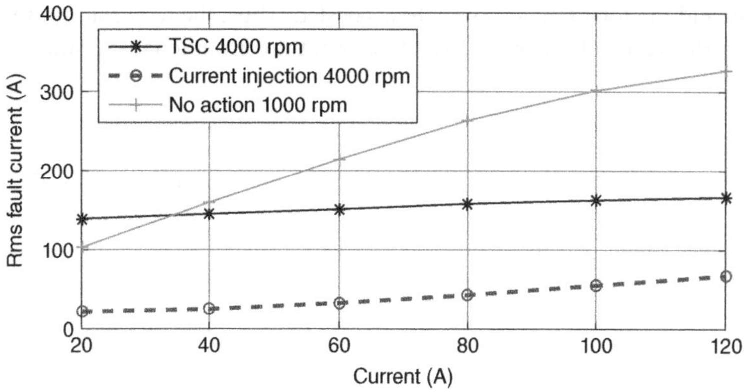

Figure 6.23 Rms turn fault current variation with healthy set currents with turn fault in coil B2 in different operation conditions.

Fig. 6.23. It is seen the turn fault currents are effectively reduced with both strategies but they are much lower with the current injection over the current operation range.

Meanwhile, the torques in healthy condition are compared in Fig. 6.24 with those under the TSC and current injection methods when the current in the healthy sets is varied from 20 to 120 A. It is seen that the output torques under current injection control are always higher than those of TSC when the fault turn is in coil B2. The torque reduction under the current injection when the current in the healthy sets reaches 120 A is 20 Nm, which is about quarter of the rated torque. At low current range, the torque under current injection is close to those in the healthy condition since the injected currents contribute some torque in the fault

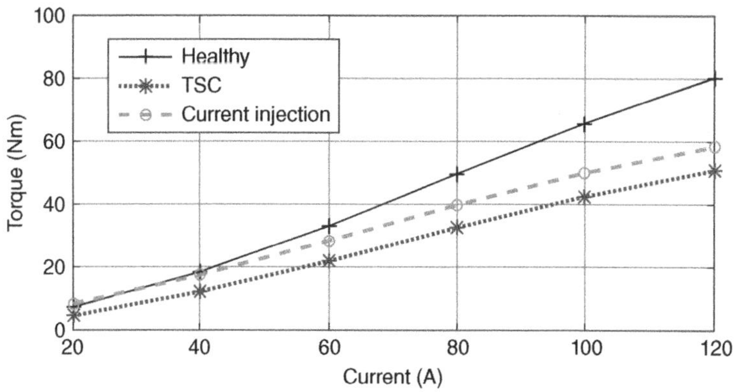

Figure 6.24 Variations of torques with the load currents at 4000 rpm in different operation conditions.

set region and the total torque is somehow boosted. Thus, the tests confirm the superior performance of the proposed method over the conventional TSC strategy.

6.6.2 Turn Fault in Coil A1

Next, similar tests are performed by rotating the machine in the reversal direction. This effectively renders the fault in coil A1 with the same fault injection setup as explained in Chapter 3. This fault scenario is first tested at 4000 rpm while the healthy sets are fed with 120 A currents in magnitude under MTPA condition. The current waveforms under the current injection control are shown in Fig. 6.25, where the rms turn fault current is evaluated to be 58 A. It is seen that the turn

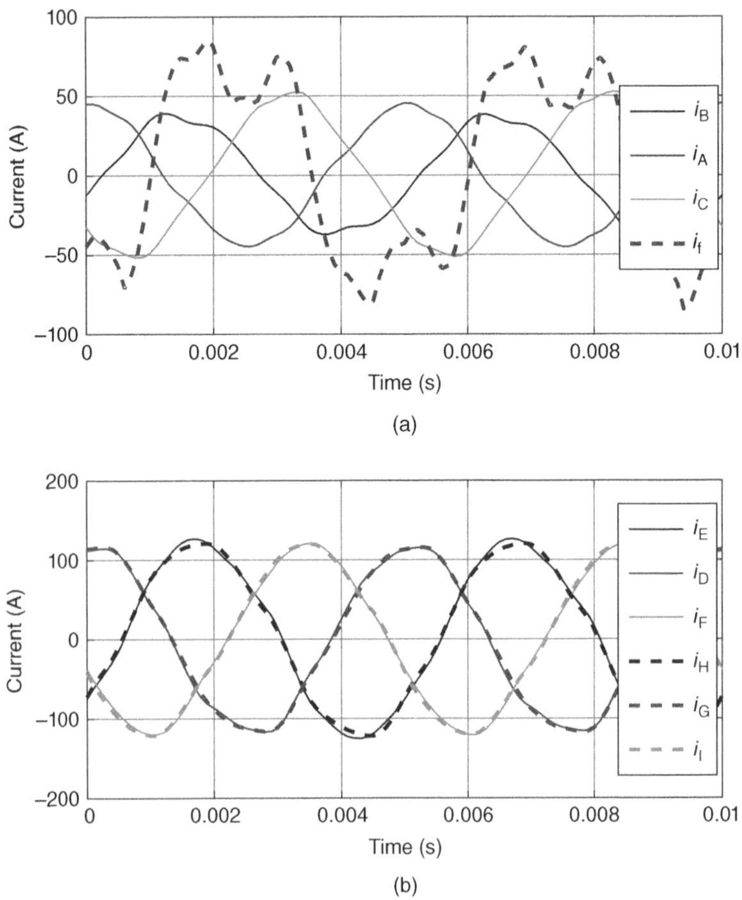

Figure 6.25 Fault current and phase currents: (a) fault current and phase currents in set ABC and (b) phase currents in healthy sets.

fault current is limited effectively, and the phase currents in both the faulted and healthy 3-phase sets are well controlled with small distortions.

The operations with the turn fault in coil A1 are further tested over the same current operation range as that stated in the previous section when the two different mitigation measures are applied. The resultant turn fault currents under the two mitigation methods are compared in Fig. 6.26. It is seen the maximum turn fault current with the current injection control is lower than that of the TSC. In the light current range, the fault current under the current injection is slightly higher than that of the TSC. However, the fault current is much lower than the rated value and will not cause any thermal issue. The resultant torques with the two mitigation methods are compared in Fig. 6.27. The output torque

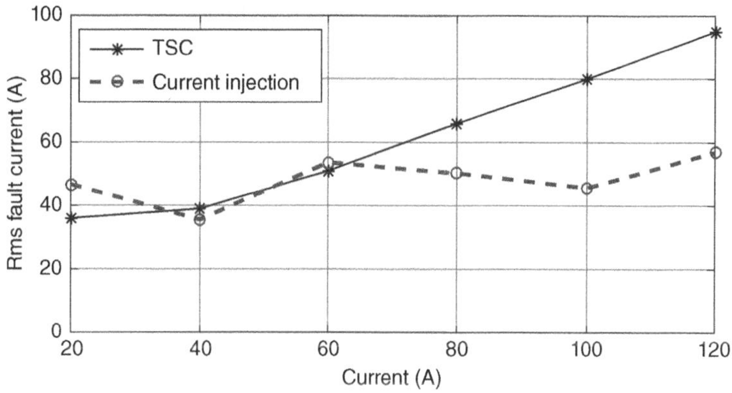

Figure 6.26 Rms turn fault current variations with healthy set currents with turn fault in coil A1 at 4000 rpm.

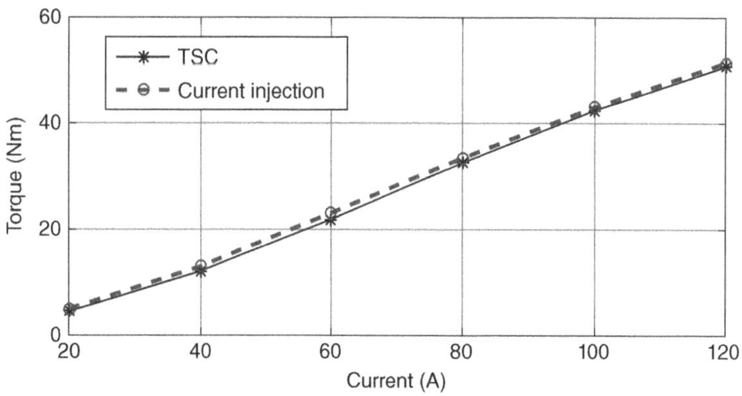

Figure 6.27 Variations of postfault torque with healthy set currents with turn fault in coil A1 at 4000 rpm.

under current injection control is close to that of the TSC when the fault turn is in coil A1. The different behaviors in the fault current and toque from those seen in Figs. 6.23 and 6.24 are mainly caused by the differences in the injected currents as listed in Table 6.2 when the fault occurs in coil A1.

6.6.3 Integrated Fault Test with Current Injection

In practical operation, the turn fault should be timely detected, and a mitigation action should be applied immediately to avoid any further damage to the machine drive. Therefore, an integrated test of turn fault injection, fault detection, and fault mitigation is performed, and the resultant fault current and phase currents in the ABC set of windings are shown in Fig. 6.28. Initially, the machine is operating at 1000 rpm with phase currents of 60 A in magnitude under MTPA condition. Then, one-turn short-circuit fault is injected in coil B2 at 0.03 second and the fault current quickly shoots to 200 A. The fault is detected, and the fault phase is identified by employing the diagnosis technique in [15]. Subsequently, the proposed current injection method is enabled to reduce the fault current. It is seen that the fault current is effectively reduced to a safe level. Meanwhile, the operation of the other two healthy 3-phase sets is not affected as seen from the current waveforms in Fig. 6.28(b). The test demonstrates that the triple redundant 3-phase machine drive can accommodate the turn fault effectively by employing the proposed current injection technique.

6.7 Summary

This chapter describes postfault control techniques that generally entail control actions to (1) manage the fault condition and (2) facilitate continuous operations with the best possible performance. To cope with potential short-circuit faults in inverter and terminal connections, a fault-tolerant PM machine must be designed such that its short-circuit current is below 1.0 pu.

For a multiphase fault-tolerant FSCW PM machine, significant torque ripple will be present under an open-circuit or short-circuit fault in one complete phase. Optimal current control strategies described in this chapter can be used to deliver smooth output torque during postfault operation. The technique is applicable in both constant torque and constant power operating regions with some adaption in the flux-weakening weighting factor. For an interturn short-circuit in a few turns, both TSC of the faulted phase via its H-bridge converter or appropriate current injection can be applied to reduce the current in the short-circuited turns to a thermally sustainable level. The optimal current control can then be employed to deliver torque-ripple free postfault operation.

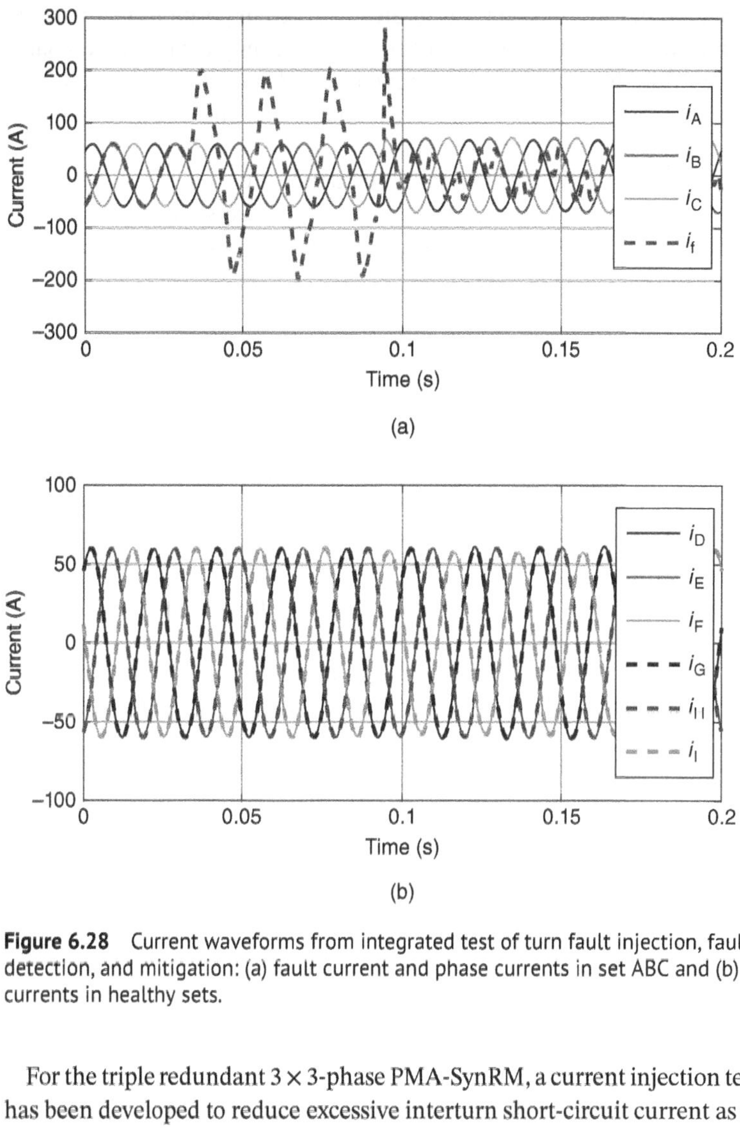

Figure 6.28 Current waveforms from integrated test of turn fault injection, fault detection, and mitigation: (a) fault current and phase currents in set ABC and (b) phase currents in healthy sets.

For the triple redundant 3 × 3-phase PMA-SynRM, a current injection technique has been developed to reduce excessive interturn short-circuit current as an alternative means of fault mitigation. The flux linkage of the fault phase is estimated by accounting for three influential sources, namely, the rotor PM field, the currents in the healthy 3-phase sets, and the currents in the fault 3-phase sets. The flux linkage and current in the faulted turn can then be reduced by injecting specific currents to the faulted 3-phase set while the healthy sets continue their operation to generate torque. The developed current injection approach is capable of limiting the fault current for all possible turn fault scenarios, including fault occurrence in different phases and subcoils, in turns located in different positions in a slot, and with different numbers of short-circuit turns. The technique has been validated by extensive FE simulations and experimental tests, demonstrating its additional benefits in the reduction of the fault current and in boosting output torque over the conventional TSC mitigation measure.

While the above two postfault control techniques are explored as examples, other control strategies, such as current harmonic suppression and copper loss minimization, can be investigated in a similar way to improve the machine drive performance during postfault operations.

References

1 B. C. Mecrow, A. G. Jack, J. A. Haylock, and J. Coles, "Fault-tolerant permanent magnet machine drives," *IEE Proceedings – Electric Power Applications*, vol. 143, no. 6, pp. 437–442, 1996.

2 A. G. Jack, B. C. Mecrow, and J. A. Haylock, "A comparative study of permanent magnet and switched reluctance motors for high-performance fault-tolerant applications," *IEEE Transactions on Industry Applications*, vol. 32, no. 4, pp. 889–895, 1996.

3 J. Wang, K. Atallah, and D. Howe, "Optimal torque control of fault-tolerant permanent magnet brushless machines," *IEEE Transactions on Magnetics*, vol. 39, no. 5, pp. 2962–2964, 2003.

4 J. D. Ede, K. Atallah, W. Jiabin, and D. Howe, "Effect of optimal torque control on rotor loss of fault-tolerant permanent-magnet brushless machines," *IEEE Transactions on Magnetics*, vol. 38, no. 5, pp. 3291–3293, 2002.

5 Z. Sun, J. Wang, G. W. Jewell, and D. Howe, "Enhanced optimal torque control of fault-tolerant PM machine under flux-weakening operation," *IEEE Transactions on Industrial Electronics*, vol. 57, no. 1, pp. 344–353, 2010.

6 J. A. Haylock, B. C. Mecrow, A. G. Jack, and D. J. Atkinson, "Operation of fault tolerant machines with winding failures," *IEEE Transactions on Energy Conversion*, vol. 14, no. 4, pp. 1490–1495, 1999.

7 Z. Sun, J. Wang, D. Howe, and G. Jewell, "Analytical prediction of the short-circuit current in fault-tolerant permanent-magnet machines," *IEEE Transactions on Industrial Electronics*, vol. 55, no. 12, pp. 4210–4217, 2008.

8 A. J. Mitcham, G. Antonopoulos, and J. J. A. Cullen, "Implications of shorted turn faults in bar wound PM machines," *IEE Proceedings – Electric Power Applications*, vol. 151, no. 6, pp. 651–657, 2004.

9 B. A. Welchko, J. Wai, T. M. Jahns, and T. A. Lipo, "Magnet-flux-ing control of interior PM Machine drives for improved steady-state response to short-circuit faults," *IEEE Transactions on Industry Applications*, vol. 42, no. 1, pp. 113–120, 2006.

10 J. G. Cintron-Rivera, S. N. Foste, and E. G. Strangas, "Mitigation of turn-to-turn faults in fault tolerant permanent magnet synchronous motors," *IEEE Transactions on Energy Conversion*, vol. 30, no. 2, pp. 465–475, 2015.

11 Y. Jiang, Z. Zhang, W. Jiang, W. Geng, and J. Huang, "Three-phase current injection method for mitigating turn-to-turn short-circuit fault in concentrated-winding permanent magnet aircraft starter generator," *IET Electric Power Applications*, vol. 12, no. 4, pp. 566–574, 2018.

12 B. Sen and J. Wang, "Stationary frame fault-tolerant current control of polyphase permanent-magnet machines under open-circuit and short-circuit faults," *IEEE Transactions on Power Electronics*, vol. 31, no. 7, pp. 4684–4696, 2016.

13 B. Sen and J. Wang, "A fast detection technique for stator inter-turn fault in multi-phase permanent magnet machines using model based approach," in *7th IET International Conference on Power Electronics, Machines and Drives (PEMD 2014)*, 2014, pp. 1–6.

14 B. Sen and J. Wang, "Stator inter-turn fault detection in SPM machines using PWM ripple current measurement," in *7th IET International Conference on Power Electronics, Machines and Drives (PEMD 2014)*, 2014, pp. 1–6.

15 B. Sen and J. Wang, "Stator interturn fault detection in permanent-magnet machines using PWM ripple current measurement," *IEEE Transactions on Industrial Electronics*, vol. 63, no. 5, pp. 3148–3157, 2016.

16 B. C. Mecrow *et al.*, "Design and testing of a four-phase fault-tolerant permanent-magnet machine for an engine fuel pump," *IEEE Transactions on Energy Conversion*, vol. 19, no. 4, pp. 671–678, 2004.

17 B. Wang, J. Wang, A. Griffo, and B. Sen, "A general modeling technique for a triple redundant 3x3-phase PMA SynRM," *IEEE Transactions on Industrial Electronics*, vol. 65, no. 11, pp. 9068–9078, 2018.

18 B. Sen, J. Wang, and P. Lazari, "A detailed transient model of interior permanent magnet motor accounting for saturation under stator turn fault," in *Energy Conversion Congress and Exposition (ECCE), 2013 IEEE*, 2013, pp. 3548–3555.

19 M. Dai, A. Keyhani, and T. Sebastian, "Fault analysis of a PM brushless DC motor using finite element method," *IEEE Transactions on Energy Conversion*, vol. 20, no. 1, pp. 1–6, 2005.

20 B. Sen and J. Wang, "Analytical modelling of stator turn fault in surface mounted permanent magnet machines," in *Energy Conversion Congress and Exposition (ECCE), 2013 IEEE*, 2013, pp. 4445–4452.

21 K. Kyung-Tae, P. Jun-Kyu, H. Jin, and K. Byeong-Woo, "Comparison of the fault characteristics of IPM-type and SPM-type BLDC motors under inter-turn fault conditions using winding function theory," *IEEE Transactions on Industry Applications*, vol. 50, no. 2, pp. 986–994, 2014.

7

Novel Segregated Windings with Enhanced Fault Tolerance for Multiple 3-Phase Machine

7.1 Introduction

In Chapter 2, it has been shown that the triple redundant 3-phase permanent magnet-assisted synchronous reluctance machine (PMA-SynRM) exhibits good performance while being capable of tolerating various electrical faults. However, in the worst-fault scenario, namely, single-turn short-circuit fault in a phase winding, the resultant fault current is about 3.2 pu after application of terminal short circuit (TSC) on the fault 3-phase set. While this fault current is thermally sustainable in the machine, it becomes a bottleneck for the further improvement of fault tolerance and limitation on the increases in power density of the machine drive. In Chapter 6, the fault current has been limited by adopting the current injection method from the perspective of postfault control, and this chapter aims to suppress the excessive turn fault current by adopting different winding configurations [1].

It is well known that winding configuration and connection (star or delta) has a profound impact on the machine's performance and fault-tolerant capability [2]. Induction machines and permanent magnet (PM) synchronous machines employ distributed winding to produce highly sinusoidal magneto-motive force (MMF) with low harmonics. However, their overlapped end windings bundled together have bulky size and are vulnerable to interphase failure due to combined mechanical, thermal, electrical, and environmental stresses. The nonoverlapped concentrated winding exhibits short-end winding and good isolation between phases. Therefore, it has been adopted in fault-tolerant applications in the form of fractional slot concentric winding (FSCW) PM machines [3], switched reluctance machines (SRMs) [4], and switched flux PM machines [5]. However, both FSCW and switched flux PM machines contain considerable low- and high-order MMF harmonics. Although the winding space harmonics in the FSCW PM machine can be reduced by employing two tooth-span coils, the interturn short-circuit current in such types of PM machines is too high to be managed safely.

The winding connection of a machine also affects its behavior and level of fault tolerance. Most PM machines adopt star winding connection to avoid zero-sequence current and the resultant copper loss. The star connection, however, makes it impossible to operate with two phases without a neutral connection. In contrast, machines with delta-connected windings are capable of two-phase operation in an event of single-phase failure, which can be utilized to enhance fault tolerance [6, 7]. In [8, 9], it has been shown that a star-delta-connected machine has superiority in both healthy performance and fault tolerance. Therefore, winding layouts, configurations, and connections could be carefully designed to achieve the best trade-off between performance and level of fault tolerance.

For the triple redundant 3-phase PMA-SynRM under consideration, although there are good electrical and thermal isolations among the three 3-phase winding sets, they are magnetically coupled via the MMF offset component [10, 11], as has been analyzed in Chapter 4. The magnetic mutual coupling causes a number of undesirable effects in the fault conditions, such as current distortion and increased torque ripple [12]. It is also responsible for high fault current in short-circuited turns [13] after TSC application and deteriorates the postfault control performance of the drive system [14]. Therefore, the mutual coupling effect should be minimized as low as possible to suppress the adverse effects.

In addition, as has been shown in Chapter 2, the fault current after TSC application is much larger if one-turn short-circuit fault occurs in the leading or trailing coil of a 3-phase winding set, depending on modes of operation. This can be attributed to the fact that the reduction of flux linkage by TSC application is less effective in these two subcoils. It shows that the asymmetry due to the magnetic coupling exists not only in different phases but also in the different coils within the same phase. Consequently, the turn fault current exhibits different magnitudes when the fault occurs in different coils of a 3-phase winding set [15–17]. Similarly, a switch failure in different legs of a 3-phase inverter will lead to different current patterns and fault behaviors [18]. For better fault management, clearly it is desirable to minimize the coil-dependent fault behaviors.

This chapter explores alternative winding configurations in order to reduce turn fault current and hence further improve the fault tolerance level of the triple redundant 3-phase PMA-SynRM. Thus, four new winding configurations, namely, delta-connected winding, star-delta-connected winding, mixed-pitch winding, and concentric winding for the machine under study are described, and their behaviors in healthy and fault conditions are analyzed and experimentally tested [19]. The reduction in the turn fault current is achieved by a change in the winding connection or in the coil layout, which has a very little impact on the manufacturing process and cost. The merits of the machines with the alternative winding configurations in reduction of the turn fault current are assessed in comparison with the classic star-connected machine. This study provides some useful insights into selection of appropriate winding configuration for safety critical applications.

7.2 Multiple 3-Phase Machines with Four Alternative Segregated Windings

As has been discussed in Chapters 2, 3, and 4, a short circuit fault across a single turn located on top of a slot results in the largest possible fault current. An important parameter that influences this current is the number of series connected turns that have a close magnetic coupling with the faulted turn. As a rule-of-thumb indication, if the number of series connected turns is N, the fault current may reach N times the rated current. It is therefore possible to reduce the worst-case fault current in pu by selecting a low N during design if there is freedom to do so. For example, for a given power rating, a combination of low voltage and high current ratings will lead to low series connected turns in a phase winding, and hence lower fault current in pu. Likewise, a split of a phase winding into a few sub-coils that are less magnetically coupled is also conducive to lower fault current in pu since the number of series connected turns that are magnetically coupled to a fault turn is reduced. In many cases, however, such design freedoms may not be available. To further reduce the turn fault current in the multiple 3-phase machine with star-connected windings when the above freedoms are not available, four alternative winding configurations are examined and the mechanism by which the turn fault current can be reduced with these winding configurations is discussed.

7.2.1 Zero-Sequence Flux Linkage in Star-Connected Windings

For the purpose of comparison, the triple redundant fault-tolerant machine with segregated, star-connected winding configuration is shown on the left side of Fig. 7.1. This winding configuration is denoted as W1 for short. As discussed in the previous chapters 2–4, magnetic coupling through the MMF offset component exists among the three 3-phase winding sets under fault conditions. This offset component can cause positive, negative, and zero-sequence flux linkages in the 3-phase winding set where a turn fault occurs. Following TSC application as the mitigation action on the faulted 3-phase set, the resultant short-circuit currents contain positive and negative sequence components but no zero-sequence component as it has no flow path in a star-connected winding. Hence, while the induced short-circuit currents have cancellation effects on the positive and negative sequence flux linkages, they have no cancellation effect on the zero-sequence flux linkages. As a result, the zero-sequence flux linkages in the coils of the 3-phase windings remain after the application of TSC.

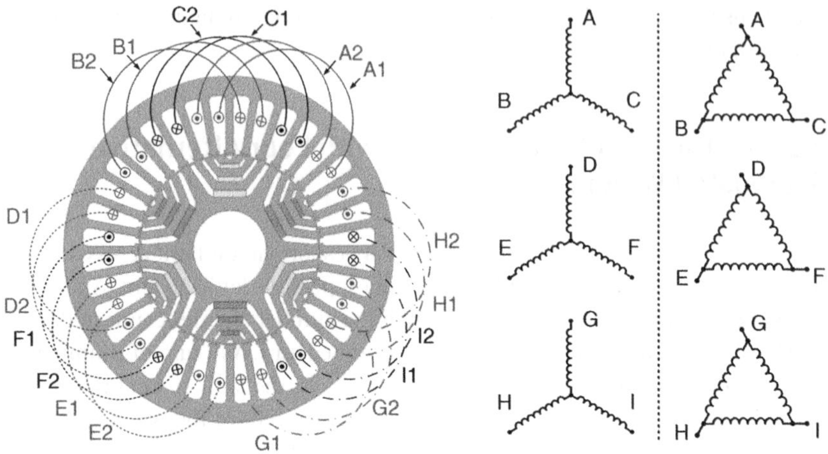

Figure 7.1 Triple redundant 3-phase PMA-SynRM with star-connected windings and delta-connected windings.

By way of example, the phase flux linkages of the ABC winding set after TSC application are plotted in Fig. 7.2 with the time-domain waveforms on the left and sequence components on the right. It is seen that the positive and negative sequence components of the phase flux linkages are very small while the major components are the zero-sequence component. The zero-sequence flux linkage is linked by the fault turns and is the cause of a relatively large fault current. This can be further understood from the flux linkages of the six subcoils shown in Fig. 7.3. The turn fault current is indeed proportional to the residual flux linkage of subcoils. It is also observed that after the mitigation action, the residual flux linkage of coil B2 is highest, which leads to the largest turn fault current. This analysis agrees with the findings in Chapter 2.

Thus, there are two possible means to further reduce the turn fault current for the multiple 3-phase machine: (1) reduce zero-sequence component of winding flux linkage and (2) reduce the mutual magnetic coupling. The first approach can be achieved by employing delta or star-delta-connected windings as they allow the zero-sequence current to flow, which has an effect on reducing zero-sequence flux linkage. The second approach can be realized by mixed-pitch windings. Further, by adopting concentric winding for each phase, the short-circuit currents induced by the mitigation action will be more effective in reducing the flux linkage of faulted turns, leading to lower fault current. The four alternative winding configurations will be introduced, and their performances in healthy and fault conditions are analyzed and compared with those of the star-connected winding configuration in the subsequent sections.

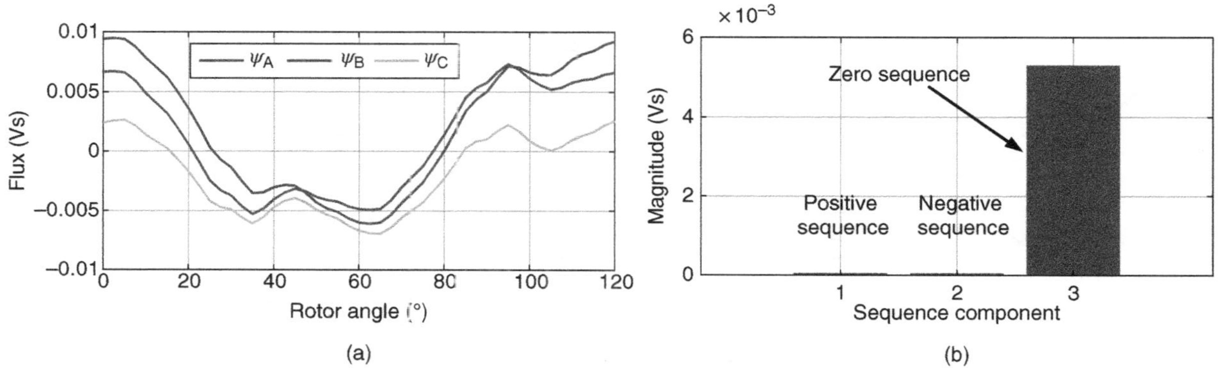

Figure 7.2 Phase flux linkages in phases A, B, and C after TSC application: (a) time-domain waveform and (b) sequence components.

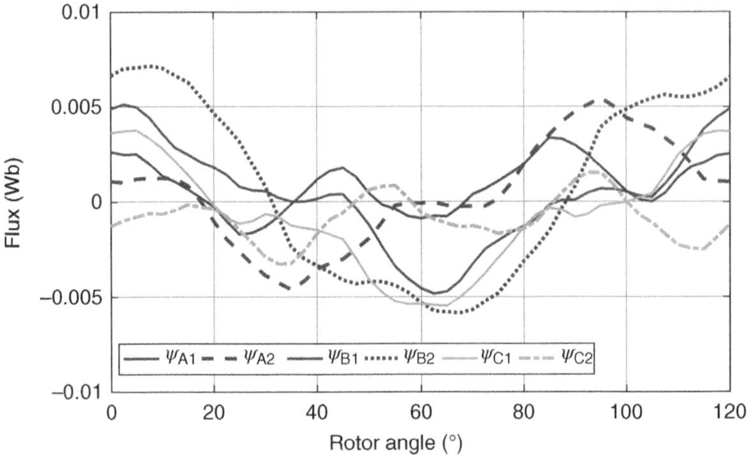

Figure 7.3 Flux linkages of six subcoils in ABC set after TSC.

7.2.2 Delta-Connected Winding Configuration

The segregated, delta-connected winding configuration for the fault-tolerant machine under study is illustrated on the right side of Fig. 7.1, and it is named as W2 for short. Compared with the star-connected winding configuration, the delta connection provides a zero-sequence current flow path. Consequently, any zero-sequence flux linkage component in the air gap, such as triplen harmonics, will induce zero-sequence current that flows through the delta-connected three phases. These currents do not produce any useful torque but incur additional copper losses. In PM-excited synchronous machines, it is difficult to eliminate triplen harmonics of the PM field, and star-connected winding configuration is more preferred to avoid the extra loss and resultant thermal loading. In the machine under study, however, the extra copper loss is negligible as the triplen harmonic PM field in the air gap is relatively weak with the multilayer rotor structure and in contrast its fault-tolerant capability is enhanced with the delta connection. In case of an open-circuit fault in one-phase winding, it does not need to shut down the whole 3-phase set as it does in the star-connected windings. The two remaining healthy phases can continue their operation to generate useful torque with modification to the drive control. Hence, the torque reduction due to the open-circuit fault is minimized and the torque pulsation is also reduced. In the event of a turn fault, by applying TSC on the fault 3-phase set, the zero-sequence current can flow in the delta-connected the phase windings, which leads to lower residual flux linkage in the faulted turn and lower fault current. Hence, the machine with delta-connected windings is expected to have improved fault-tolerant performance than the star-connected machine.

It is worth noting that the above merits are achieved without introducing any additional cost. Only the number of turns and the winding connections are changed. For like-with-like comparison, the machines with delta- or star-connected winding configuration will have the same rated voltage and current. Hence, the number of turns of each coil in the delta winding configuration is 1.73 times of that in the star configuration while the cross-sectional area of each turn is reduced proportionally. As a result, the machine exhibits the same line voltage and line current at the rated operation.

7.2.3 Star-Delta Winding Configuration

Star-delta winding connection, shown in Fig. 7.4, is commonly used in high-power machines. With proper design, the machine with such a winding configuration can achieve an increased fundamental winding factor and reduced spatial harmonics, leading to increased average torque and reduced torque ripple and iron loss, etc. In addition, there is also a zero-sequence current path in the delta-connected winding part, which is conducive to a reduction of the fault current in an event of turn-to-turn short circuit. Thus, star-delta winding configuration shown in Fig. 7.4 and denoted as W3 is adopted for the triple redundant 3-phase PMA-SynRM. As can be seen, the two coils in each phase are split into the star- and delta-connected sections. The coils with potentially large turn fault currents, namely, B2, E2, and H2, are arranged in the delta section, which provides a zero-sequence current path. Therefore, the residual flux linkages of these coils could be more effectively reduced by the mitigation action in an event of turn fault. The turn fault current is expected to be further reduced. Since the circulating zero-sequence currents only flow in the inner delta-connected winding section, the extra copper loss introduced is also lower than that with delta-connected winding configuration.

As is well known, the star-delta-connected winding has a higher fundamental winding factor and lower spatial harmonics than those of W1 and W2 configurations shown in Table 7.1. The fundamental winding factor calculated based on the method described in [9] is increased by 3.5%, which will result in a torque

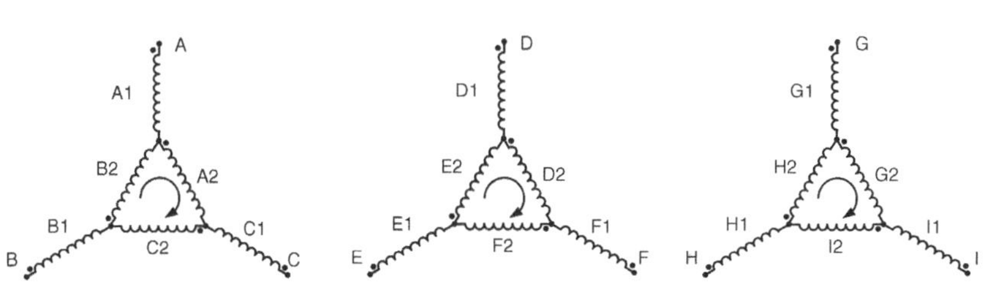

Figure 7.4 Star-delta winding connection, W3.

Table 7.1 Comparisons of winding factors with star, delta, and star-delta configurations.

Harmonic order	Star or delta	Star-delta
1st	0.966	1
5th	0.259	0
7th	0.259	0
11th	0.966	1
13th	0.966	1
17th	0.259	0
19th	0.259	0

increase by the same amount. Further, the harmonic factors for the 5th, 7th, 17th, 19th,…, harmonics are zero, and hence these harmonics are totally cancelled in the flux linkages and back electromotive force (back-emf), while the 11th and 13th harmonics are increased by 3.5%. As a result, the total harmonic distortion in the machine with this winding configuration is reduced, leading to lower torque ripple and iron loss. Thus, the machine with the star-delta winding configuration is expected to have improved performance and enhanced fault-tolerant capability. A full comparison of the winding factors up to 19th harmonics for the three winding configurations is given in Table 7.1.

It should be noted from Fig. 7.4 that the currents in the delta-connected coils are $(1/\sqrt{3})$ times the currents in star-connected coils. To produce the same MMF by both star- and delta-connected coils, the number of turns in the delta-connected coils should be increased by a factor of $\sqrt{3}$ while the cross-sectional area of each turn is reduced by the same factor. This complicates the manufacturing process and may cause marginal cost increases.

7.2.4 Mixed-Pitch Winding

The W2 and W3 winding configurations are employed to facilitate zero-sequence current in the delta-connected windings or coils following TSC application. This current will reduce the residual zero-sequence flux linkage in the faulted turns and hence the fault current. Since the zero-sequence flux linkage is caused by the MMF offset component of 3-phase windings, it will be logical to explore possible means to reduce the MMF offset component directly. For this purpose, the MMF offset component, $F^{11}_{\mathrm{os\_W1}}$, caused by the currents in star-connected ABC windings, which has been derived in Chapter 2, is rewritten in (7-1) [20] together with the expressions for the phase currents that generate this component. It is a time-varying sinusoidal whose magnitude is proportional to $\frac{4}{3}I_m$ where N is the number of turns per each coil and I_m is the phase current amplitude.

$$
\begin{aligned}
F^{11}_{os\_W1} &= \frac{2}{3}N(i_A + i_B - i_C) = \frac{4}{3}NI_m \cos(\omega t - 60) \\
i_A &= I_m \cos(\omega t) \\
i_B &= I_m \cos(\omega t - 120) \\
i_C &= I_m \cos(\omega t + 120)
\end{aligned}
\tag{7.1}
$$

In healthy conditions, the sum of the MMF offset components of all three 3-phase winding sets is zero since the currents in each 3-phase winding set are identical and balanced [21]. However, it will be nonzero in fault or unbalanced operation conditions. Therefore, by reducing this offset component, it is possible to further reduce the fault current in case of a turn-to-turn short-circuit fault.

A segregated, mixed-pitch winding configuration (named as W4) is devised as shown in Fig. 7.5 to reduce the magnitude of the MMF offset component. Similar to the W1 configuration, the three 3-phase distributed windings are segregated into three nonoverlapped winding sets. Each module forms a balanced 3-phase set, which can be driven by individual inverters. Compared with the W1 configuration, only the coil layouts of phases B, E, and H in the three 3-phase sets are changed while the coils of phases A and C, phases D and F, and phases G and I remain the same. In the W1 configuration, all the winding coils are full-pitched with a coil span of six slots over the space of one electrical pole. In the W4 configuration, by contrast, the coils in phases B, E, and H are short-pitched with a span of five slots, and the two coils in these phases have an opposite go-return direction, while the coils of phases A and C, D and F, and G and I are full-pitched. Therefore, this winding configuration is termed as mixed-pitch segregated windings. It should be noted that the winding factors of phases B, E, and H with short-pitched

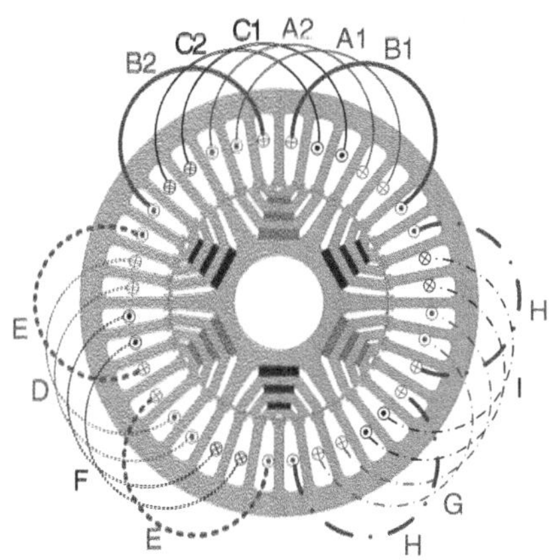

Figure 7.5 Triple redundant 3-phase PMA-SynRM with mixed-pitch windings, W4.

coils are not affected because the distribution factor becomes 1.0 while the pitch factor is reduced to 0.966. In contrast, in the full-pitched phases, the distribution factor is 0.966 but the pitch factor is 1.0. Thus, each of the 3-phase windings in the W4 configuration forms a balanced set for the fundamental electric and magnetic quantities.

If all three 3-phase sets of the machine with the W4 configuration are excited with the same 3-phase sinusoidal currents as that of the machine with the W1 configuration, the current distributions in the slots of the former are the same as those in the latter. Thus, the resultant MMFs in healthy operations of the two machines have no difference. It indicates that the two machines have the same performance in healthy conditions. Thus, all the merits of the PMA-SynRM, and the measures and processes to achieve the desired fault tolerance are maintained for the machine with the W4 configuration.

However, the W4 configuration will yield a decreased mutual coupling effect between the three 3-phase sets than that of the W1 configuration, as will be shown subsequently.

The reduced mutual coupling effect can be quantified by derivation of the MMF distribution of one 3-phase set with the W4 configuration, as described in Chapter 2. The turn functions of the six mixed-pitch coils in set ABC are plotted in Fig. 7.6. It is seen that the turn function of coil B1 is negative while it is positive for coil B2 as a result of the modified coil layouts. The resultant MMF distribution in space at a given instant of time is shown in Fig. 7.7. It is seen that the MMF has a similar distribution to that of the W1 configuration. The MMF over the ABC winding region consists of an AC component and an offset component $F^{11}_{os\_W4}$ while only offset components $F^{21}_{os\_W4}$ and $F^{31}_{os\_W4}$ exist over the other two regions occupied by the DEF and GHI sets, respectively. It can be shown that the MMF AC component of the W4 configuration is exactly the same as that of the W1, while the offset components over the three 3-phase sets can be derived and given in (7.2) and further simplified for $F^{11}_{os\_W4}$ in (7.3). The overall MMF offset components in

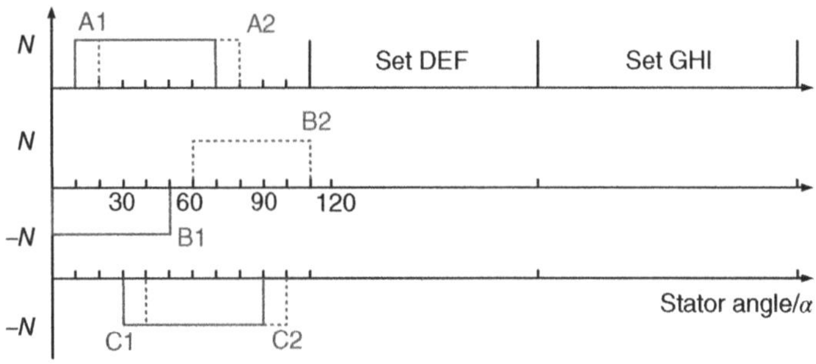

Figure 7.6 Turn functions of the six coils in the ABC set of W4.

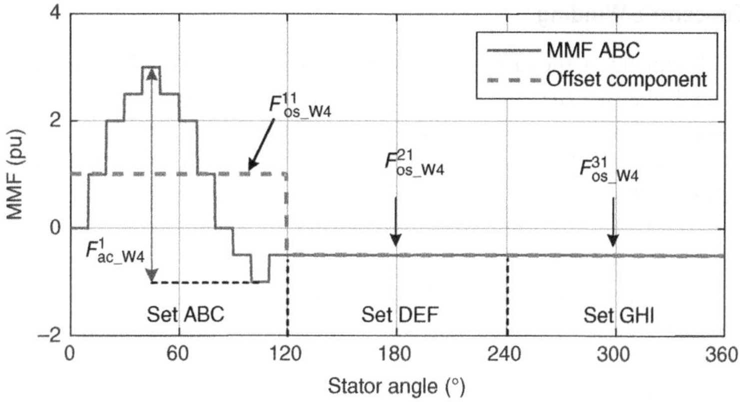

Figure 7.7 MMF profile produced by ABC set of W4.

each 3-phase region can be obtained by summing the offset components produced by all the three 3-phase winding sets. Thus, under healthy conditions, they are zero when all the three 3-phase sets are fed with the same 3-phase currents. They will not be zero in fault or unbalanced operation conditions.

$$F^{11}_{os\_W4} = \frac{2}{3}N(i_A - i_C)$$
$$F^{21}_{os\_W4} = -\frac{1}{3}N(i_A - i_C)$$
$$F^{31}_{os\_W4} = -\frac{1}{3}N(i_A - i_C) \tag{7.2}$$

$$F^{11}_{os-W4} = \frac{2\sqrt{3}}{3}N(i_d\cos(\omega t - 30) - i_q\sin(\omega t - 30))$$
$$= \frac{2\sqrt{3}}{3}NI_m\cos(\omega t - 30 + \delta), I_m = \sqrt{i_d^2 + i_q^2} \tag{7.3}$$

Comparing equations (7.1) and (7.2), it is seen that the magnitude of the offset component of $F^{11}_{os\_W4}$ is 14% lower than that of $F^{11}_{os\_W1}$. It confirms that the mutual coupling between the different 3-phase sets is reduced and so are the undesirable effects associated with the mutual coupling, such as current distortion and relatively large turn fault current due to the residual zero-sequence flux linkage in the faulted turn.

It should be noted that the mixed-pitch winding configuration is achieved simply by rearranging the coils. It will cause very little additional cost. The proposed winding configuration is applicable to other machines with segregated, multiple 3-phase distributed windings to obtain fault tolerance with reduced mutual coupling, such as synchronous reluctance machines, surface-mounted PM machines, synchronous wound field machines, and induction machines.

7.2.5 Concentric Winding

As previously explained, the turn fault current in the machine with W1 configuration is unequal when a turn fault occurs in a different coil within a 3-phase winding set. This is because the residual flux linkages are different when the coil is located in a different position as illustrated in Fig. 7.3. By employing a concentric winding configuration as illustrated in Fig. 7.8, it is possible to reduce the fault current in the leading or trailing coil of the W1 configuration. This winding configuration is denoted as W5. As can be seen, each 3-phase winding set also has no contact with others to guarantee segregation. It is observed that the current distribution in the slots is also the same as that of the W1 configuration, which implies that the machines with the W5 and W1 configurations will have the same performances in healthy operations.

The key characteristic of the W5 distribution is the fact that the two coils in a phase are concentric as shown in Fig. 7.9, meaning that the axes of the two coils

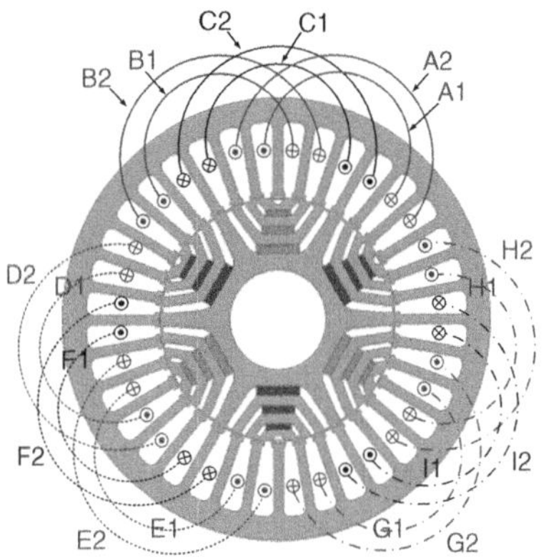

Figure 7.8 Triple 3-phase PMA-SynRM with concentric windings, W5.

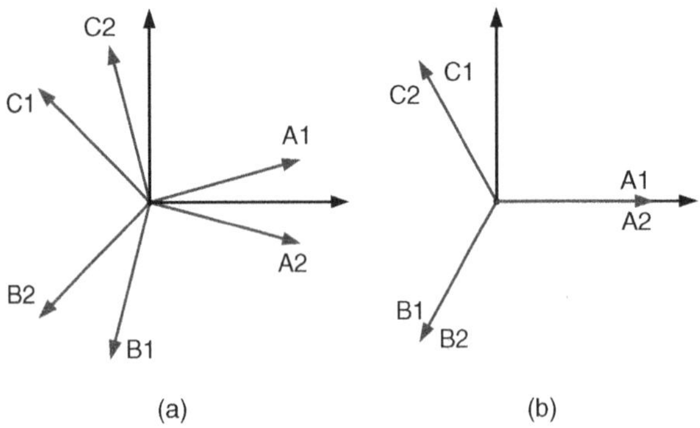

Figure 7.9 Space vectors of the six coils in phases ABC (a) W1 and (b) W5.

coincide with each other. Hence, the winding distribution factor is 1. However, the coil spans of A1 and A2 are 7 and 5 slots, respectively, which corresponds to 210 and 150 electrical degrees. Thus, the pitch factor of the winding is 0.966. The overall winding factor of the W5 configuration is the same as that of the W1 configuration, although the flux linkage of each coil is 4% lower, which is beneficial to limit the turn fault current.

Further, the no-load flux linkages of the ABC phases are plotted in Fig. 7.10 for the two machines with the W5 and W1 configurations. Obviously, the phase flux linkages of the two machines coincide with each other. Further, the flux linkages of the two coils in phase A of the two winding configurations are illustrated in Fig. 7.11. It is seen that the two coil flux linkages of the W1 configuration have a phase shift of 30 electric degrees while those of the W5 overlap with each other. This is the main feature of the difference of the two machines. As a result, the residual flux linkages of the two coils in one phase after TSC will be closer and may lead to a lower fault current. The 4% lower coil flux linkage of the W5 configuration is caused by its lower pitch factor while its distribution factor is 1.0, higher than that of the W1 counterpart. Thus, the overall winding factors of the two configurations are the same. However, the lower coil flux linkage helps reduce the turn fault current under the fault operating conditions.

It is worth noting that the four alternative winding configurations are known in literature and have been employed in machines for various industry applications. However, only their operations in healthy conditions are analyzed in the existing literature. Their behaviors with multiple 3-phase segregated windings for fault-tolerant applications are not fully understood. Hence, detailed FE simulations and experimental testing will be performed to fill this knowledge gap in the subsequent sections.

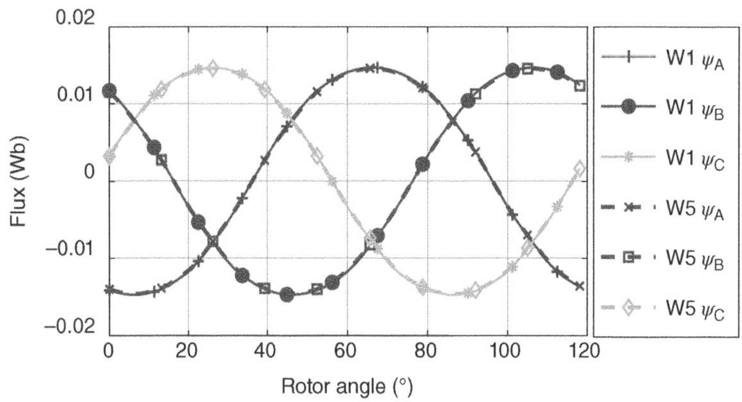

Figure 7.10 Comparison of no-load phase flux linkages of W1 and W5 configurations.

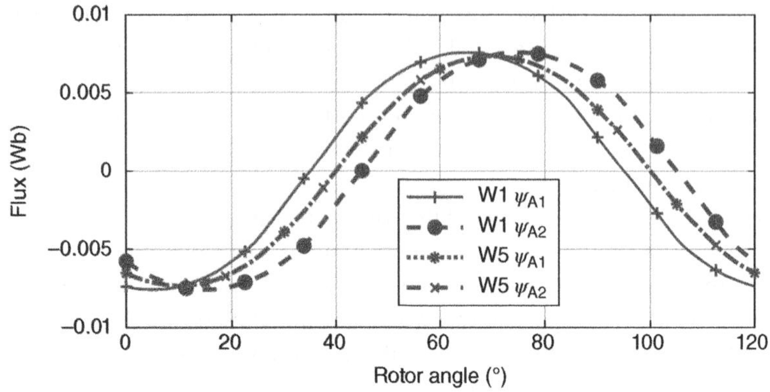

Figure 7.11 Comparison of no-load coil flux linkages of W1 and W5 configurations.

7.3 FE Analysis of Machines with Five Segregated Winding Configurations

In this section, five PMA-SynRM drives with different segregated winding configurations are modeled in FE software based on the leading specification and geometry parameters provided in Table 7.2. Different from the previous 40 kW machine prototype, it is naturally cooled with 5 kW rated output power at 2000 rpm. It has a shorter axial length but with similar stator and rotor geometries for high rotor saliency. For like-with-like comparison of the five winding configurations, the number of turns of each coil is 8 for the W1, W4, and W5 configurations while it is 14 for the delta-connected configuration (W2). The number of turns in the W3 configuration is 8 for the star-connected part and 14 for the delta part. The cross-sectional area of each turn is adjusted accordingly to

Table 7.2 Machine specification and leading parameters.

Specification	Symbol	Value
Rated speed	n_b	2000 rpm
Rated power	P_r	5 kW
Rated phase current (magnitude) and gamma angle	$I_{ratedY}/I_{rated\Delta}$	67 A(49°)/38.7 A(49°)
Turn number of star-connected coil	N_Y	8
Turn number of delta-connected coil	N_Δ	14
Axial length	LA	86 mm
Stator diameter	R_s	180 mm

yield the same effective slot-fill factor for all five configurations. Consequently, the five machines share the same electrical, magnetic, and thermal loads and can be operated with the same inverter and control in both healthy and fault conditions. The machines are evaluated under healthy and several common fault conditions, and their performances and characteristic behaviors are compared. For the simplicity of discussion, the fault is always assumed to occur in set ABC.

7.3.1 Healthy Performance Evaluation

Although the machines are targeted for fault-tolerant applications, the performance in healthy operations is also of paramount importance as they operate in this condition most of the time. Thus, the five machines are simulated under the rated healthy condition. They are all excited with the rated line current of 67 A. The presence of the zero-sequence current path in the W2 and W3 configurations results in zero-sequence currents of triplen harmonics in the delta-connected part of the windings, as shown in Fig. 7.12. It is seen that the triplen harmonic currents are very small in magnitude compared with the rate current. The resultant overall copper losses have been compared in Table 7.3 for the five machines. It can be seen that the zero-sequence currents have negligible influence on the copper loss at the rated operation albeit the increase in the copper loss will be more pronounced at lower load conditions.

The predicted torques for the five machines are compared in Fig. 7.13, and their average values and absolute torque ripples, denoted as ΔT, are listed in Table 7.3. As can be seen, the machines with the W1, W4, and W5 configurations exhibit the same torque characteristics due to the same current distributions in the slots. The machine with the W3 configuration exhibits the highest torque of 25 N m among the five candidate designs while its torque ripple is the lowest. The higher torque

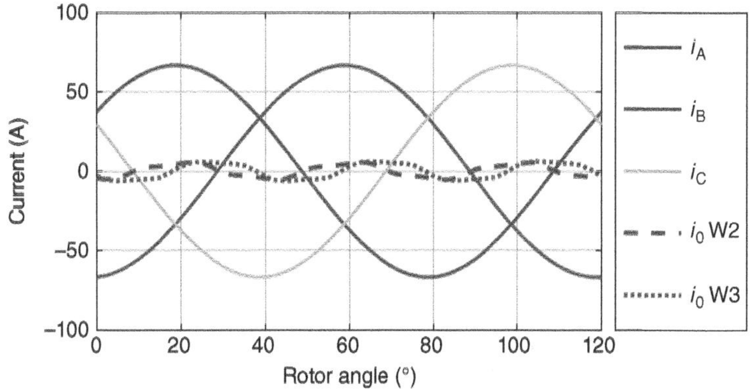

Figure 7.12 Line currents and induced zero-sequence currents.

Table 7.3 Performance comparison at rated healthy operation.

Items	W1	W2	W3	W4	W5
Torque (N m)	24	23.5	25	24	24
ΔT ($T_{max} - T_{min}$)	2.97	2.57	2.37	2.97	2.97
Copper loss (W)	302	307	309.6	302	302
Iron loss (W)	68.3	66.45	66.84	68.3	68.3
Efficiency (%)	93.1	93.0	93.3	93.1	93.1

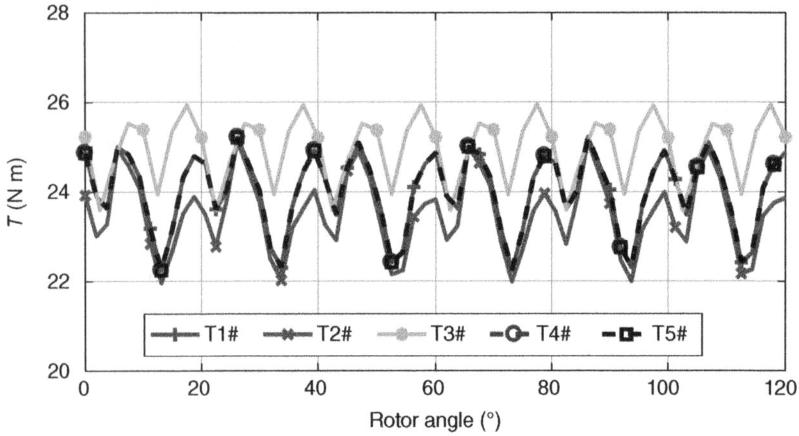

Figure 7.13 Comparison of torque waveforms of the five machines in rated healthy operation.

and low torque ripple are attributed to the increased fundamental winding factor and reduced harmonic winding factors because of the star-delta connection. The machine with the W2 configurations shows slight torque reduction due to the zero-sequence currents in the delta connection, which induce additional eddy current loss in the rotor magnets and give rise to drag torque. The torque ripple is also marginally smaller.

In addition, the five machines' copper loss, iron loss, and electromagnetic efficiency are quantified and given in Table 7.3. The copper loss of the machines with the W2 or W3 configuration is marginally higher due to the induced zero-sequence currents. Thus, their influence on the efficiency at the rated operation is negligible. As expected, the machine with the W3 configuration exhibits the highest efficiency due to the higher fundamental winding factor while the efficiencies of the other four machines are more or less the same.

7.3.2 Open-Circuit Fault

Open-circuit failure in one phase is very common in electrical machine drives because of a failure in a switching device or connection. For the machine under study, the whole inverter of the faulted 3-phase set will be deactivated while the other two 3-phase modules continue their operation to provide torque at the rated speed with the rated current. This mode is simulated with the ABC phases open circuited while the other two modules maintain the same operation without interruption. The resultant torque waveforms of the five machines are shown in Fig. 7.14. Again, the machine with the W3 configuration shows the highest torque while the torques of the remaining four machines are almost the same with an average of about 2/3 of the healthy value. The torque waveforms of the machines with the W1 and W5 configurations coincide with each other as the currents in the slots of the two machines are still the same under this open-circuit mode. Further, the torque ripples are also provided in Table 7.4. It is seen that they are only slightly increased compared with those in the rated healthy operation.

It is worth noting that there are zero-sequence currents flowing in the delta-connected winding sections of the open-circuited ABC phases in the machines with the W2 or W3 configuration. Their amplitudes are very small with negligible influence on the postfault operation.

With the delta configuration (W2), it is possible to operate the remaining two phases in case of an open-circuit fault in one-phase winding [6]. This will provide additional torque under the open-circuit condition. By way of example, the schematic of postfault two-phase operation of the delta-connected ABC winding set, where phase A is open-circuited, is shown in Fig. 7.15. The phase current vectors of phases A, B, and C in the healthy condition are shown in Fig. 7.16 in

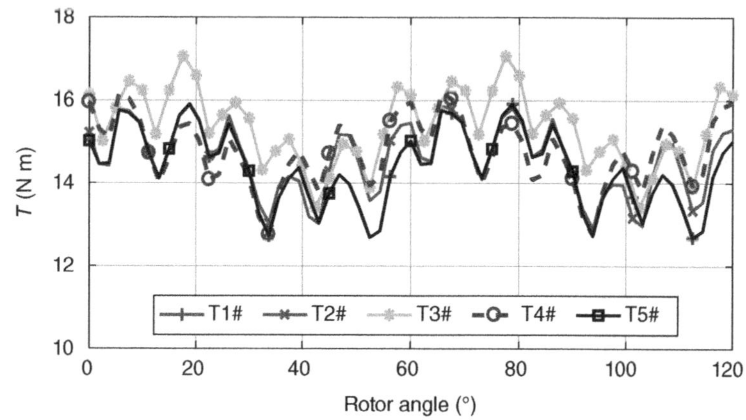

Figure 7.14 Comparison of torque waveforms of the five machines in open-circuit condition.

Table 7.4 Torque comparison under fault conditions

Items	W1	W2	W3	W4	W5
OC torque (N m)	14.37	14.58	15.34	14.73	14.37
OC ΔT ($T_{max} - T_{min}$)	3.22	2.97	3.66	3.41	3.22
SC torque (N m)	15.90	15.62	16.58	15.85	15.90
SC ΔT ($T_{max} - T_{min}$)	4.42	4.24	4.62	4.56	4.42

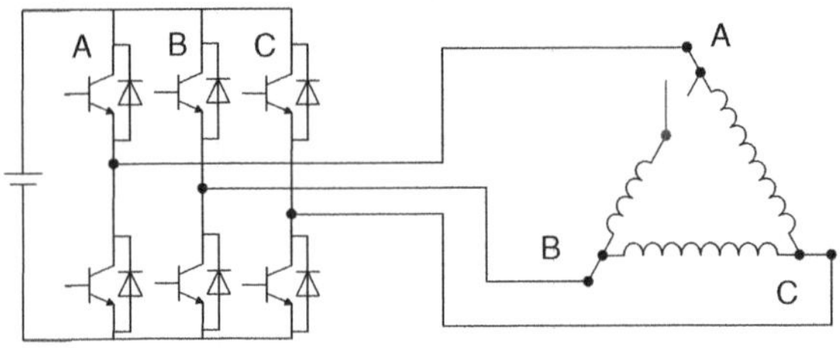

Figure 7.15 Open-circuit fault in phase A winding for delta configuration (W2).

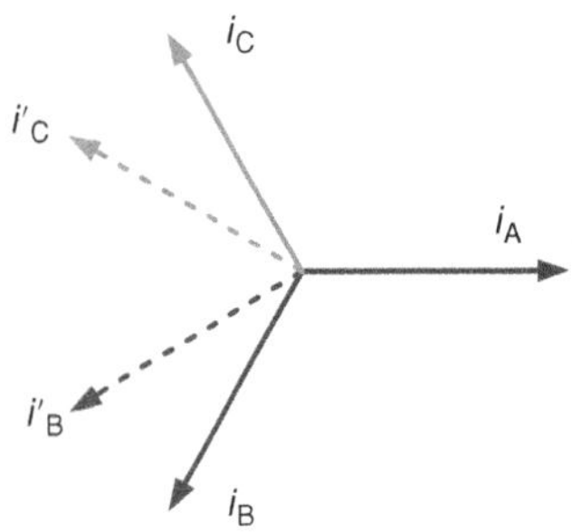

Figure 7.16 Phase current vectors in healthy and phase A open-circuited conditions for W2.

the solid arrow lines. After the open-circuit fault in phase A, the phase A current is forced to 0, but the currents in phases B and C can be regulated as the current vectors, i'_B and i'_C, shown as the dashed arrow lines with the inverter to generate positive sequence MMF and torque. Under this operation mode, extra 4.1 Nm of torque, or 17% of the rated torque, can be produced with the rated current in phases B and C of the faulted ABC windings.

7.3.3 Short-Circuit Fault

As has been discussed in Chapter 2, upon detection of a short-circuit fault in a phase winding, TSC is applied to the fault 3-phase windings to mitigate the fault impact. This fault condition is simulated for the machines with the five different winding configurations by applying TSC on set ABC with sets DEF and GHI fed with the rated current at the rated speed. The postfault torque waveforms of the

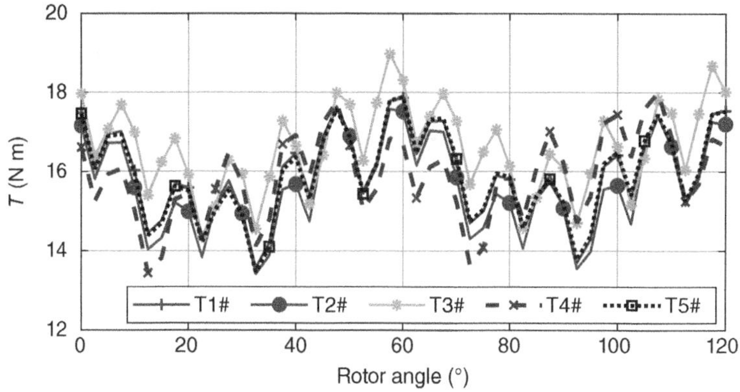

Figure 7.17 Comparison of torque waveforms of the five machines in short-circuit condition.

five machines under this mode are compared in Fig. 7.17, where similar trends are seen. The average toques and torque ripples of the five machines are given in Table 7.4. It is seen that both the average torque and torque ripple of a given machine under the short-circuit condition are slightly larger than those of the machine with the same winding configuration under the open-circuit condition. Again, it is observed that the torque of the machine with the W3 configuration is the highest among the five candidates mainly due to its advantage on the winding factor.

Due to the low PM flux linkages, all the five machines have low short-circuit (line) currents as shown in Fig. 7.18. For the machine with the W3 configuration, both the line currents and the currents in the delta-connected windings are shown. Although some differences in waveform shape and phase shift are seen, the RMS currents of all the five machines, including those in the delta-connected coils are below the rated. Thus, they are all capable of sustained operation in this failure mode. The asymmetric short-circuit currents seen are attributed to the mutual coupling effect between the three 3-phase modules under the fault condition, as has been discussed previously.

7.3.4 Single-Turn Short-Circuit Fault

The fault current in a single short-circuited turn after TSC application is the focus of the investigation on the four alternative winding configurations. Thus, simulations have been performed to assess their relative merits on this. The residual flux linkage of the faulted turn, which has a key influence on the fault current, is dependent on fault location and the currents in the remaining healthy 3-phase sets. Therefore, single-turn fault under TSC application is simulated for the five machines in motoring mode by varying fault locations in all possible coils

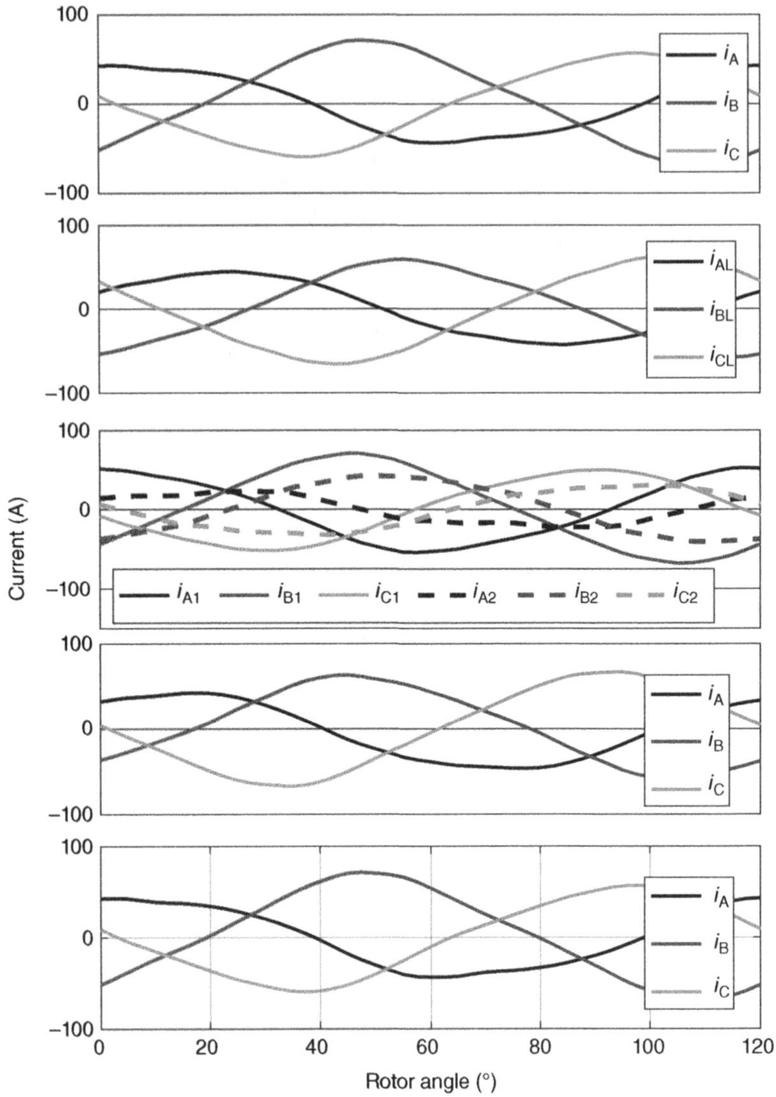

Figure 7.18 Induced short-circuit currents for the five machines, from top to bottom, W1, W2, W3, W4, and W5.

and the currents in the healthy sets. For a fault located in each coil, it is assumed that the turn closest to the slot opening is short circuited. The resultant variations of the RMS fault currents and the normalized values to the rated coil current are shown from Figs. 7.19 to 7.24 with the coil designations given in Figs. 7.1, 7.4, 7.5

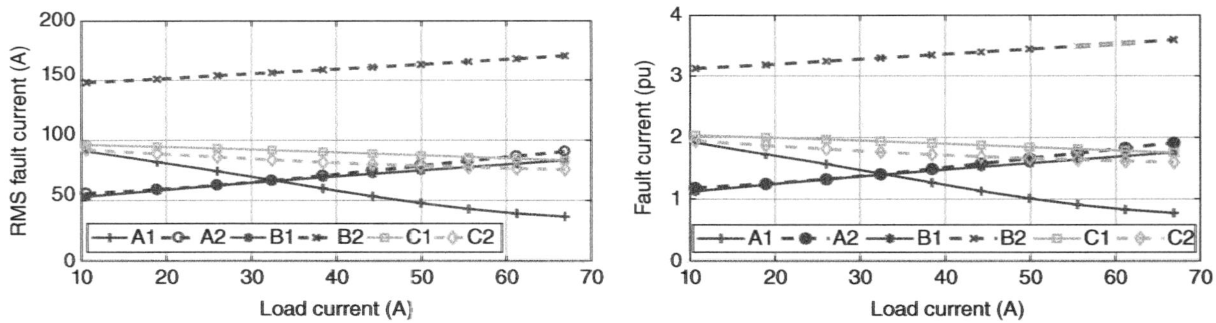

Figure 7.19 Variations of turn fault current with load current and fault location in 6 coils of W1 winding configuration.

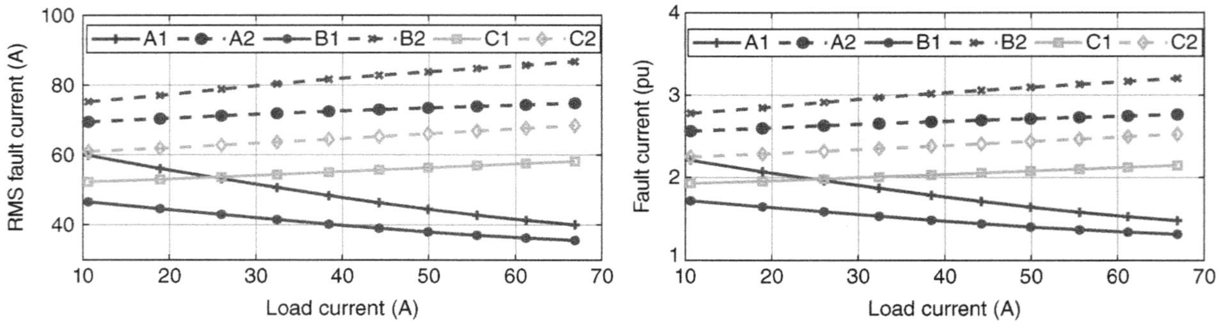

Figure 7.20 Variations of turn fault current with load current and fault location in 6 coils of W2 winding configuration.

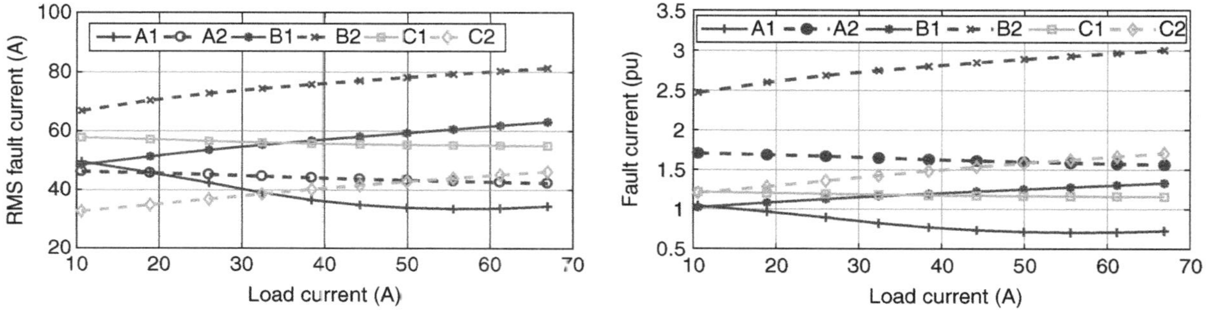

Figure 7.21 Variations of turn fault current with load current and fault location in 6 coils of W3 winding configuration.

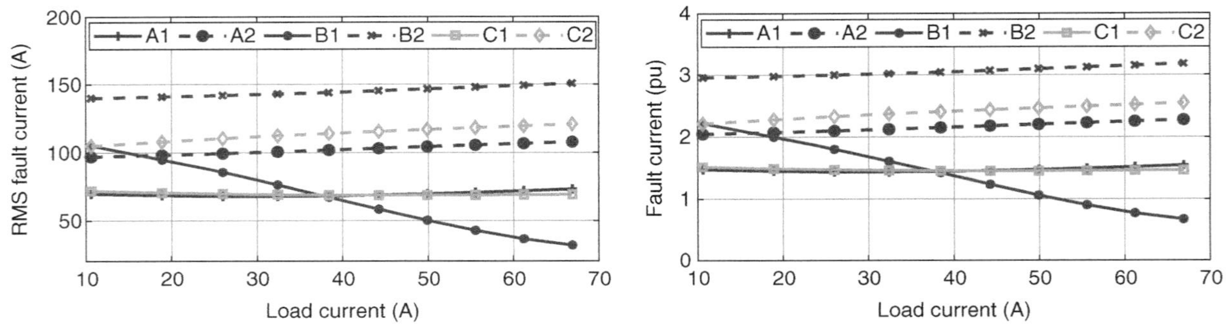

Figure 7.22 Variations of turn fault current with load current and fault location in 6 coils of W4 winding configuration.

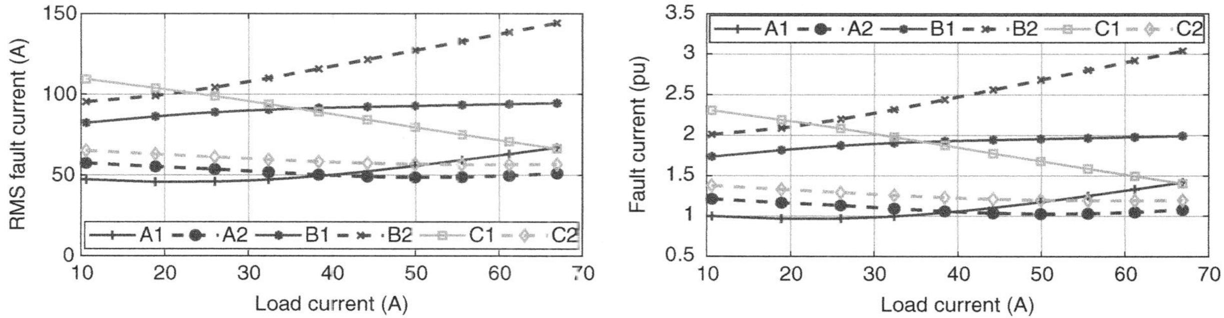

Figure 7.23 Variations of turn fault current with load current and fault location in 6 coils of W5 winding configuration.

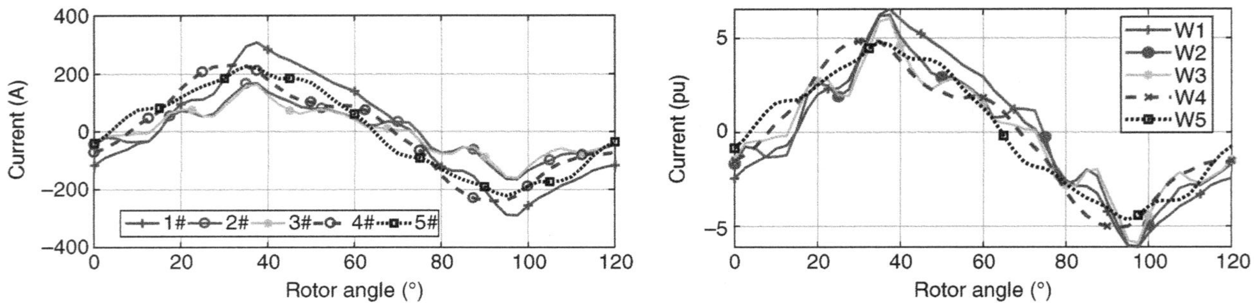

Figure 7.24 Turn fault current waveforms of the worst cases for the five winding machines.

and 7.8. It is shown that in all five machines a single-turn fault in B2 coil yields the largest fault current when the currents in the healthy 3-phase sets reach the rated. For this reason, coil B2 is arranged in the inner delta part of the W3 configuration.

The fault current waveforms together with the normalized values in the worst case of each winding configuration are plotted in Fig. 7.24, and their RMS values are given in Table 7.5. It should be noted that since the fault turn occurs in the delta-connected coil in the W2 and W3 configurations, their fault currents are normalized to $I_{\text{rated}\Delta}$. It is seen that compared with the W1 configuration, the fault currents in all four alternative winding configurations have been reduced. The reduction in the machines with the W2 and W3 configurations is more significant since the residual flux linkage in the faulted turn is reduced by the zero-sequence current in the delta-connected coils. The reduction in the fault current with the W4 and W5 configurations is very similar, ~16%. Further, the copper losses in the faulted turn and healthy part of the B2 coil for each winding configuration have been quantified and compared in Table 7.5. The resultant temperature distributions under the worst case for each machine are predicted in Ansys Fluent and shown in Fig. 7.25, from which the hot-spot temperature in the fault turn of each machine is identified and the results are listed in Table 7.5.

It is seen that the hot-spot temperature in the machine with the W3 configuration is the lowest and this is followed by the machine with the W2 configuration. This confirms the effect of the zero-sequence current on the reduction of the turn fault current and hence the resultant temperature. For the machine with the W3 configuration, further benefit is derived from the placement of coil B2 in the delta-connected part. In the machine with the W4 configuration, the reduction

Table 7.5 Turn fault current, copper loss, and hotspot temperature.

Items	W1	W2	W3	W4	W5
RMS fault current (A) Its normalized values (pu)	173 (3.65 pu)	86.5 (3.19 pu)	78.7 (2.91 pu)	145.8 (3.07 pu)	144 (3.04 pu)
Copper loss of faulted turn (W)	30.2	13.2	11	21.7	21.2
Copper loss of healthy turns (W)	10.2	14.8	14.4	8.3	11.4
Maximum temperature in faulted turn (°)	178	164	157	168	170

in the fault current is roughly proportional to the reduction of the MMF offset component, which is the cause of the mutual coupling effect. In the machine with the W5 configuration, the residual flux linkage in the faulted turn is more effectively reduced by the short-circuit currents in the healthy part of the concentric windings. The overall effect appears to be similar to that of the W4 configuration. The benefits of the alternative winding configurations are also evident in the reduced hot-spot temperatures in the worst-case fault condition.

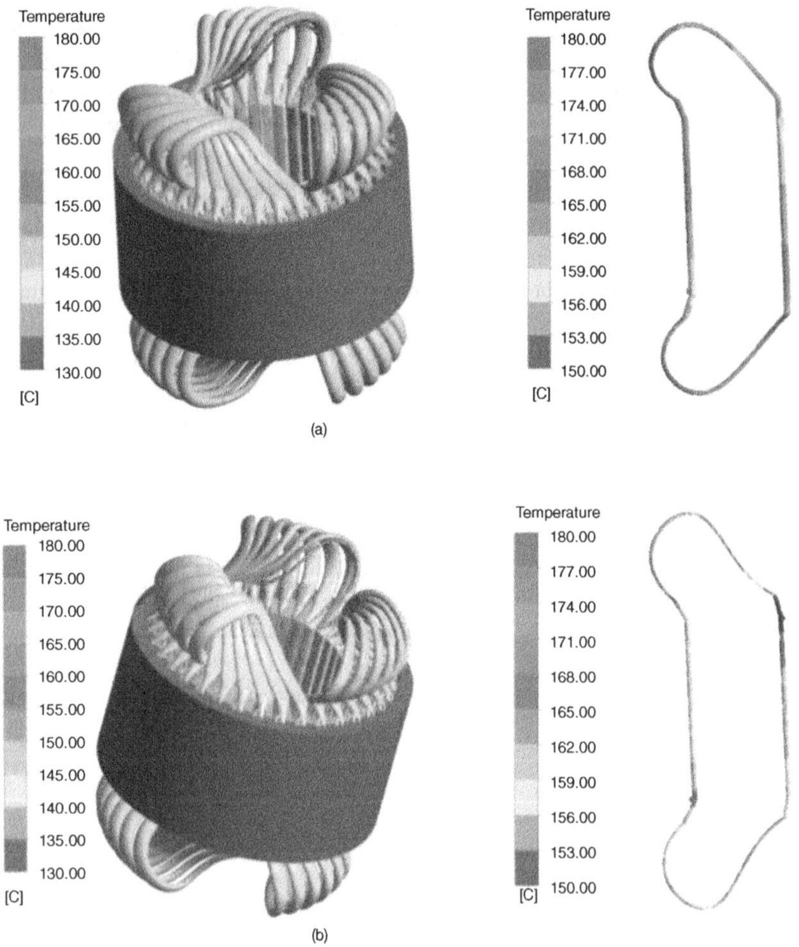

Figure 7.25 Thermal distribution prediction of the five machines: (a) W1, (b) W2, (c) W3, (d) W4, and (e) W5.

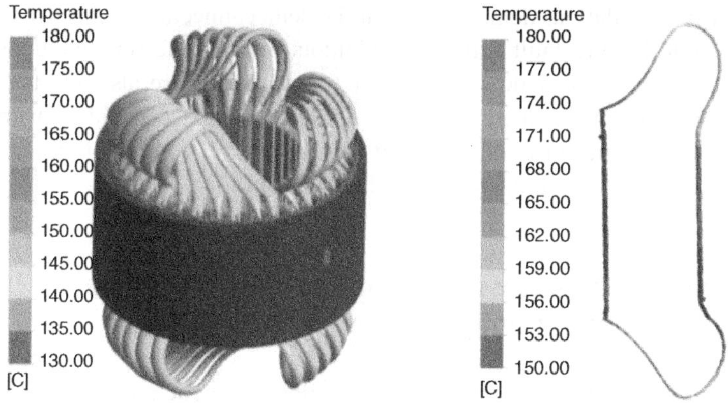

(c)

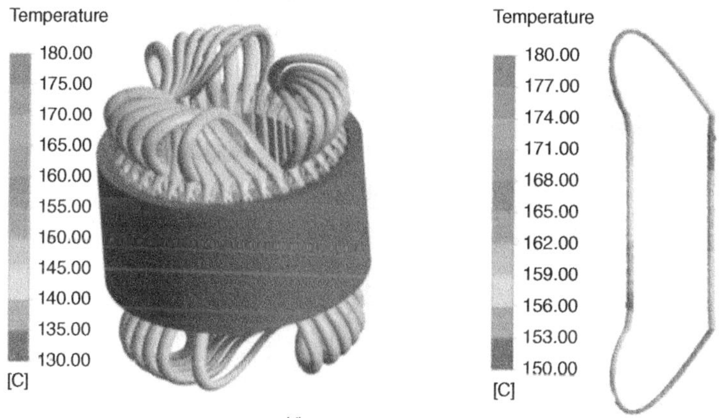

(d)

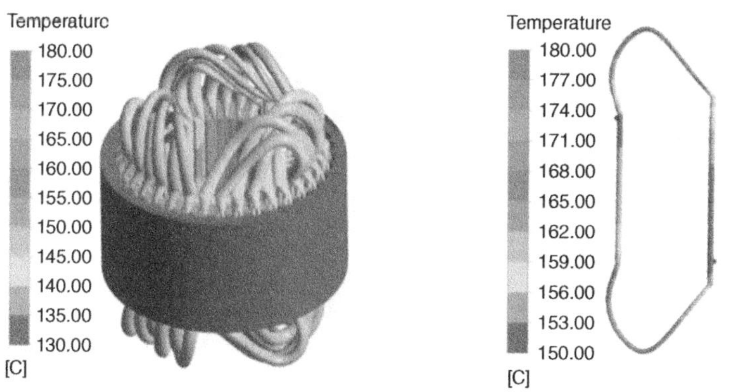

(e)

Figure 7.25 (*Continued*)

The extensive simulations demonstrate that the delta connection is effective in the reduction of the turn fault current. In addition, the winding configurations with reduced mutual magnetic coupling or concentric coils are also effective. Therefore, combinations of these concepts, e.g. delta-connected windings with the mixed-pitch or concentric coils shown in the W4 and W5 configurations, are expected to achieve further reduction in the fault current.

7.4 Experimental Assessments of Machines with Five Segregated Winding Configurations

The merits of the alternative winding configurations in reduction of the turn fault current are further assessed by experiments. The machines with the five winding configurations are designed according to the specifications in Table 7.2,

Figure 7.26 Stators and winding ptototypes of the five machines: (a) W1 and W2, (b) W3, (c) W4, and (d) W5. Wang et al. [1]/with permission of IEEE.

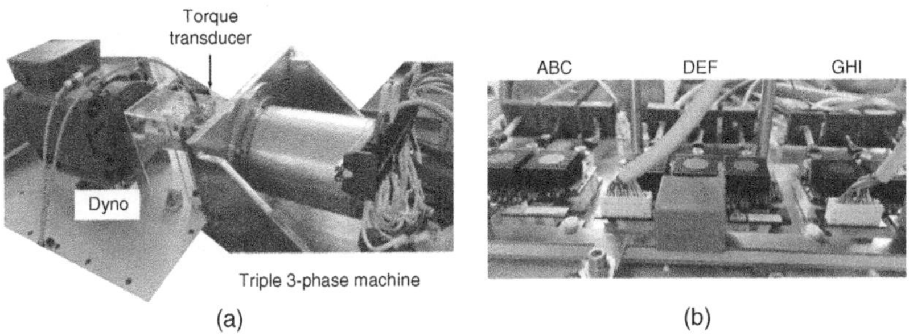

Figure 7.27 Established experimental test platform of the five machines: (a) machine and dynamometer designated as dyno for short and (b) triple 3-phase inverters. Wang et al. [1]/with permission of IEEE.

and prototyped as shown in Fig. 7.26. They are built with the same stator stack and rotor assembly and hence have the same thermal dissipation capability. To reduce the number of stators to be built, the machine with the W2 configuration is obtained by changing the stator windings of the W1 configuration to the delta connection. Thus, the number of turns in each phase in these two machines is the same, which implies the rated line current of the delta-connected machine is 1.732 times of the rated 67 A in the star-connected phase windings, i.e. 117 A.

An experimental test platform similar to that described in Chapter 3 has been established as shown in Fig. 7.27. Each prototype machine is fed by three 3-phase inverters operating at 10 kHz switching frequency and mechanically coupled, via an inline torque transducer, to the dynamometer, which acts as a load and controls the speed of the test operations. The currents in each inverter phase are measured by the integrated inverter current sensors and recorded in the DSP unit. All other important signals, such as voltages, speed, torque, and currents in delta-connected windings or in faulted turns, are measured and recorded during tests.

7.4.1 No-Load Test

First, the five machines are tested with the same rotor assembly under no-load condition. They are driven by the dynamometer at the rated speed of 2000 rpm, and the back-emfs of each machine are measured. The measured phase back-emfs are shown in Fig. 7.28 for all machines except for the machine with the W3 configuration. Since there is no neutral point in the star-delta-connected windings, direct measurements of the phase back-emfs are not possible. Instead, its line back-emfs are measured and shown in Fig. 7.28(c).

It is seen that the phase back-emf waveforms of the machines with W1, W4, and W5 configurations are essentially the same with a peak value around 11.2 V. Small differences in harmonic contents are not visible.

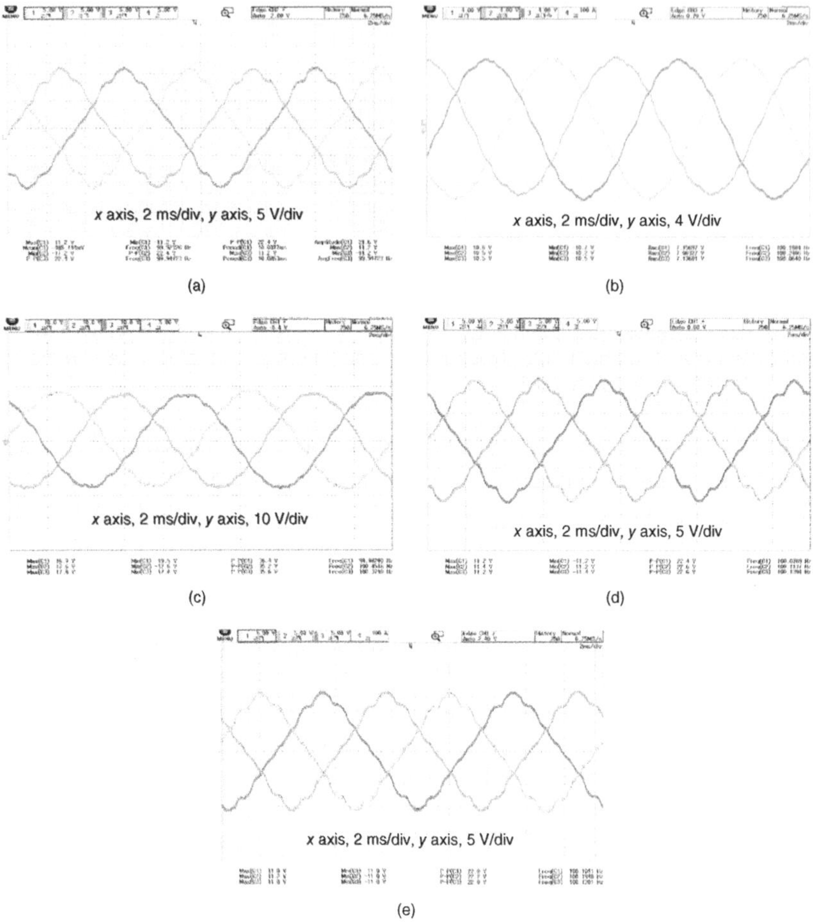

Figure 7.28 Measured phase back-emfs of the machines with winding configurations of (a) W1, (b) W2, (d) W4, and (e) W5. Measured line back-emfs of the machine with W3 configuration shown in (c).

For the delta-connected machine, the measured phase back-emfs are also the line quantities, and they differ from those of the star-connected machines in both waveform shape and magnitude. This is because zero-sequence currents can flow in the delta-connected windings even under the no-load condition. Consequently, the triplen harmonic voltages are significantly reduced by the zero-sequence current and the measured back-emfs exhibit a lower harmonic distortion with a peak voltage of 10.6 V. The rated line voltage of this machine will be ~1.73 times lower than those of the star-connected machines.

A similar effect of the zero-sequence current in the delta-connected coils is also seen in the measured line back-emfs of the machine with the W3 configurations.

The waveforms are less-distorted and highly symmetrical with a peak value of 17.86 V, close to the line back-emfs of the star-connected machines. Their rated line voltage will also be similar.

7.4.2 Load Test Under Healthy Condition

Load tests are performed on the five machines to evaluate their torque capability in healthy operations. First, tests are carried out at 500 rpm by varying the magnitude and phase angle (denoted as gamma angle) of the stator currents, and the results are plotted in Fig. 7.29 for the four machines with the W1, W3, W4, and W5 configurations, from which the optimal current angles for the maximum torque per Ampere (MTPA) operation of each machine can be determined. It can be seen that the four machines have similar MTPA characteristics and their optimal gamma angle at the rated current of 67 A is about 44°. The machine with the W2 configuration is the delta-connected version and, hence, has the same MTPA characteristic of the machine with the W1 configuration. Thus, the separate test is not performed. However, the inverter-fed currents for this machine should be adjusted in both magnitude and phase angle based on the governing relation between the delta and star connections.

The measured output torque variations with the load currents under MTPA operations are shown in Fig. 7.30, from which it is evident that the machine with the W3 configuration exhibits the highest torque as expected while that with the W2 configuration of the lowest torque. This is consistent with the FE predictions in Fig. 7.13. Further, the torque waveforms are measured at a very low speed of 10 rpm for the five machines at the rated current and shown in Fig. 7.31. The characteristic of the torque ripple observed in each machine are close to the FE results, and any difference is likely caused by the torque ripple of the dynamometer and mechanical unbalance of the test setup. As predicted, the torque ripple of the W3 machine is the lowest while those of the rest are very close.

The measured line currents of the five machines at the rated operation of 2000 rpm are shown in Fig. 7.32. They are well tracked by three independent inverter controllers and exhibit good symmetry. The currents in three 3-phase sets of each machine coincide with each other. It is worth noting that the line current magnitude of the W2 machine is 117 A, which is 1.732 times that of the W1 machine, owing to its delta connection.

The currents in the delta-connected phase windings and coils are also measured by current probes and they are shown in Fig. 7.33. The currents are slightly more distorted due to zero-sequence triplen harmonics circulating in the delta-connected phases or coils, which are not under inverter control. Switching harmonics, captured by the high bandwidth probes, are also visible.

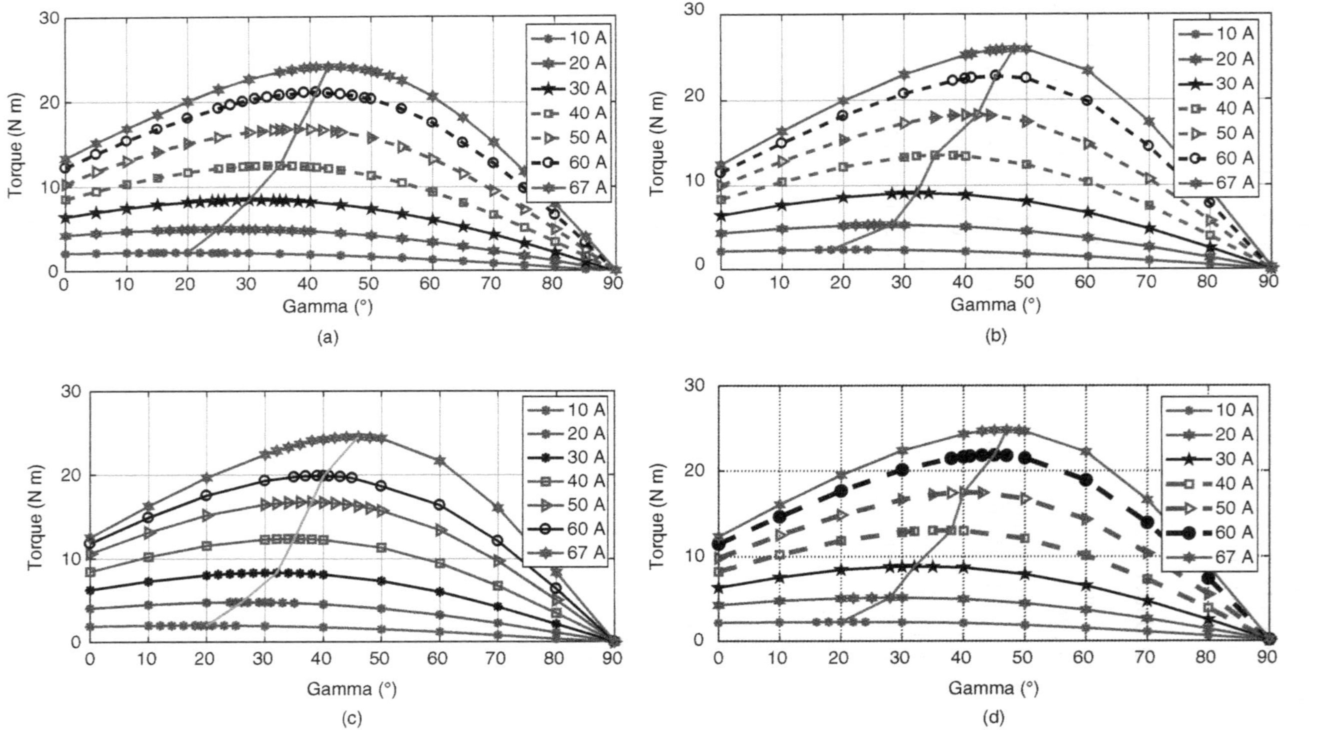

Figure 7.29 Measured torque variations with current magnitude and gamma angle of four machines with winding configurations of (a) W1, (b) W3, (c) W4, and (d) W5.

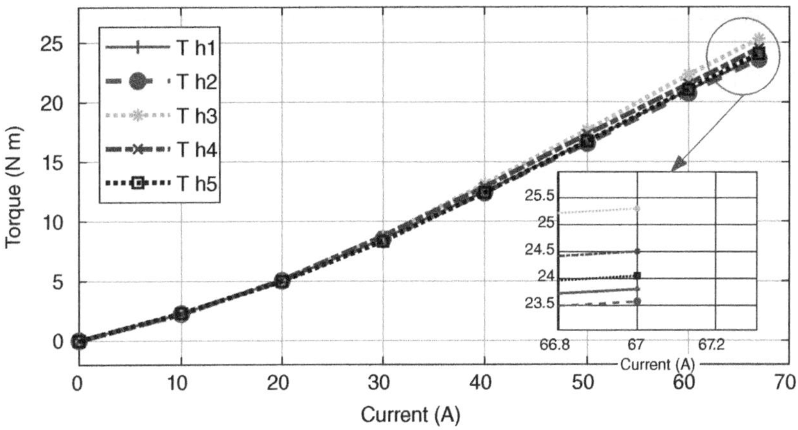

Figure 7.30 Output torque variations with load current under MTPA operation in healthy mode.

The load tests show that all five machines exhibit similar torque characteristics in healthy operations as predicated with slightly better performance from the machine with the W3 configuration.

7.4.3 Load Test with Open Circuit in One 3-Phase Set

Load tests are also performed for each machine at 2000 rpm with the ABC set open circuited while the other two 3-phase sets are fed with the rated currents shown in Fig. 7.34. With feedback control in the inverters, the currents in the healthy sets of the five machines are essentially kept the same as those in the healthy operation with very small harmonic distortion. However, circulating current is presented in the delta-connected ABC phases or coils even in the open-circuit condition. The measured circulating currents in the delta-connected phases of the W2 configuration and in the delta-connected coils of the W3 configuration are shown in Fig. 7.35. The RMS circulating currents are 13.8 and 9.6 A in the machines with the W2 and W3 configurations, respectively. They are below 35% of the rated and the resultant heating effect is insignificant.

The measured torque variations with current magnitude for the five machines under the open-circuit mode are compared in Fig. 7.36. Compared with the healthy operations, the torque capability of all five machines is reduced by 40%, which is consistent with the trend observed in FE predictions. The tests confirm that the five machines can cope with the open-circuit fault with reduced torque capability.

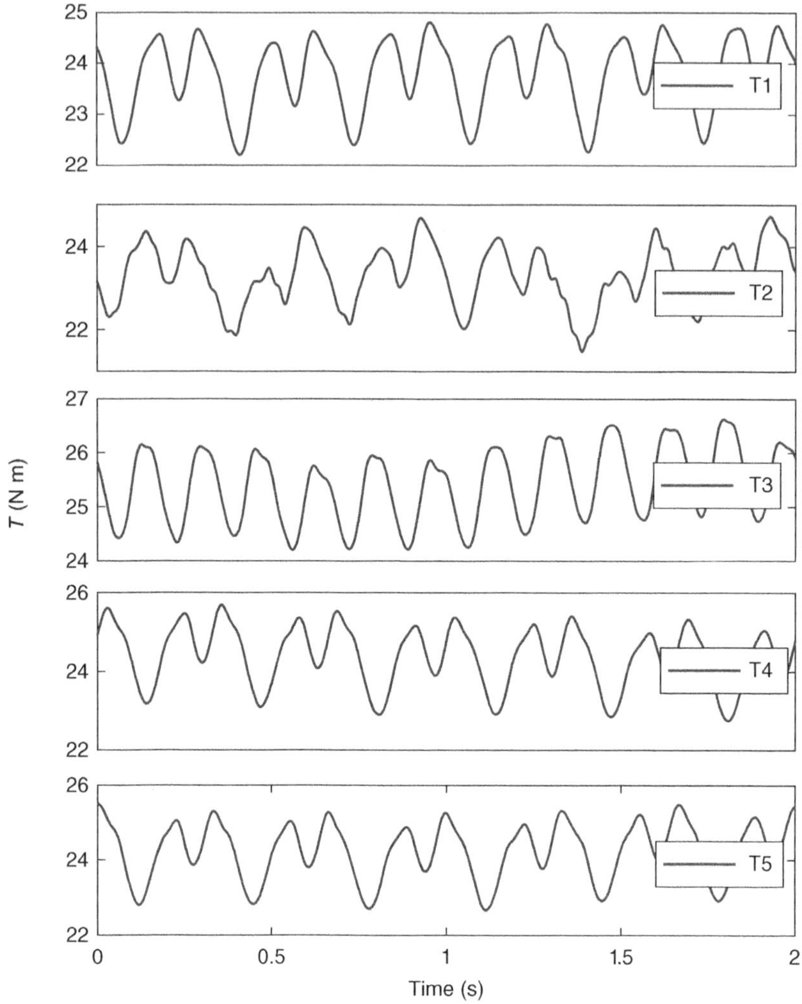

Figure 7.31 Measured torque waveforms of five machines at 10 rpm with rated current, T1, T2, T3, T4, and T5 denote measurements of machines with W1, W2, W3, W4, and W5 configurations, respectively.

7.4.4 Load Test with Short Circuit in One 3-Phase Set

Load tests with one 3-phase set under TSC mode via the inverter are undertaken with the five machines operating at the rated speed. The tests in this mode are of particular interest since upon detection of a short circuit in one phase or a turn fault in any coil, TSC of the 3-phase set where the fault is located is often adopted as a mitigation measure, which reduces the fault current to a sustainable level and enables continuous operation of the remaining healthy 3-phase sets. Figure 7.37

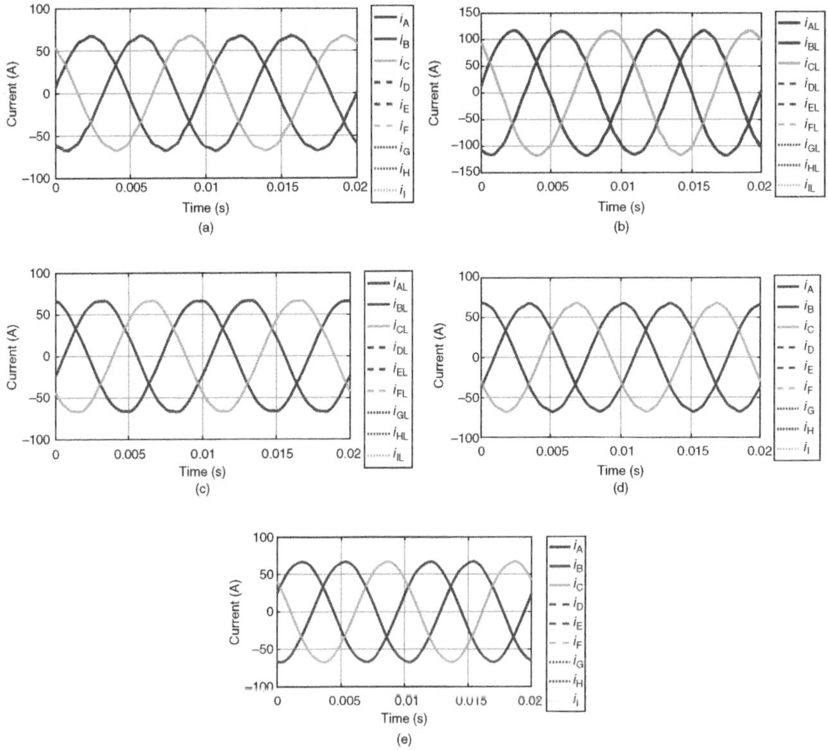

Figure 7.32 Measured line currents of the five machines: (a) W1, (b) W2, (c) W3, (d) W4, and (e) W5.

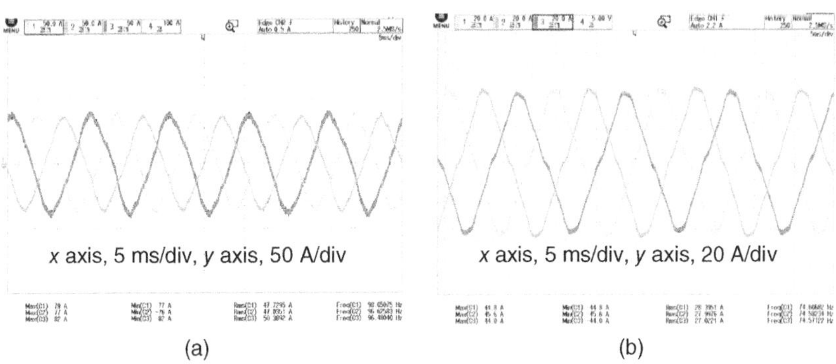

Figure 7.33 Measured currents under rated healthy operation: (a) phase currents of W2 configuration and (b) currents in delta-connected coils of W3 configuration.

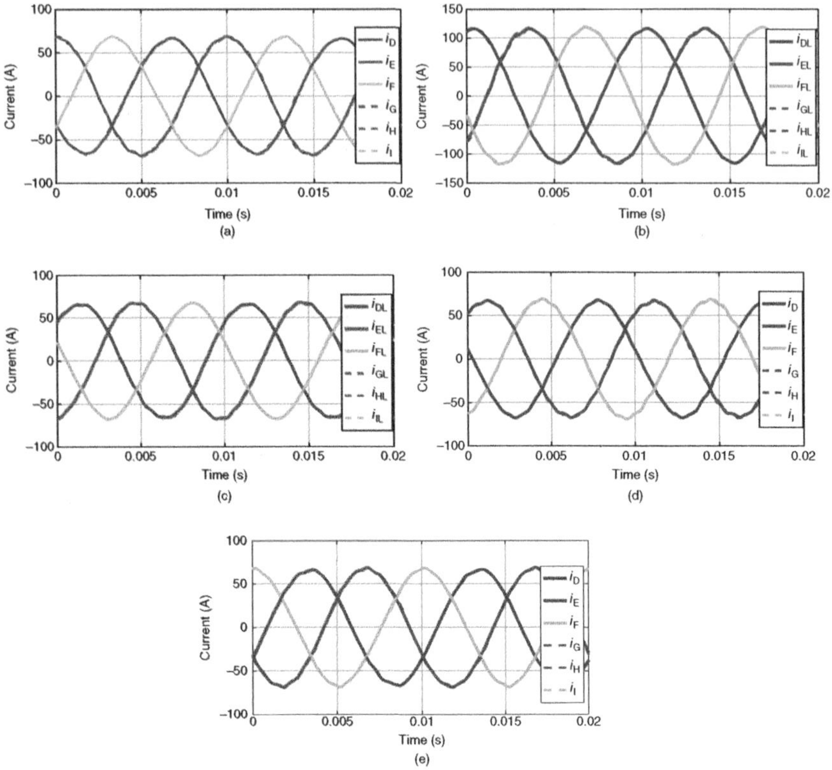

Figure 7.34 Measured line currents of the five machines with winding configuration of (a) W1, (b) W2, (c) W3, (d) W4, and (e) W5, when ABC phases are open circuited.

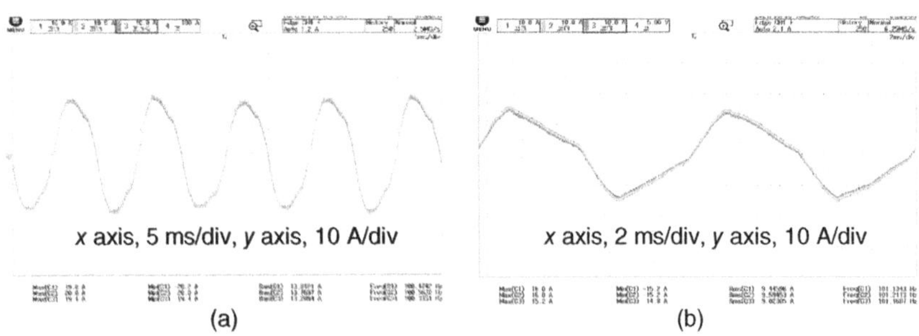

Figure 7.35 Measured currents when phases ABC are open circuited: (a) currents in delta-connected phases of W2 configuration and (b) currents in delta-connected coils of W3 configuration.

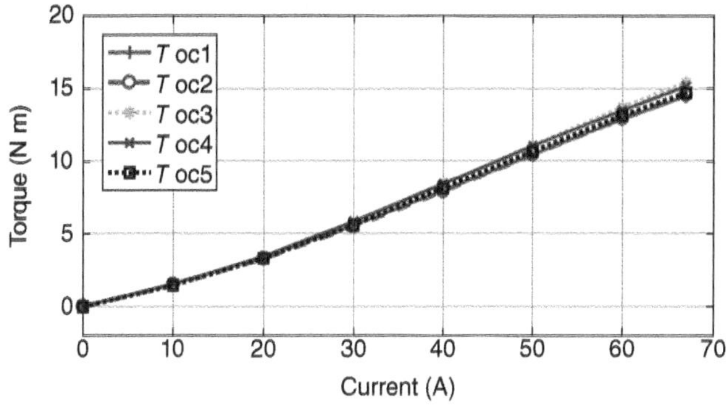

Figure 7.36 Variations of the output torques with the load currents for the five machines in open-circuit mode.

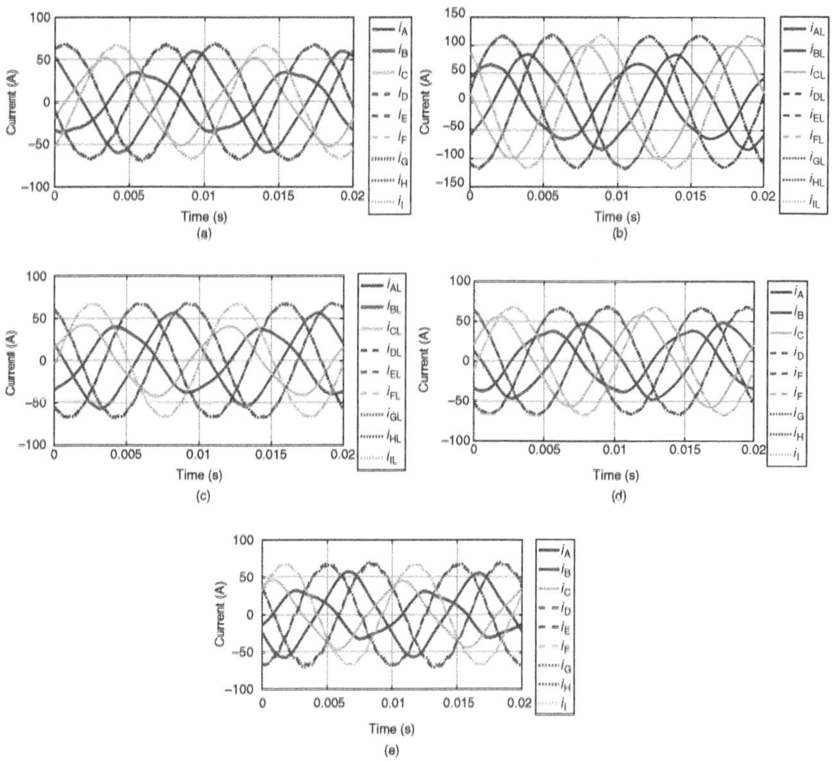

Figure 7.37 Measured line currents of five machines with winding configuration of (a) W1, (b) W2, (c) W3, (d) W4, and (e) W5, when ABC phases are short circuited.

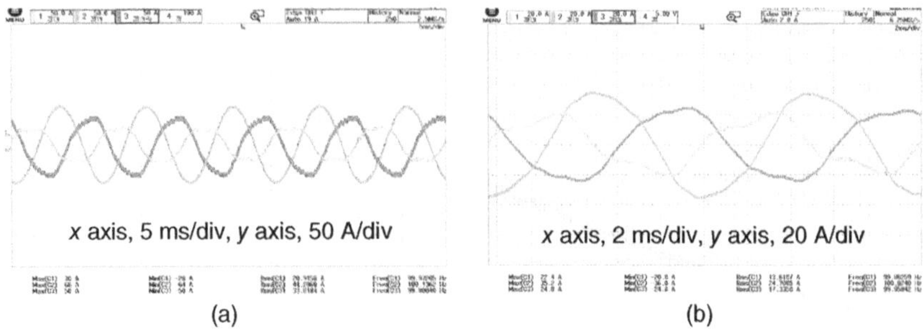

Figure 7.38 Measured currents when ABC phases are short circuited: (a) phase currents in ABC set of W2 configuration and (b) currents of delta-connected coils in ABC set of W3 configuration.

shows the measured currents in the healthy sets and short-circuit currents in the faulted ABC set of the five machines under the tests. As observed, the currents in the two healthy 3-phase sets of each machine are well controlled and overlapped as those seen in the open-circuit mode while the short-circuit currents are asymmetric and distorted because of the mutual magnetic coupling. However, the RMS values of the short-circuit currents are much lower than the rated in all the five machines, which provides a thermal margin to accommodate a relatively large turn fault current in case such a fault occurs in any part of a 3-phase winding set. Because of the delta connection, relatively larger currents by a factor of 1.732 in the machine with the W2 configuration are seen.

In addition, the currents in the delta-connected phases or coils of the short-circuited ABC phases in the machines with the W2 and W3 configurations are also measured and shown in Fig. 7.38. Similar distortion is seen with the average RMS value being slightly larger than that of the line quantities divided by 1.732 because of the circulating zero-sequence components in the delta-connected loop. Nevertheless, they are all below the rated and within the safe region of operation under this fault mode.

The torque variations with current magnitude in the healthy 3-phase sets under this mode are shown in Fig. 7.39 for the five machines, where a reduction in torque capability by ~1/3 is observed for all the five machines compared with their operations in the healthy condition.

7.4.5 Single-Turn Short-Circuit Test

Finally, the merits of the alternative winding configurations in the reduction of turn fault current following TSC application are experimentally evaluated. A single turn is split from the B2 coil in each machine and placed close to the slot opening at which the fault current is the largest. Tests are performed with this turn

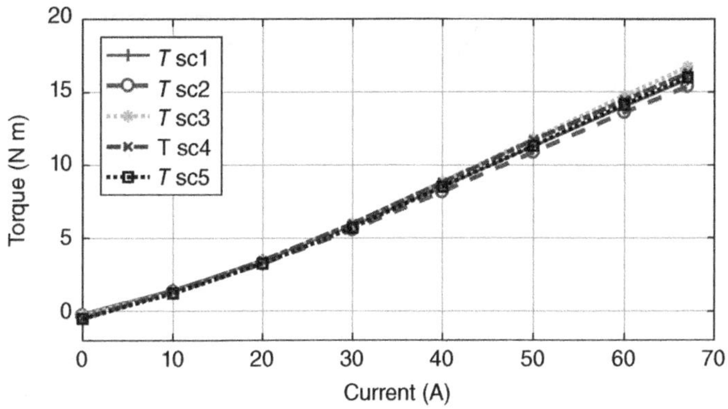

Figure 7.39 Variations of the output torques with the load currents for the five machines in short-circuit mode.

being short-circuited externally. Meanwhile, TSC is applied to the ABC set while the currents in the other two 3-phase sets are controlled by their inverters. To avoid excessive current in the short-circuited turn that may cause permanent damage to the windings, the tests are carried out by gradual increases in the speed and load current until both the currents in the healthy sets and the speed reach the rated. The measured short-circuit currents in the ABC set and in the fault turn are shown in Fig. 7.40 for the five machines under the rated operation point. The waveforms match well with the FE calculation. Their RMS currents in the short-circuited turn are 151, 113, 72, 122, and 128 A in the machines with the W1, W2, W3, W4, and W5 configurations, respectively. They are slightly lower than the FE prediction in all the machines except for that with the W2 configuration, which are caused by the additional resistance and inductance of the cable and connections to form the external short circuit. The larger fault current than the prediction given in Table 7.5 in the machine with the W2 configuration is because the number of turns in a phase winding of the prototype is kept to that of the star-connected machine and not scaled for the same line current and voltage ratings as in the simulations. Nevertheless, if the RMS fault currents are normalized to the rated value, the measurement agrees reasonably well with the prediction. In addition, the currents in the healthy sets are very similar to those observed in the load tests under the short-circuit mode, and hence are not repeated. Likewise, the torque capability of the five machines under this failure mode is very similar to that under one 3-phase set being short-circuited, shown in Fig. 7.39.

The turn fault currents and torque capabilities in the rated healthy and typical fault conditions of the five machines are summarized in Table 7.6. Compared with the classical star configuration, the normalized turn fault currents are reduced by 25%, 17%, 19%, and 15% with the W2, W3, W4, and W5 configurations, respectively.

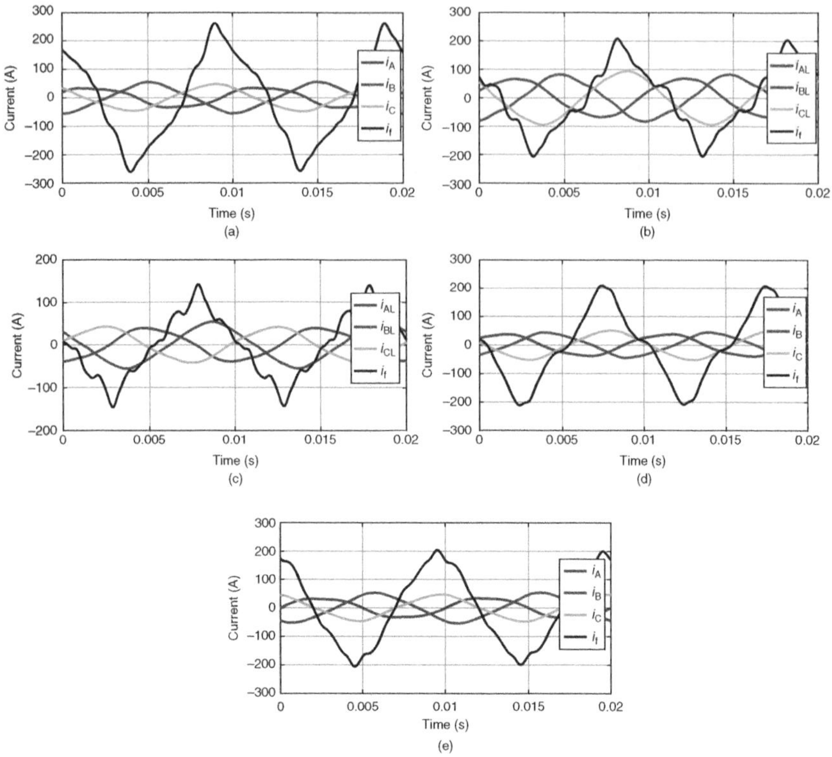

Figure 7.40 Measured line currents and turn fault current in five machines with winding configuration of (a) W1, (b) W2, (c) W3, (d) W4, and (e) W5, when TSC is applied to fault ABC set.

Table 7.6 Overall performance comparison.

Items	W1	W2	W3	W4	W5
Healthy torque (N m)	23.8	23.55	25.3	24.4	24.0
OC torque (N m)	14.55	14.47	15.38	15.16	14.7
SC torque (N m)	15.87	14.4	16.72	16.3	16.0
RMS turn fault current (A)	151	113	72	122	128
Normalized RMS turn fault current (pu)	3.19	2.38	2.65	2.57	2.70

These reductions are obtained without much affecting their torque capabilities in both healthy and fault conditions. Compared with the FE prediction for the rms turn fault currents shown in Table 7.5, the measured fault currents in pu are lower except for that of the W2 configuration. This is mainly caused by the additional impedance of the connection cables and relay introduced to generate the short-circuit fault in the experiments. Since this parasitic impedance is difficult to control during experiments, its influence on the measured fault currents differs. It is evident that the relative reduction due to the parasitic impedance in the measured fault current in the W3 configuration is much less than that in the other configurations. Consequently, the reduction in the measured fault current with the W3 configuration appears lower than the other configurations. Further, since the machine with the W2 configuration in the experiment uses the same phase windings of the W1 configuration with 8 turns per coil but with a delta connection, the number of series connected turns across the line voltage is lower, leading to 1.732 times lower voltage rating than the other configurations. This is conducive to a lower fault current in pu as discussed at the beginning of section 7.2. For this reason, the experimentally measured fault current of the W2 configuration is not directly comparable to the FE prediction, though the largest reduction in the fault current is obtained with the W2 configuration. While the measured and predicted fault currents differ to some extent, their trends are still very close. In conclusion, the reduction in the turn fault current with these alternative winding configurations will lead to lower hot-spot temperatures in the faulted turns, and hence enhance fault-tolerant capability of the machine drive under the worst-case fault condition.

Other faults, like the demagnetization fault and DC-link capacitor fault, can also be accommodated by the machines with the alternative winding configurations in a similar manner to the star-connected machine. These faults can be analyzed with the methods described in Chapter 2 and, therefore, are not repeated.

7.5 Summary

In this chapter, the cause of the relatively large turn fault current in the worst-case fault condition of the triple redundant PMA-SynRM with star-connected 3-phase windings have been identified and four alternative winding configurations have been presented to address this weakness. The zero-sequence residual flux linkage in the faulted turn due to mutual magnetic coupling in the fault condition is the main cause of the turn fault current. Thus, four alternative winding configurations have been devised to reduce the residual flux linkage by the zero-sequence currents in the delta-connected windings or coils or directly reduce the mutual coupling effect. Both simulations and tests show that, compared with the classic star-connected machine, the machines with the alternative winding configurations exhibit similar torque characteristics under both healthy and fault conditions, while the turn fault current in the worst-case fault condition is significantly reduced. The machine with the W3 configuration is most effective in reduction of the turn fault current and it will lead to the lowest temperature rise in the fault turn. Meanwhile, its torque capability is also slightly improved. These improvements are obtained with a few additional steps in the stator winding manufacture process, which may lead to a slight increase in cost. In contrast, 13% reduction in the normalized turn fault current with the delta winding configuration is gained at the expense of a slight reduction in the torque capability, mainly due to the circulating currents in the delta-connected phases. This disadvantage is indeed the main reason why the star winding configuration is more commonly used in PM machines. The other two alternative winding configurations are also effective in the reduction of the turn fault current, while there is very little difference in the stator winding manufacturing process compared to the classical winding configuration. Finally, the combination of the delta or star-delta winding configurations with the mixed-pitch or concentric coils in each phase is likely to have further reduction in the turn fault current under the worst-case fault condition.

References

1 Wang, B., Zha, C., Xu, Y. et al. (2022). Comparative study on fault tolerant triple 3-phase PM machine drive with five modular windings. *IEEE Transactions on Industrial Electronics* 70 (10): 1–11.

2 Liu, C. (2018). Emerging electric machines and drives — an overview. *IEEE Transactions on Energy Conversion* 33 (4): 2270–2280.

3 Mecrow, B.C. et al. (2004). Design and testing of a four-phase fault-tolerant permanent-magnet machine for an engine fuel pump. *IEEE Transactions on Energy Conversion* 19 (4): 671–678.

4 Radun, A.V., Ferreira, C.A., and Richter, E. (1998). Two-channel switched reluctance starter/generator results. *IEEE Transactions on Industry Applications* 34 (5): 1026–1034.

5 W. Zhao, M. Cheng, W. Hua, and H. Jia, "A redundant flux-switching permanent magnet motor drive for fault-tolerant applications," in *IEEE Vehicle Power and Propulsion Conference, 2008. VPPC '08*, 2008, pp. 1-6.

6 Sayed-Ahmed, A. and Demerdash, N.A.O. (2012). Fault-tolerant operation of delta-connected scalar- and vector-controlled AC motor drives. *IEEE Transactions on Power Electronics* 27 (6): 9.

7 Sayed-Ahmed, A., Mirafzal, B., and Demerdash, N.A.O. (2011). Fault-tolerant technique for Δ-connected AC-motor drives. *IEEE Transactions on Energy Conversion* 26 (2): 646–653.

8 Ibrahim, M.N.F., Abdel-Khalik, A.S., Rashad, E.M., and Sergeant, P. (2018). An improved torque density synchronous reluctance machine with a combined star–delta winding layout. *IEEE Transactions on Energy Conversion* 33 (3): 1015–1024.

9 Lei, Y., Zhao, Z., Wang, S. et al. (2011). Design and analysis of star–delta hybrid windings for high-voltage induction motors. *IEEE Transactions on Industrial Electronics* 58 (9): 3758–3767.

10 Alberti, L. and Bianchi, N. (2012). Experimental tests of dual three-phase induction motor under faulty operating condition. *IEEE Transactions on Industrial Electronics* 59 (5): 2041–2048.

11 J. Chai, J. Wang, K. Atallah, and D. Howe, "Performance comparison and winding fault detection of duplex 2-phase and 3-phase fault-tolerant permanent magnet brushless machines," in *Industry Applications Conference, 2007. 42nd IAS Annual Meeting. Conference Record of the 2007 IEEE*, 2007, pp. 566-572.

12 Fu, J.-R. and Lipo, T.A. (1994). Disturbance-free operation of a multiphase current-regulated motor drive with an opened phase. *IEEE Transactions on Industry Applications* 30 (5): 1267–1274.

13 Wang, B., Wang, J., Griffo, A., and Sen, B. (2018). Stator turn fault detection by 2nd harmonic in instantaneous power for a triple redundant fault-tolerant PM drive. *IEEE Transactions on Industrial Electronics* 65 (9): 7279–7289.

14 Bianchi, N., Bolognani, S., and Pre, M.D. (2007). Strategies for the fault-tolerant current control of a five-phase permanent-magnet motor. *IEEE Transactions on Industry Applications* 43 (4): 960–970.

15 Wang, B., Wang, J., Griffo, A., and Sen, B. (2018). Experimental assessments of a triple redundant 9-phase fault tolerant PMA SynRM drive. *IEEE Transactions on Industrial Electronics* 66 (1): 772–783.

16 Dusek, J., Arumugam, P., Brunson, C. et al. (2016). Impact of slot/pole combination on inter-turn short-circuit current in fault-tolerant permanent magnet machines. *IEEE Transactions on Magnetics* 52 (4): 1–9.

17 Arumugam, P. (2016). Design optimization on conductor placement in the slot of permanent magnet machines to restrict turn-turn short-circuit fault current. *IEEE Transactions on Magnetics* 52 (99): 1–8.

18 S. Dwari, L. Parsa, and T. A. Lipo, "Optimum control of a five-phase integrated modular permanent magnet motor under normal and open-circuit fault conditions," in *IEEE Power Electronics Specialists Conference, 2007. PESC 2007*, 2007, pp. 1639-1644.

19 Park, J.K., Babetto, C., Berardi, G. et al. (2021). Comparison of fault characteristics according to winding configurations for dual three-phase synchronous reluctance motor. *IEEE Transactions on Industry Applications* 57 (3): 2398–2406.

20 Faiz, J. and Tabatabaei, I. (2002). Extension of winding function theory for nonuniform air gap in electric machinery. *IEEE Transactions on Magnetics* 38 (6): 3654–3657.

21 Wang, B., Wang, J., Griffo, A., and Sen, B. (2018). A general modeling technique for a triple redundant 3×3-phase PMA SynRM. *IEEE Transactions on Industrial Electronics* 65 (11): 9068–9078.

Index

Multiple 3-Phase Fault Tolerant Permanent Magnet Machine Drives: Design and Control, First Edition.
Bo Wang and Jiabin Wang.

Printed and bound by CPI Group (UK) Ltd, Croydon, CR0 4YY
17/09/2026
14964329-0003